T0182019

Communications
in Computer and Information Science 1489

More information about this series at http://www.springer.com/series/7899

Azlinah Mohamed · Bee Wah Yap ·
Jasni Mohamad Zain · Michael W. Berry (Eds.)

Soft Computing
in Data Science

6th International Conference, SCDS 2021
Virtual Event, November 2–3, 2021
Proceedings

 Springer

Editors
Azlinah Mohamed
Universiti Teknologi MARA
Shah Alam, Malaysia

Bee Wah Yap
Universiti Teknologi MARA
Shah Alam, Selangor, Malaysia

Jasni Mohamad Zain
Universiti Teknologi MARA
Shah Alam, Malaysia

Michael W. Berry
University of Tennessee
Knoxville, TN, USA

ISSN 1865-0929   ISSN 1865-0937   (electronic)
Communications in Computer and Information Science
ISBN 978-981-16-7333-7   ISBN 978-981-16-7334-4   (eBook)
https://doi.org/10.1007/978-981-16-7334-4

This Springer imprint is published by the registered company Springer Nature Singapore Pte Ltd.
The registered company address is: 152 Beach Road, #21-01/04 Gateway East, Singapore 189721, Singapore

Preface

We are pleased to present the proceedings of the Sixth International Conference on Soft Computing in Data Science 2021 (SCDS 2021). SCDS 2021 was held as a virtual conference in collaboration with Institut Teknologi Sepuluh Nopember, Indonesia. The theme of the conference was 'Science in Analytics: Harnessing Data and Simplifying Solutions'. SCDS 2021 aimed to provide a platform for knowledge sharing on theory and applications of big data analytics and artificial intelligence (AI). Big data analytics empowers enterprises to leverage data for better data-driven decisions and AI for optimal performance. The world is moving towards automation and innovation using AI, and industry experts are capitalizing on emerging technologies in artificial intelligence, big data, and cloud computing.

The papers in this proceeding covers issues, challenges, theory, and innovative applications of big data analytics and artificial intelligence including, but not limited to, AI techniques and applications, computing and optimization, data mining and image processing, and machine and statistical learning.

For the advancement of society in the 21st century, there is a need to transfer knowledge and technology on big data and AI to industrial applications, and to solve real-world problems that benefit the global community. Research collaborations involving academia, government, industry, and community can lead to novel innovative computing applications for real-world problems and facilitate near real-time insights and solutions.

We are delighted that this year we received paper submissions from a diverse group of national and international researchers. We received 79 paper submissions, among which 31 were accepted. SCDS 2021 utilized a double-blind review procedure. All accepted submissions were assigned to at least two independent reviewers in order to ensure a rigorous, thorough, and convincing evaluation process. A total of 42 international and 68 local reviewers were involved in the review process. The conference proceeding volume editors and the Springer CCIS Editorial Board made the final decisions on acceptance, with 31 of the 79 submissions (39%) being published in the conference proceedings.

We would like to thank the authors who submitted manuscripts to SCDS 2021. We thank the reviewers for voluntarily spending time to review the papers. We thank all conference committee members for their tremendous time, ideas, and efforts in ensuring the success of SCDS 2021. We also wish to thank the Springer CCIS Editorial Board, cooperating organizations, and sponsors for their continuous support. We would like to dedicate this proceedings to the late Associate Professor Dr Suhatono (Institut Teknologi Sepuluh Nopember, Indonesia) for his dedicated support.

We sincerely hope that SCDS 2021 provided a venue for knowledge sharing, publication of good research findings, and new research collaborations. Last but not least, we hope everyone benefited from the keynote and parallel sessions, and enjoyed engaging with other researchers through this virtual conference.

November 2021

Azlinah Mohamed
Bee Wah Yap
Jasni Mohamad Zain
Michael W. Berry

Organization

Patrons

Roziah Mohd Janor (Vice-chancellor)	Universiti Teknologi MARA, Malaysia
Mochamad Ashari (Rektor)	Institut Teknologi Sepuluh Nopember, Indonesia

Honorary Chairs

Mohd Nazip Suratman	Universiti Teknologi MARA, Malaysia
Jasni Mohamad Zain	Universiti Teknologi MARA, Malaysia
Haryani Haron	Universiti Teknologi MARA, Malaysia
Hamzah Fansuri	Institut Teknologi Sepuluh Nopember, Indonesia
Azlinah Mohamed	Universiti Teknologi MARA, Malaysia
Michael W. Berry	University of Tennessee, USA
Mahadhir Bin Aziz	Malaysia Digital Economy Corporation, Malaysia

Conference Chairs

Yap Bee Wah	Universiti Teknologi MARA, Malaysia
Jerry D. T. Purnomo	Institut Teknologi Sepuluh Nopember, Indonesia

Secretaries

Nurain Ibrahim	Universiti Teknologi MARA, Malaysia
Achmad Choiruddin	Institut Teknologi Sepuluh Nopember, Indonesia

Secretariat

Santi Puteri Rahayu	Institut Teknologi Sepuluh Nopember, Indonesia
Wibawati	Institut Teknologi Sepuluh Nopember, Indonesia
Adatul Mukarromah	Institut Teknologi Sepuluh Nopember, Indonesia
Khairil Anuar Md Isa	Universiti Teknologi MARA, Malaysia

Finance Committee

Nurbaizura Borhan	Universiti Teknologi MARA, Malaysia
Mohd Fikri Hadrawi	Universiti Teknologi MARA, Malaysia
Nur Asyikin Abdullah	Universiti Teknologi MARA, Malaysia
Vita Ratnasari	Institut Teknologi Sepuluh Nopember, Indonesia
Azira Mohamed Amin	Universiti Teknologi MARA, Malaysia

Technical Program Committee

Jasni Mohamad Zain	Universiti Teknologi MARA, Malaysia
Azlinah Mohamed	Universiti Teknologi MARA, Malaysia
Azlan Ismail	Universiti Teknologi MARA, Malaysia
Kartika Fithriasari	Institut Teknologi Sepuluh Nopember, Indonesia
Santi Wulan Purnami	Institut Teknologi Sepuluh Nopember, Indonesia
Dedy D. Prastyo	Institut Teknologi Sepuluh Nopember, Indonesia
Rizauddin Saian	Universiti Teknologi MARA, Malaysia

Publicity and Sponsorship Committee

Ahmad Nazim Aimran	Universiti Teknologi MARA, Malaysia
Nooritawati Md Tahir	Universiti Teknologi MARA, Malaysia
Siti Haslinda Mohd Din	Jabatan Perangkaan Malaysia, Malaysia
Sharifah Sakinah Syed Ahmad	Universiti Teknikal Malaysia, Melaka, Malaysia
Muhammad Asmu'i Abdul Rahim	Universiti Teknologi MARA, Malaysia
Fairoza Amira Hamzah	Women in AI Ambassador/STMicroelectronics, Malaysia
Syed Muslim Jameel	Sir Syed University of Engineering and Technology, Pakistan

Publication Committee (Program Book)

Marina Yusoff	Universiti Teknologi MARA, Malaysia
Saiful Farik Mat Yatin	Universiti Teknologi MARA, Malaysia
Sayang Mohd Deni	Universiti Teknologi MARA, Malaysia
Muhammad Ahsan	Institut Teknologi Sepuluh Nopember, Indonesia

Website Committee

Mohamad Asyraf Abdul Latif	Universiti Teknologi MARA, Malaysia
Muhammad Ridhwan Muhamad Razali	Universiti Teknologi MARA, Malaysia

International Scientific Committee

Nur Iriawan	Institut Teknologi Sepuluh Nopember, Indonesia
Mario Köppen	Kyushu Institute of Technology, Japan
Sri Hartati	Universitas Gadjah Mada, Indonesia
Chidchanok Lursinsap	Chulalongkorn University, Thailand
Aiden Doherty	University of Oxford, UK
Richard Millham	Durban University of Technology, South Africa

Simon Fong	Durban University of Technology, South Africa
Layth Sliman	Efrei Paris, France
Hizir Sofyan	Syiah Kuala University, Indonesia
Dhiya Al-Jumeily	Liverpool John Moores University
Adel Al-Jumaily	Charles Sturt University, Australia
Min Chen	University of Oxford, UK
Mohammed Bennamoun	University of Western Australia, Australia
Agus Harjoko	Universitas Gadjah Mada, Indonesia
Yasue Mitsukura	Keio University, Japan
Dariusz Krol	Wroclaw University of Science and Technology, Poland
Richard Weber	University of Chile, Chile
Jose Maria Pena	Technical University of Madrid, Spain
Yusuke Nojima	Osaka Perfecture University, Japan
Siddhivinayak Kulkarni	University of Ballarat, Australia
Tahir Ahmad	Universiti Teknologi Malaysia, Malaysia
Daud Mohamed	Universiti Teknologi MARA, Malaysia
Ku Ruhana Ku Mahamud	Universiti Utara Malaysia, Malaysia
Ubydul Haque	University of North Texas Health Care Centre, USA
Rajalida Lipikorn	Chulalongkorn University, Thailand
Suhartono	Insititut Teknologi Sepuluh Nopember, Indonesia
Wahyu Wibowo	Insititut Teknologi Sepuluh Nopember, Indonesia
Edi Winarko	Universitas Gadjah Mada, Indonesia
Retantyo Wardoyo	Universitas Gadjah Mada, Indonesia
Rayner Alfred	Universiti Malaysia Sabah, Malaysia
Faiz Ahmed Mohamed Elfaki	Qatar University, Qatar
Mohamed Chaouch	Qatar University, Qatar
Abdul Haris Rangkuti	BINUS University, Indonesia
Sapto Wahyu Indratno	Institut Teknologi Bandung, Indonesia
Angela Kim	Women in AI Ambassador/Teacher's Health, Australia

Reviewers

Achmad Choiruddin	Institut Teknologi Sepuluh Nopember, Indonesia
Adam Shariff Adli Aminuddin	Universiti Malaysia Pahang, Malaysia
Ahmad Farid Abidin	Universiti Teknologi MARA, Malaysia
Ahmad Nazim Aimran	Universiti Teknologi MARA, Malaysia
Ahmad Taufek Abdul Rahman	Universiti Teknologi MARA, Malaysia
Aida Mustapha	Universiti Tun Hussein Onn Malaysia, Malaysia
Albert Guvenis	Boğaziçi University, Turkey
Alexander Bolotov	University of Westminster, UK
Ali Seman	Universiti Teknologi MARA, Malaysia
Angela Lee	Sunway University, Malaysia

Anis Farihan Mat Raffei   Universiti Malaysia Pahang, Malaysia
Azlan Iqbal   Universiti Tenaga Nasional, Malaysia
Azlin Ahmad   Universiti Teknologi MARA, Malaysia
Azlinah Mohamed   Universiti Teknologi MARA, Malaysia
Azman Taa   Universiti Utara Malaysia, Malaysia
Bong Chih How   Universiti Malaysia Sarawak, Malaysia
Chidchanok Lursinsap   Chulalongkorn University, Thailand
Chin Kim On   Universiti Malaysia Sabah, Malaysia
Choong-Yeun Liong   Universiti Kebangsaan Malaysia, Malaysia
Christoph Friedrich   University of Applied Sciences and Arts Dortmund,
                              Germany
Dariusz Krol   Wroclaw University of Science and Technology,
                              Poland
Daud Mohamad   Universiti Teknologi MARA, Malaysia
Dedy Prastyo   Institut Teknologi Sepuluh Nopember, Indonesia
Deepti Theng   G. H. Raisoni College of Engineering, India
Dittaya Wanvarie   Chulalongkorn University, Thailand
Harish Kumar   King Khalid University, Saudi Arabia
Edi Winarko   Universitas Gadjah Mada, Indonesia
Ensar Gul   Maltepe University, Turkey
Ezzatul Akmal Kamaru   Universiti Teknologi MARA, Malaysia
  Zaman
Faroudja Abid   CDTA, Algeria
Foued Saâdaoui   King Abdulaziz University, Saudi Arabia
Hamidah Jantan   Universiti Teknologi MARA, Malaysia
Hamzah Abdul Hamid   Universiti Malaysia Perlis, Malaysia
Hanaa Ali   Zagazig University, Egypt
Haris Rangkuti   Bina Nusantara University, Indonesia
Hasan Kahtan   University of Malaya, Malaysia
Heri Kuswanto   Institut Teknologi Sepuluh Nopember, Indonesia
Hizir Sofyan   Syiah Kuala University, Indonesia
Jasni Mohamad Zain   Universiti Teknologi MARA, Malaysia
Karim Al-Saedi   Mustansiriyah University, Iraq
Khairul Anwar Sedek   Universiti Teknologi MARA, Malaysia
Khatijah Omar   Universiti Malaysia Terengganu, Malaysia
Khyrina Airin Fariza Abu   Universiti Teknologi MARA, Malaysia
  Samah
Krung Sinapiromsaran   Chulalongkorn University, Thailand
Ku Ruhana Ku-Mahamud   Universiti Utara Malaysia, Malaysia
Lee Chinh How   Universiti Tunku Abdul Rahman, Malaysia
Marina Yusoff   Universiti Teknologi MARA, Malaysia
Mario Koeppen   Kyushu Institute of Technology, Japan
Marshima Mohd Rosli   Universiti Teknologi MARA, Malaysia
Maslina Abdul Aziz   Universiti Teknologi MARA, Malaysia
Michael W. Berry   University of Tennessee, USA
Min Chen   University of Oxford, UK

Mohamad Johari Ibrahim	Universiti Teknologi MARA, Malaysia
Mohamed Imran Mohamed Ariff	Universiti Teknologi MARA, Malaysia
Mohd Sulaiman	Universiti Teknologi MARA, Malaysia
Muhammad Ahsan	Institut Teknologi Sepuluh Nopember, Indonesia
Muhammad Firdaus Mustapha	Universiti Teknologi MARA, Malaysia
Muhammad Izzad Ramli	Universiti Teknologi MARA, Malaysia
Muhammad Syafiq Mohd Pozi	Universiti Utara Malaysia, Malaysia
Mustafa Abuzaraida	Universiti Utara Malaysia, Malaysia
Nasiroh Omar	Universiti Teknologi MARA, Malaysia
Ng Kok Haur	University of Malaya, Malaysia
Noor Asiah Ramli	Universiti Teknologi MARA, Malaysia
Noor Azilah Muda	Universiti Teknikal Malaysia Melaka
Nooritawati Md Tahir	Universiti Teknologi MARA, Malaysia
Norazliani Md Lazam	Universiti Teknologi MARA, Malaysia
Norhaslinda Kamaruddin	Universiti Teknologi MARA, Malaysia
Noriko Etani	All Nippon Airways Co., Ltd., Japan
Noryanti Muhammad	Universiti Malaysia Pahang, Malaysia
Norzehan Sakamat	Universiti Teknologi MARA, Malaysia
Nurazzah Abd Rahman	Universiti Teknologi MARA, Malaysia
Nurin Asmuni	Universiti Teknologi MARA, Malaysia
Omar Moussaoui	Université Mohammed Premier, Morocco
Pakawan Pugsee	Chulalongkorn University, Thailand
Peraphon Sophatsathit	Chulalongkorn University, Thailand
Rajalida Lipikorn	Chulalongkorn University, Thailand
Raseeda Hamzah	Universiti Teknologi MARA, Malaysia
Rayner Alfred	Universiti Malaysia Sabah, Malaysia
Retantyo Wardoyo	Universitas Gadjah Mada, Indonesia
Richard Millham	Durban University of Technology, South Africa
Rizauddin Saian	Universiti Teknologi MARA, Malaysia
Rizwan Aslam Butt	NED University of Engineering and Technology, Pakistan
Rodrigo Campos Bortoletto	Instituto Federal de São Paulo, Brazil
Rohit Gupta	IIT Delhi, India
Roselina Sallehuddin	Universiti Teknologi Malaysia, Malaysia
Saiful Akbar	Institut Teknologi Bandung, Indonesia
Sakhinah Abu Bakar	Universiti Kebangsaan Malaysia, Malaysia
Sayang Mohd Deni	Universiti Teknologi MARA, Malaysia
Seng Ong	Universiti Malaya, Malaysia
Shukor Sanim Mohd Fauzi	Universiti Teknologi MARA, Malaysia
Siow Hoo Leong	Universiti Teknologi MARA, Malaysia
Siti Sakira Kamaruddin	Universiti Utara Malaysia, Malaysia
Sofianita Mutalib	Universiti Teknologi MARA, Malaysia
Sri Hartati	Universitas Gadjah Mada, Indonesia

Suraya Masrom	Universiti Teknologi MARA, Malaysia
Syaripah Ruzaini Syed Aris	Universiti Teknologi MARA, Malaysia
Syazreen Niza Shair	Universiti Teknologi MARA, Malaysia
Tahir Ahmad	Universiti Teknologi Malaysia, Malaysia
Tajul Rosli Razak	Universiti Teknologi MARA, Malaysia
Tri Priyambodo	Universitas Gadjah Mada, Indonesia
Wahyu Wibowo	Institut Teknologi Sepuluh Nopember, Indonesia
Waidah Ismail	Universiti Sains Islam Malaysia, Malaysia
Wan Fairos Wan Yaacob	Universiti Teknologi MARA, Malaysia
Weng Siew Lam	Universiti Tunku Abdul Rahman, Malaysia
Xiaolong Jin	Institute of Computing Technology, China
Yasue Mitsukura	Keio University, Japan
Yap Bee Wah	Universiti Teknologi MARA, Malaysia
Yusuke Nojima	Osaka Prefecture University, Japan
Zainab Othman	University of Basrah, Iraq
Zainura Idrus	Universiti Teknologi MARA, Malaysia

Organized by

IBDAAI

Institute for Big Data Analytics
and Artificial Intelligence

Fakulti
Sains Komputer
Dan Matematik

ReNeU
Research
Nexus
UiTM

Hosted by

Technical Co-sponsor

In Cooperation with

Supported by

Contents

Data Mining and Image Processing

Machine and Statistical Learning

AI Techniques and Applications

Comparison Performance of Long Short-Term Memory and Convolution Neural Network Variants on Online Learning Tweet Sentiment Analysis

Muhammad Syamil Ali[1] and Marina Yusoff[1,2(✉)]

[1] Faculty of Computer and Mathematical Sciences, Universiti Teknologi MARA, Shah Alam, Selangor, Malaysia
marina998@uitm.edu.my
[2] Institute for Big Data Analytics and Artificial Intelligence (IBDAAI), Universiti Teknologi MARA, Al-Khawarizmi Complex, Shah Alam, Selangor, Malaysia

Abstract. Sentiment analysis can be act as an assisted tool in improving the quality of online teaching and learning between teachers and students. Twitter social media platform currently more than 500 million tweets sent each day which is equal to 5787 tweets per second. Therefore, it is hard to track users' overall opinions on the topics contained in social media. To catch up with the feedback on online learning, it is crucial to detect the topic being discussed and classify users' sentiments towards those topics. Even though there are many approaches in developing sentiment analysis models, DL models prove to provide the best performance in the sentiment analysis field. Convolutional Neural Network (CNN) and Long Short-Term Memory (LSTM) are two mainstream models in DL used for sentiment analysis classification. Therefore, we evaluate CNN, LSTM, and its hybrids to classify sentiment or an online learning tweet from 2020 until 2021 of 23168 tweets. CNN-LSTM, LSTM-CNN, Bidirectional LSTM, CNN-Bidirectional LSTM models were designed and evaluated based on random hyperparameter tuning. We explain the proposed methodology and model design illustration. The outcome assesses the superiority of all models with a remarkable improvement of accuracy and a reduction loss when applying the random oversampling technique. Specifically, the LSTM-CNN model with random oversampling technique outperformed the other six models with an accuracy of 87.40% and loss value of 0.3432. However, the computational time has resulted increased when with random oversampling technique. Thus, in the future, the performance can be improved on computational time and hyperparameter selection with the employment of nature-inspired computing for fast and optimal results.

Keywords: CNN · LSTM · Online learning · Random oversampling · Sentiment analysis

1 Introduction

In the new era of pandemic COVID-19, online learning is becoming more popular and relevant with the rapid development of technology as it can provide more flexibility

© Springer Nature Singapore Pte Ltd. 2021
A. Mohamed et al. (Eds.): SCDS 2021, CCIS 1489, pp. 3–17, 2021.
https://doi.org/10.1007/978-981-16-7334-4_1

to learn [1]. Even though it is appropriate, various kinds of feedback from people, especially students who give different perceptions [2–4]. It could be positive and negative Uncovering all these kinds of perception are very useful to help various stakeholders include organization, university, schools, and related agencies to improve the learning and teaching experience [5]. Many researchers have started investigating the perception of online learning in the teaching and learning process to understand better the people on the online class system in the past years. Some use WhatsApp calls and thematic analysis [4], online surveys [2, 6, 7].

Sentiment analysis is viable to assist in improving teaching and learning processes. For instance, using an online learning platform that accommodates sentiment analysis allows monitoring the positive and negative feelings [12]. With the use of social media, sentiment analysis can be act as an assisted tool in improving the quality of teaching and learning between teachers and students [13, 14]. Past researchers have applied sentiment analysis to the online learning domain. Recent research used sentiment analysis to investigate the public opinion on online learning in the COVID-19 pandemic was reported as an advantage in the education towards the satisfaction of teaching and learning processes of the relevant parties [14] (Bhagat et al., 2021). Previously, the work on sentiment analysis systems has demonstrated good initiative for course improvement based on student feedback [13].

DL techniques are proven to have the upper hand in obtaining higher performance in the sentiment analysis field compared to the other machine learning techniques [9–11, 15]. Out of all DL models that have been developed in the sentiment analysis field, CNN and LSTM networks have been proved to be more dominant in the area [16]. They also add that each of the DL techniques is entirely different in terms of the model structure itself, and it may produce different performance. DL classification techniques have been widely used in sentiment classification problems [17, 18]. However, in the case of tweet sentiment analysis for online learning, comparatively little work has been done on DL models. Most of the research only classifies the tweets into positive and negative perceptions [19, 20], and more research is required on features selection, pre-process refinement and apply a better approach for NLP [21]. Therefore, evaluating multiple DL models to find the best online learning tweets is prime of importance. Thus, a comparison of different DL approaches and selection of the best model is required. Also, more investigation on online learning perception is expected to be established in the pandemic COVID-19 era. However, investigating public opinion and measuring its effectiveness in online learning, especially during the COVID-19 pandemic, remains challenging.

Therefore, the present work aims at evaluating DL methods to see the capability of classifying sentiment or perception of an online learning tweet. In this way, we hope to assist online educators and training organizations in teaching and training students or customers by providing an excellent online learning sentiment analysis mechanism. We choose CNN and LSTM because of their online Twitter sentiment analysis [9–11, 22]. CNN and LSTM have performed well can be due to deep feature extraction and sequence learning pattern capabilities. This paper is organized as follows. Section 2 describes the proposed methodology. Section 3 presents result of the experiments conducted to determine the best DL models. Section 4 focuses on summary of the result and analysis. Finally, we present the conclusion and future works in Sect. 5.

2 Proposed Methodology

The proposed approach captures the steps to see the performance of DL models on online learning tweet sentiment analysis. The approach includes data collection, pre-processing, model design, and evaluation. We used seven DL models include hybrid CNN-LSTM, LSTM-CNN, and CNN-Bidirectional LSTM models, Single Layer CNN model, Single Layer LSTM model, Bidirectional LSTM, and 2 Layer CNN model used in the evaluation. Figure 1 demonstrates the overview of the proposed methodology. In addition, we add a random oversampling technique to balance the minority class. Details steps is elaborated in the following sub-section.

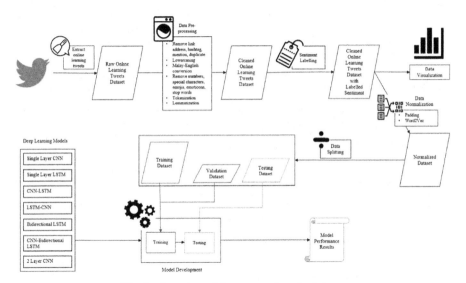

Fig. 1. Overview of the proposed methodology

2.1 Data Collection

The tweet datasets are ranging from the year 2010 to 2021, with about 23168 tweets. It was an online tweet throughout Malaysia. Data collection was conducted by scrapping a collection of tweets from Twitter using the snscrape python library. Extensive research to find suitable keywords related to online learning has been done using past papers, websites, and brainstorming. A total of 23 Twitter keywords related to online learning have been identified.

2.2 Data Pre-processing

In data pre-processing, a thorough cleaning process has been done towards the dataset to make it as clean as possible. The activities associated are as shown in Fig. 1. First, link addresses, hashtags, mentions, and duplicates are removed from the tweets. After that, all the tweets were converted into lowercase. Then, all kinds of numbers and unwanted characters, stop words such as 'me', 'it', and 'be' removed from the tweets. Then, the tokenization process and lemmatization process were done to the tweets where tokenization split the tweets into a set of tokens and lemmatization converted each word token into their meaningful root form.

2.3 Sentiment Labelling

Sentiment labeling is a process where all of the tweets in the dataset annotate with appropriate sentiment according to the tweets' meaning. The labels involved in this process are 'Positive', 'Negative', and 'Neutral'. This process has been done by implementing an automatic tagging technique using a DL model. The DL model used is an LSTM type model trained with 1.6 million Twitter tweets and achieved about 80% accuracy.

2.4 Sentiment Visualization

A visualization technique is applied to visualize the result of the model's sentiment output. Many visualization techniques are chosen to visualize the total sentiment for each class sentiment to identify the comparison of the sentiments on online learning.

2.5 Dataset Normalization

Before the dataset is applied to the DL model, it is required to normalize all the tweets in the dataset into the same format. There are two steps applied to the dataset to normalize it, which are padding and word embedding.

2.6 Online Learning Dataset Analysis

The result found from all the previously discussed processes is applied and visualized using Power BI tools for further analysis. From the visualization that has been made, there is much analysis that can be extracted. The first information that can be gained is the total of the tweets that have been posted on Twitter which is about 10.78 thousand tweets. Next, out of the total tweets, negative sentiment is the highest, which is about 4.85 thousand tweets, followed by neutral sentiment, which is about 3.41 thousand tweets, and positive sentiment, which is about 2.51 thousand tweets.

2.7 Splitting Dataset

The basic principle to develop a DL model is to have a training set to train the model and a testing set to evaluate the model's performance. Therefore, the dataset was divided into two sets: the training set and testing set with a probability of 80% of the dataset to be training set and the rest of it, which is 20% of the dataset, to be a testing set. The training set will be divided into two sets: the training set used to train the model and the validation set, which is used to adjust the hyperparameter during the model training phase with a 90% training set probability and 10% validation set.

2.8 Deep Learning Model Design

This section explains the model design by firstly illustrating the architecture of CNN and LSTM. CNN and LSTM are chosen due to their performance in sentiment analysis and good performance in deep feature extraction and sequence learning pattern, as mentioned above. A hybrid model of CNN and LSTM was derived, looking at the capabilities in sentiment analysis to find a good solution on online learning tweet sentiment. A total of seven DL models were designed. The Single Layer CNN model, Single Layer LSTM model, CNN-LSTM model, LSTM-CNN model, Bidirectional LSTM model, CNN-Bidirectional LSTM model, and 2 Layer CNN model. The proposed hybrid CNN-LSTM, LSTM-CNN, and CNN-Bidirectional LSTM models are shown in Fig. 2, 3, and 4, respectively. The other models are also briefly described.

The architecture of the CNN-LSTM model utilizes a convolution layer in extracting the local features from the word embedding and LSTM layers to learn and store information based on the local features and make sentiment classification from it. As shown in Fig. 2, the process of the model started with the embedding layer converting the input into a set of embedding vectors and passed it to the convolution layer. In the convolution layer, the embedding vectors undergo a convolution process that results in feature maps. The pooling layer then pools the output to reduce the dimension. Then, the output feeds into the LSTM layer to produce a new set of encoding vectors based on meaningful information stored. The same setting is imposed for dense layers and dropouts.

The LSTM-CNN model's architecture uses the LSTM layer to learn and store information from the sequence of words and the convolution layer to extract the local information from the LSTM output and use it to classify the sentiment. Fig. 3 demonstrates the process of the model. It starts with the embedding layer converting the input sentences into embedding vectors. The LSTM layer then received each embedding vector to learn the words sequentially, store and produce a new encoding vector. The output is then fed into the convolution layer, which produces a set of features map and the features map pooled by the pooling layer. The same setting for the output was used as in the previous models.

Hybrid CNN-Bidirectional LSTM model applied convolution layer in extracting the local features from the word embedding and the bidirectional LSTM to learn the local features in bidirectional sequence as illustrated in Fig. 4. In brief, the process started with the embedding layer converting the input sentence into a set of embedding vectors and passing it to the convolution layer to extract the local features and produce feature maps. The pooling layer then pools all the feature maps. The output from the pooling layer then forwarded to the bidirectional LSTM layer to learn the sequence of its input and provide a new set of encoding output.

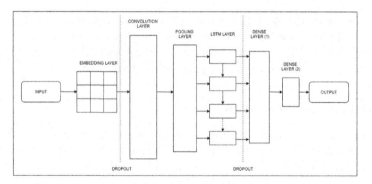

Fig. 2. CNN-LSTM model architecture

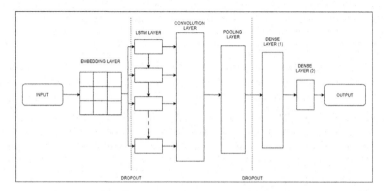

Fig. 3. LSTM-CNN model architecture

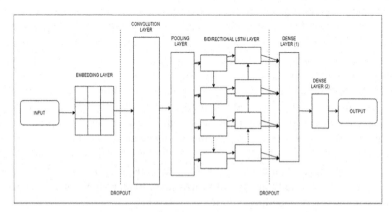

Fig. 4. CNN-Bidirectional LSTM model architecture

2.9 Deep Learning Model Evaluation

The first experiment was conducted to evaluate the performance of the deep learning models based on using hyperparameter setting as shown in Table 1.

Table 1. Hyperparameter setting for all experiments

Hyperparameter	Description
CNN	Size = 3, Num of filter = 100
LSTM	Number of hidden cells = 100
Dropout (1)	0.5, 0.55
Dropout (2)	0.3
Pooling	Max Pooling
Dense (1)	Activation = 'relu', dimension = 64
Dense (2)	Activation = 'softmax', dimension = 3
Learning rate	0.0001
Optimizer	'Adam'
Batch Size	32, 64
Epochs	10, 30, 50
Additional Applied Technique	Random Oversampling

The experiments have been conducted on the seven DL models: Single Layer CNN model, Single Layer LSTM model, CNN-LSTM model, LSTM-CNN model, Bidirectional LSTM model, CNN-Bidirectional LSTM model, and 2 Layer CNN model. To find the highest possible performance that the DL models can achieve. The metrics evaluated from the models are the accuracy and loss that the models can reach from the unseen dataset and the computational time to train the model using the training and validation datasets. Loss is a distance between the actual value of the problem and the value predicted by the model. Accuracy is the number of correct predictions made by the model divided by the total number of predictions [24, 25]. The higher the accuracy, the better the performance of the model would be.

3   Results

This section explains, in brief, the pre-processed datasets result from five experiments conducted for the seven models, and summary results of all experiments.

3.1   Pre-processed Datasets

The dataset scrapped has been pre-processed. All of the methods have been listed and described in the previous section. The comparison table before and after pre-processing the dataset is in Table 2. To add, the size of the dataset has become a lot smaller compared to its original size, which has become about 10775 tweets from 23168 tweets. We used the dataset that has been labelled using a pre-trained DL model. Based on the result, it is found that the number for the negative sentiments in the dataset is the highest, followed by neutral sentiment and positive sentiment, which are 4853, 3410, and 2510 respectively. The dataset has been labelled using a pre-trained DL model. Based on the result, it is found that the number for the negative sentiments in the dataset is the highest, followed by neutral sentiment and positive sentiment, which is 4853, 3410, and 2510, respectively.

Table 2. Sample of pre-processing results

Before	After
pisangggg......3 orng anak esok ade online class.mcm ner nk layan ni.1 talefon tok 3 orng.kene kite kerja x belajar lah...	['pisangggg', 'child', 'tomorrow', 'online', 'classmcm', 'ner', 'want', 'serve', 'phone', 'tok', 'peoplekene', 'work', 'study', 'lah']
Done online class for form 3 science subject, idk y i love teaching so much https://t.co/5wh V8KEB6w	['do', 'online', 'class', 'form', 'science', 'subject', 'idk', 'love', 'teach', 'much']

3.2   Results from Experiment 1: 64 Batch Size and 10 Epoch

The first experiment was conducted to evaluate the performance of the DL models based on batch size, which is the number of samples that work in one iteration, is set to 64, and the number of epochs set for experiment 1 is set to 10. We demonstrate in Fig. 5 (a) accuracy for CNN-LSTM (b) loss for CNN-LSTM (c) accuracy for LSTM-CNN (d) loss for LSTM-CNN. All the models show a decent performance when using the hyperparameter setting described in the early part of the experiment since all the models achieved accuracy in a range of 70% to 75%. Out of the models, LSTM-CNN has been found to provide the best performance for this hyperparameter setting since the model has the highest accuracy, which is 74.57% and has the lowest loss, which is 0.5570. The model with the shortest time to train is the Single Layer CNN model, which is about 21 s.

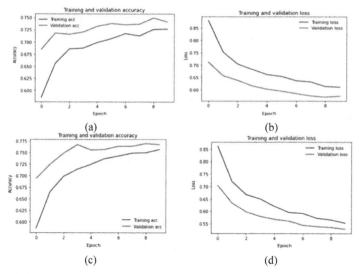

Fig. 5. Result for Experiment 1 (a) accuracy for CNN-LSTM (b) loss for CNN-LSTM (c) accuracy for LSTM-CNN (d) loss for LSTM-CNN.

3.3 Results from Experiment 2: 64 Batch Size and 30 Epochs

The second experiment was conducted to evaluate the performance of the DL models by adding a number of 20 epochs to the experiment 1 hyperparameter setting. Figures 6 (a) and (b) illustrate the results of Single Layer LSTM. The training and validation accuracy shows an increasing pattern over the 30 epochs and has a minimal generalization gap with each other at the last epoch meanwhile training and validation loss show a decreasing pattern over the 30 epochs. The Single Layer LSTM needs around 92 s to train using the training and validation dataset. To add, by using the testing dataset, the model achieved an accuracy of 76.84% and a loss value of 0.5272. Figures 6 (c) and (d) illustrate the results of Bidirectional LSTM. It shows that training accuracy shows an increasing pattern over the 30 epochs meanwhile the validation accuracy increases and decreases at the 27th epoch. Training loss shows a decreasing pattern over the 30 epochs meanwhile the validation loss decreases and increases at the 26th epoch. Both training and validation loss have a minimal generalization gap with each other at the last epoch. Bidirectional LSTM needs around 156 s to train using the training and validation dataset. To add, by using the testing dataset, the model achieved an accuracy of 77.96% and a loss value of 0. 5322.

All the models show a decent performance when using the hyperparameter setting described in the early part of the experiment to achieve accuracy in a range of 71% and 78%. Out of the models, the Bidirectional LSTM model has been found to have the best performance when using this hyperparameter setting as the model has the highest accuracy, which is 77.96%, and has the lowest loss is 0.5322. The model that has the shortest time to train is the Single Layer CNN model, which is about 61 s.

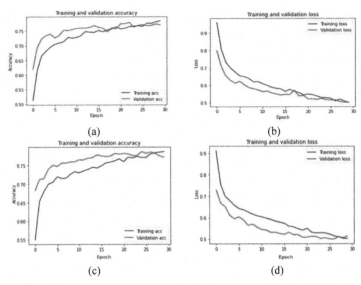

Fig. 6. Result for Experiment 2 (a) accuracy for Single Layer LSTM (b) loss for Single Layer LSTM (c) accuracy for Bidirectional LSTM (d) loss for Bidirectional LSTM

3.4   Results from Experiment 3: 32 Batch Size and 30 Epoch

The third experiment evaluates the performance of the DL models using a similar hyper-parameter setting with experiment 2 but with a change in batch size, which is the batch size of 64 configured to 32. All the models achieved accuracy in a range of 72% to 78%. Out of the models, the LSTM-CNN model has the highest accuracy which is 78.19% while the Bidirectional LSTM model has the lowest loss which is 0.5146. To add, the model that has the shortest time to train is the Single Layer CNN model which is about 60 s.

3.5   Results from Experiment 4: 32 Batch Size and 30 Epochs with Random Oversampling

The fourth experiment was conducted to evaluate the performance of the DL models using a similar hyperparameter setting with experiment 3 but with an oversampling technique applied to the dataset since the dataset used did not have the same distribution classes. Oversampling technique is a technique to balance a dataset by copying and pasting some of their respective class data randomly until it has a similar distribution with other classes. We used the dataset of about 4853 negative class, 3410 neutral class and 2510 negative class. Therefore, the dataset will contain 4853 negative class, 4853 neutral class, and 4853 positive class by applying an oversampling technique to the dataset. All the models achieved accuracy in a range of 84% to 87%. Out of the models, LSTM-CNN has the highest accuracy which is 87.40% and has the lowest loss which is 0.3432. The model with the shortest time to train is the 2 Layer CNN model, which is about 91 s.

3.6 Results from Experiment 5: 64 Batch Size 64 and 50 Epoch with Random Oversampling

The fifth experiment was conducted to evaluate the performance of the DL models by adding a number of 30 epochs and 0.05 first dropout value in the experiment using four hyperparameter settings. As a summary, all the models show good performance when using the hyperparameter setting that has been described in the early part of the experiment since all the models achieved accuracy in a range of 85% to 87%. Out of the models, CNN-Bidirectional LSTM model has the highest accuracy which is 87.39%, the LSTM-CNN model has the lowest loss which is 0.3458- and the Single-Layer CNN model has the shortest time to train the model which is about 101 s.

4 Summary of the Results and Analysis

As a summary, all the models show good performance when using the hyperparameter setting that has been described in the early part of the experiment since all the models achieved accuracy in a range of 85% to 87%. Out of the models, the LSTM-CNN model has the highest accuracy, which is 87.40%, the LSTM-CNN model has the lowest loss which is 0.3432 meanwhile Single Layer CNN model obtained the shortest time to train the model which is about 101 s. Table 3 shows the summary table for all five experiments that have been conducted on Single Layer CNN model, Single Layer LSTM model, CNN-LSTM model, LSTM-CNN model, Bidirectional LSTM model CNN-Bidirectional LSTM model and 2 Layer CNN model. There are a few analyses that can be made from the table. First, experiment 4 and experiment 5 demonstrate a significant difference in accuracy and loss compared to experiment 1, experiment 2 and experiment 3. The major difference between experiment 4 and 5 with experiment 1, experiment 2 and experiment 3 is experiment 4 and 5 trained with a dataset applied with oversampling technique meanwhile models in experiment 1, 2 and 3 trained with a dataset before applying the random oversampling technique. This finding proves that using the random oversampling technique to an imbalanced dataset can improve models' performance.

As seen from the table, the computational time trend for all DL models increases by each experiment. The increasing training time for experiment 2 from experiment 1 is increasing the number of epochs by 20. The increasing of models training time for experiment 3 from experiment 2 is a change in the decreasing batch size value. The increasing of models training time for experiment 4 from experiment 3 is by applying oversampling to the dataset, which increases the size of the dataset used. Meanwhile, the increase of model training time for experiment 5 from experiment 4 is because of the dropout value and epoch increase. Besides that, all models experimented did not provide a consistent result when using different hyperparameter settings. For example, the Single Layer CNN model and 2 Layer CNN model in experiment 2 did show a decrease in terms of accuracy and loss when adding 20 epochs meanwhile the rest of the models show improvement in terms of accuracy and loss when adding 20 epochs. Last but not least, out of the models experimented, it is found out that the LSTM-CNN model with implementation of hyperparameter in experiment 4 provides the best accuracy and loss compared to all models in experiment 1, experiment 2, experiment 3, experiment

Table 3. Summary of the results for all experiments

Model	Experiment 1			Experiment 2			Experiment 3			Experiment 4			Experiment 5		
	Acc (%)	Loss	T(s)	Acc (%)	Loss	T(s)	Acc (%)	Loss	T(s)	Acc (%)	Loss	T(s)	Acc (%)	Loss	T(s)
Single Layer CNN	72.81	0.6073	21	72.02	0.6201	61	73.04	0.6237	60	85.78	0.3968	97	85.61	0.3818	101
Single Layer LSTM	72.11	0.6271	31	76.84	0.5272	92	77.54	0.5178	126	84.31	0.3925	185	85.13	0.3839	277
CNN-LSTM	74.01	0.5837	22	76.61	0.5352	63	75.92	0.5492	93	85.44	0.3686	123	86.02	0.3459	203
LSTM-CNN	74.57	0.557	33	76.75	0.5336	93	78.19	0.5337	152	**87.40**	**0.3432**	111	87.09	0.3458	303
Bidirectional LSTM	74.48	0.5839	54	77.96	0.5322	156	77.77	0.5146	216	84.31	0.4032	306	84.89	0.3761	502
CNN-Bidirectional LSTM	74.11	0.5807	35	75.78	0.5559	95	76.29	0.5468	154	84.89	0.3901	184	87.39	0.3548	305
2 Layer CNN	71.93	0.6172	22	71.74	0.6426	62	72.62	0.6429	61	84.03	0.4063	91	85.47	0.3794	151

*Acc –Accuracy
*T– Computational time during training

4, and experiment 5. The LSTM-CNN model successfully shows high accuracy and minimal loss value in every experiment compared to the other model.

CNN capability in feature extraction and its association among the features (Liao et al., 2017). Hyperparameter tuning is typically considered to determine the best CNN model as mentioned in many literature. However, different batch sizes and epochs used do not significantly affect Single Layer CNN results, but with the additional oversampling technique, the accuracy has increased more than 10%. Generally, in terms of accuracy measure, a random oversampling technique absolutely gives a significant remark in sentiment analysis, mainly using online learning tweet datasets. All DL models obtained a significant improvement of about an additional 10% in accuracy and a loss of about -0.02. We adapt the oversampling technique to balance a dataset by copying and pasting some of their respective class data randomly until it has a similar distribution with other classes (Brownlee, 2021).

5 Conclusion

Sentiment analysis is a popular tool in natural language processing and has been used in various online tweets datasets. Past research highlights that CNN and LSTM provide improvement in the detection and classification of tweet data. In this paper, we evaluate seven models based on CNN and LSTM on online learning tweets. The tweets datasets were scrapped online using the identified keyword. A detailed proposed methodology for model constructions is highlighted. This study also shows that hyperparameter tuning and random oversampling technique leads to an increase in model performance. The random oversampling idea has an impact on balancing the minority class towards a balanced class. Evaluation of other oversampling methods could give better results, and hyperparameter tuning can be enhanced with nature-inspired optimization algorithms such as particle swarm optimization and cuckoo search.

Acknowledgement. The authors would like to thank Institute for Big Data Analytics and Artificial Intelligence (IBDAAI) and Research Managemet Center, Universiti Teknologi MARA, Shah Alam, Malaysia for providing essential support and knowledge for the work.

References

1. 5 Reasons Why Online Learning is the Future of Education. https://www.educations.com/art icles-and-advice/5-reasons-online-learning-is-future-of-education-17146. Accessed 30 Apr 2021
2. Adnan, M.: Online learning amid the COVID-19 pandemic: students perspectives. J. Pedagogical Res. 1(2), 45–51 (2020)
3. Dhawan, S.: Online learning: a panacea in the time of COVID-19 crisis. J. Educ. Technol. Syst. 49(1), 5–22 (2020)
4. Allo, M.D.G.: Is the online learning good in the midst of Covid-19 Pandemic? The case of EFL learners. Jurnal Sinestesia 10(1), 1–10 (2020)

5. Dessí, D., Dragoni, M., Fenu, G., Marras, M., Reforgiato Recupero, D.: Deep learning adaptation with word embeddings for sentiment analysis on online course reviews. In: Agarwal, B., Nayak, R., Mittal, N., Patnaik, S. (eds.) Deep Learning-Based Approaches for Sentiment Analysis. AIS, pp. 57–83. Springer, Singapore (2020). https://doi.org/10.1007/978-981-15-1216-2_3
6. Rojabi, A.R.: Exploring EFL students' perception of online learning via Microsoft teams: university level in Indonesia. Engl. Lang. Teach. Educ. J. 3(2), 163 (2020)
7. Kalyanasundaram, P., Madhavi, C.: Students' perception on e-learning with regard to online value-added courses. Int. J. Manag. 11(3), 89–96 (2020)
8. Yang, L., Li, Y., Wang, J., Sherratt, R.S.: Sentiment analysis for e-commerce product reviews in Chinese based on sentiment lexicon and deep learning. IEEE Access 8, 23522–23530 (2020)
9. Behera, R.K., Jena, M., Rath, S.K., Misra, S.: Co-LSTM: convolutional LSTM model for sentiment analysis in social big data. Inf. Process. Manag. 58(1), 102435 (2021)
10. Tam, S., Said, R.B., Tanriöver, Ö.Ö.: A ConvBiLSTM deep learning model-based approach for Twitter sentiment classification. IEEE Access 9, 41283–41293 (2021)
11. Gandhi, U.D., Kumar, P.M., Babu, G.C., Karthick, G.: Sentiment analysis on Twitter data by using convolutional neural network (CNN) and long short term memory (LSTM) (2021)
12. Rani, S., Kumar, P.: Cover feature advances in learning technologies. Adv. Learn. Technol. 36–43 (2017)
13. Barron-Estrada, M.L., Zatarain-Cabada, R., Oramas-Bustillos, R., Gonzalez-Hernandez, F.: Sentiment analysis in an affective intelligent tutoring system. In: Proceedings - IEEE 17th International Conference on Advanced Learning Technologies, ICALT 2017, pp. 394–397 (2017)
14. Bhagat, K.K., Mishra, S., Dixit, A., Chang, C.Y.: Public opinions about online learning during covid-19: a sentiment analysis approach. Sustainability (Switzerland) 13(6) (2021)
15. Jain, K., Kaushal, S.: A comparative study of machine learning and deep learning techniques for sentiment analysis. In: 2018 7th International Conference on Reliability, Infocom Technologies and Optimization: Trends and Future Directions, ICRITO 2018, pp. 483–487 (2018)
16. Kamiş, S., Goularas, D.: Evaluation of deep learning techniques in sentiment analysis from Twitter data. In: Proceedings - 2019 International Conference on Deep Learning and Machine Learning in Emerging Applications, Deep-ML 2019, pp. 12–17 (2019)
17. Huang, Q., Chen, R., Zheng, X., Dong, Z.: Deep sentiment representation based on CNN and LSTM. In: Proceedings - 2017 International Conference on Green Informatics, ICGI 2017, pp. 30–33 (2017)
18. Colón-Ruiz, C., Segura-Bedmar, I.: Comparing deep learning architectures for sentiment analysis on drug reviews. J. Biomed. Inform. 110(February), 103539 (2020)
19. Ankit, Saleena, N.: An ensemble classification system for twitter sentiment analysis. Procedia Comput. Sci. 132(Iccids), 937–946 (2018)
20. Persada, S.F., Oktavianto, A., Miraja, B.A., Nadlifatin, R., Belgiawan, P.F., Redi, A.A.N.P.: Public perceptions of online learning in developing countries: a study using the ELK stack for sentiment analysis on twitter. Int. J. Emerg. Technol. Learn. 15(9), 94–109 (2020)
21. Rajesh, P., Suseendran, G.: Prediction of n-gram language models using sentiment analysis on e-learning reviews. In: Proceedings of International Conference on Intelligent Engineering and Management, ICIEM 2020, pp. 510–514 (2020)
22. Priyan, M.K., Babu, G.C., Karthick, G.: Sentiment analysis on Twitter data by using convolutional neural network (CNN) and long short-term memory (LSTM) (2021)
23. Basiri, M.E., Nemati, S., Abdar, M., Cambria, E., Acharya, U.R.: ABCDM: an attention-based bidirectional CNN-RNN deep model for sentiment analysis. Futur. Gener. Comput. Syst. 115, 279–294 (2021)

24. Classification Accuracy is Not Enough: More Performance Measures You Can Use. https:// machinelearningmastery.com/classification-accuracy-is-not-enough-more-performance-measures-you-can-use/. Accessed 20 June 2019
25. A Gentle Introduction to Dropout for Regularizing Deep Neural Networks. https://machinele arningmastery.com/dropout-for-regularizing-deep-neural-networks/. Accessed 6 Aug 2019

Performance Analysis of Hybrid Architectures of Deep Learning for Indonesian Sentiment Analysis

Theresia Gowandi, Hendri Murfi$^{(\boxtimes)}$, and Siti Nurrohmah

Department of Mathematics, Faculty of Mathematics and Natural Sciences, Universitas Indonesia, Depok 16424, Indonesia
hendri@ui.ac.id

Abstract. Sentiment analysis is one of the fields of Natural Language Processing that builds a system to recognize and extract opinions in the form of text into positive or negative sentiment. Nowadays, many researchers have developed methods that yield the best accuracy in performing analysis sentiment. Three particular models are Convolutional Neural Network (CNN), Long Short-Term Memory (LSTM), and Gated Recurrent Unit (GRU), which have deep learning architectures. CNN is used because of its ability to extract essential features from each sentence fragment, while LSTM and GRU are used because of their ability to memorize prior inputs. GRU has a more straightforward and more practical structure compared to LSTM. These models have been combined into hybrid architectures of LSTM-CNN, CNN-LSTM, and CNN-GRU. In this paper, we analyze the performance of the hybrid architectures for Indonesian sentiment analysis in e-commerce reviews. Besides all three combined models mentioned above, we consider one more combined model, which is GRU-CNN. We evaluate the performance of each model, then compare the accuracy of the standard models with the combined models to see if the combined models can improve the performance of the standard. Our simulations show that almost all of the hybrid architectures give better accuracies than the standard models. Moreover, the hybrid architecture of LSTM-CNN reaches slightly better accuracies than other hybrid architectures.

Keywords: CNN · Deep learning · GRU · LSTM

1 Introduction

In September 2019, GlobalWebIndex reported that 79% of Indonesia's internet users aged 16 to 64 made online purchases via mobile phones. The top mobile e-commerce applications are Tokopedia, Shopee, and Lazada [1]. Also, thousands of online opinions were written by customers every day on many platforms, e.g., Google Play. They posted about products they paid for, services that they used, and items they took into consideration. Also, customers will use these reviews before they make a decision. These applications developers got a new challenge to disclose what is being said about their application or the products and services they advertise. The standard technique would be

© Springer Nature Singapore Pte Ltd. 2021
A. Mohamed et al. (Eds.): SCDS 2021, CCIS 1489, pp. 18–27, 2021.
https://doi.org/10.1007/978-981-16-7334-4_2

checking thousands of reviews manually. However, nowadays, there is a popular technique to manage this huge amount of opinionated text data by using sentiment analysis. Sentiment analysis, also called opinion mining, is a field that analyzes people's opinions and sentiments toward certain entities, like products or services, to find its polarity between positive or negative sentiment [2].

Sentiment analysis can be done manually by looking through every single text one by one and decide its polarity on our own. But this approach is time-consuming and prone to human error. The solution to this problem is the machine learning approach for sentiment analysis. Machine learning develops methods that can automatically detect patterns in data and use them to predict future data [3]. Sentiment analysis is a classification problem in machine learning that needs to classify text into positive or negative sentiment. There are many studies of machine learning for text classification like Support Vector Machine and Naïve Bayes [4, 5]. However, deep learning is currently commonly used in sentiment analysis problems because it can extract features from unstructured data very well. Popular deep learning techniques that have been widely used in sentiment analysis are Convolutional Neural Network [6], Long Short-Term Memory [7], and Gated Recurrent Unit [8].

Previous researches have shown that CNN, LSTM, and GRU have a good performance for sentiment analysis. Also, Wang, Jiang, Luo [9] have shown that combined architecture CNN-LSTM and CNN-GRU models perform better than the CNN and LSTM models alone. They combined the model because they want to take advantage of both the local feature extraction from CNN and long-distance dependencies from RNN. Then, Sosa [10] proved that the combined LSTM-CNN model achieved better performance than the standard models and CNN-LSTM model.

In this paper, we analyze the performance of the hybrid architectures for Indonesian sentiment analysis in e-commerce reviews. Besides all three combined models mentioned above, which are LSTM-CNN, CNN-LSTM, CNN-GRU, we consider one more combined model, GRU-CNN. We evaluate the performance of each model, then compare the accuracy of the standard models with the combined models to see if the combined models can improve the performance of the standard. Our simulations show that almost all of the hybrid architectures give better accuracies than the standard models. Moreover, the hybrid architecture of LSTM-CNN reaches slightly better accuracies than other hybrid architectures.

The rest of the paper is organized as follows: In Sect. 2, we present materials and methods. In Sect. 3, we discuss the results of the simulations. Finally, we give a conclusion of this research in Sect. 4.

2 Materials and Method

2.1 Dataset

There are three data sets used in this study. Three of them are Indonesian e-commerce application reviews extracted from the Google Play Store website, which are Tokopedia, Shopee, and Lazada. The detail of the data sets is listed in Table 1.

Table 1. Datasets used in this study.

Dataset	Positive sentiment	Negative sentiment	Total data
Tokopedia	1697	2663	4360
Shopee	2562	2435	4997
Lazada	2976	3304	6280

2.2 Preprocessing

In this process, the raw data set is processed from unstructured textual data to structured textual data ready for the model implementation. Firstly, the data need to be cleaned from unused punctuations and emoticons. Second, convert all alphabets to lowercase. Third, remove stop words or commonly used words. And last, the text needs to be modified into a vector of representation by tokenizing, sequencing, padding, and word embedding. Tokenizing is the process of dividing text in sentence form into tokens or parts per word. Sequencing is the process of creating an index for each unique word with an integer sorted based on the most used words in a dataset. Padding is the process of making all the lengths of each document in the dataset the same. Word embedding is the process of representing word indexes using vector representations. Last, one processing process will be carried out on label/sentiment, namely one-hot encoding. One hot encoding is the process of converting categorical variables into numeric variables. Table 2 shows the steps of preprocessing text data. Table 3 shows the preprocessing of labels.

Table 2. Preprocessing text data.

Steps	Processing Results
Data	Josss.. tingkatkan pelayanannya... dan perketat sailler yang curang/nakal... barang gak layak dijual... di jual.... yg di utamakan pelanggan puas... gak kecewa... 👍👍👍cek barang sebelum di kirim... Terimakasih
Data cleaning	tingkat layan ketat jual curang nakal barang layak jual jual utama langgan puas kecewa cek barang kirim terima kasih
Data tokenizing	tingkat, layan, ketat, jual, curang, nakal, barang, layak, jual, jual, utama, langgan, puas, kecewa, cek, barang, kirim, terima, kasih
Data sequencing	147, 51, 718, 61, 428, 329, 2, 682, 9, 9, 333, 107, 127, 22, 88, 2, 4, 12, 8
Data padding	0, 0, 0, 0, 0, 147, 51, 718, 61, 428, 329, 2, 682, 9, 9, 333, 107, 127, 22, 88, 2, 4, 12, 8

2.3 Convolutional Neural Network (CNN)

CNN is a deep learning model that performs well in text data classification using feature extraction layers. Feature extraction consists of convolutional layers and pooling layers

Table 3. Preprocessing labels.

Sentiment	Processing results
0	[0, 1]
1	[1, 0]

that will choose the best features. The first step is processing the input matrix with a convolutional layer which consists of several filters **W**. Each row in the input matrix $X_{i:n}$ is an embedding vector with d dimension from the i-th word of a sentence that has n words. Filters play a role in learning the essential features of sentence fragments. The length of a sentence fragment is called the region size. Every region size has the same number of filters. The region size and number of filters will be the hyperparameter of the model and will be adjusted to get an optimal accuracy. The convolution operation involves a filter **W** that will be applied to a window of s words (region size) to produce a feature. Suppose, input matrix $X_{i:i+s-1}$ is the concatenation of words $x_i, \ldots, x_{i+s-1}$. Then, a feature c_i is defined as

$$c_i = f(X_{i:i+s-1} \cdot \mathbf{W}) \tag{1}$$

with f as the activation function. The filter will be applied to all possible sentence fragments, $X_{1:s}, X_{2:s+1}, \ldots, X_{n-s+1:n}$ to produce a feature map $\mathbf{c} = [c_1, c_2, \ldots, c_{n-s+1}]$. Then, feature map **c** will enter the pooling layer that will choose the representative value on each feature map. Max pooling layer will take the most important feature with the maximum value $\hat{c} = \max\{\mathbf{c}\}$ from each feature map [6].

2.4 Long Short-Term Memory (LSTM)

LSTM proposed by Hochreiter & Schmidhuber in 1997 [11] is a particular type of Recurrent Neural Network because it can carry out learning on long-term dependencies. It means that LSTM can remember information in the long run. The cell state is where memory is stored in LSTM. LSTM can delete or add information to the cell state using the gate mechanisms, which are forget gate ($\mathbf{f}_t$), input gate ($\mathbf{i}_t$), dan output gate ($\mathbf{o}_t$). The gate is composed of one layer with a sigmoid activation function and a simple linear operation. The sigmoid activation function will map the results into a range of 0 and 1 values that control how much information can pass. Value 0 means to forget the information, while value 1 means to remember the information.

At time t, the first step taken in LSTM's is to determine the information that should be removed from the previous cell state ($\mathbf{c}_{t-1}$) and pass the necessary information. This decision was made by forget gate ($\mathbf{f}_t$). Then, LSTM will determine the new information that will be stored in the cell state ($\mathbf{c}_t$). This stage consists of 2 steps. First, we will determine the elements in the cell state to be updated. The input gate performs this step ($\mathbf{i}_t$). Second, the tanh function will form a candidate vector for cell state ($\tilde{\mathbf{c}}_t$) which may be added to the cell state. LSTM combines these two steps to update the cell state.

After processing the information with the previous steps, the old cell state ($\mathbf{c}_{t-1}$) will then be updated to the new state cell ($\mathbf{c}_t$). Finally, LSTM will determine the output

(h_t) based on the updated cell state. The output gate carries out this step (o_t) using the sigmoid function will determine which part of the cell state will be used as the output [12]. The information of h_t then will be passed to the next unit. Figure 1 shows an illustration of LSTM's unit.

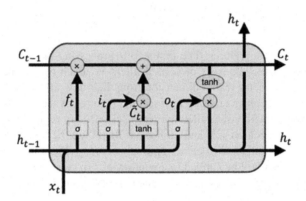

Fig. 1. Illustration of LSTM's unit [15]

$$f_t = \sigma(\mathbf{w}^f \times \mathbf{x}_t + \mathbf{u}^f \times \mathbf{h}_{t-1} + \mathbf{b}_f) \tag{2}$$

$$i_t = \sigma(\mathbf{w}^i \times \mathbf{x}_t + \mathbf{u}^i \times \mathbf{h}_{t-1} + \mathbf{b}_i) \tag{3}$$

$$\tilde{c}_t = \tanh(\mathbf{w}^c \times \mathbf{x}_t + \mathbf{u}^c \times \mathbf{h}_{t-1} + \mathbf{b}_c) \tag{4}$$

$$c_t = f_t \odot c_{t-1} + i_t \odot \tilde{c}_t \tag{5}$$

$$o_t = \sigma(\mathbf{w}^o \times \mathbf{x}_t + \mathbf{u}^o \times \mathbf{h}_{t-1} + \mathbf{b}_o) \tag{6}$$

$$h_t = o_t \odot \tanh(c_t) \tag{7}$$

2.5 Gated Recurrent Unit (GRU)

The GRU proposed by Cho et al. in 2014 [13] was another modification of the Recurrent Neural Network but simpler than the LSTM model, offered almost comparable performance to LSTM, was more computationally efficient, and was quite popular. The gate mechanism that GRU uses is the update gate (z_t) and reset gate (r_t), which are used to determine the output. The gate is composed of one layer with a sigmoid activation function that will map the results into a range of 0 and 1 values that control how much information can pass. Value 0 means to forget the information, while value 1 means to remember the information.

At time t, the first step taken in GRU is to determine how much past information still needs to be passed on to the next cell. The update gate made this decision (z_t). Then, the reset gate (r_t) is responsible for deciding how much past information should be forgotten. Next, we will calculate the candidate hidden state (h'_t) using the information from the reset gate to remember previous information that is useful. As r_t approaches 0, h'_t will only process the current input. As r_t approaches 1, h'_t will store the current input and the previous time output. This process is still a candidate because the results will be forwarded to calculate the final output values in the hidden state h_t.

The final calculation is for the hidden state (h_t), the vector that holds the information from the current unit will pass the information on to the next unit. The hidden state (h_t) will use the information from the update gate (z_t) to determine the required information from the previous hidden state (h_{t-1}) and the candidate hidden state (h'_t). As z_t approaches 0, the value of h_t approaches the value of h'_t, in other words, there will be a change in the information on the output. As z_t approaches 1, the information from x_t is ignored and effectively ignores the time step t in the process so it will only retain the old information [12]. The information of h_t then will be passed to the next unit. Figure 2 shows an illustration of GRU's unit.

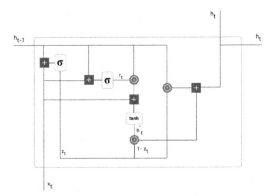

Fig. 2. Illustration of GRU's unit [16]

$$z_t = \sigma(w^z \times x_t + u^z \times h_{t-1} + b_z) \tag{8}$$

$$r_t = \sigma(w^r \times x_t + u^r \times h_{t-1} + b_r) \tag{9}$$

$$h'_t = \tanh(w^h \times x_t + r_t \odot (u^h \times h_{t-1})) \tag{10}$$

$$h_t = z_t \odot h_{t-1} + (1 - z_t) \odot h'_t \tag{11}$$

2.6 RNN-CNN

The RNN-CNN model is a combined model of the RNN and CNN models. The combined RNN-CNN model architecture refers to both LSTM-CNN and GRU-CNN models. The first step is processing the input with the RNN layer to learn the feature representation and sequence on the data. Then, the output from the RNN layer will be used as input for the CNN layer, which will look for pairs of essential features in the data. Each of the standard models has its advantages. RNN pays attention to word order, while CNN can select important features of ordered word phrases. So, these advantages are expected to maximize the data learning process. Figure 3 shows an illustration of the combined RNN-CNN model's architecture.

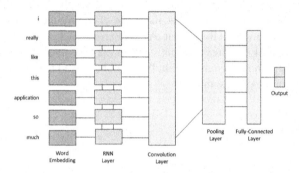

Fig. 3. The architecture of the combined RNN-CNN model [10]

2.7 CNN-RNN

In general, the combined CNN-RNN model has a similar concept to the combined RNN-CNN model. The difference lies in the orders of the model. The first step is processing the input with the CNN layer to select word phrases. Then, the output from the CNN layer will be used as input for the RNN layer, which will create a new representation for the data

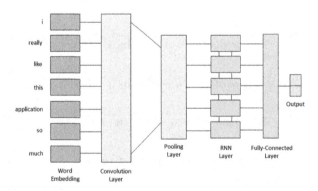

Fig. 4. The architecture of the combined CNN-RNN model [10]

sequence. The combined CNN-RNN model architecture refers to both CNN-LSTM and CNN-GRU models. Figure 4 shows an illustration of the combined CNN-RNN model's architecture.

3   Result and Discussion

3.1   Experimental Setup

After the data is being preprocessed, the following process is to implement each model using the prepared dataset to find the accuracy of each model. For each simulation, the dataset is divided into 80% training data and 20% testing data. Training data is used to build the model, and test data is used to find the model's accuracy. The training process will be done in 25 epochs and 32 batch sizes using the Adam optimization technique with a learning rate of 0.0001. The list of hyperparameters will be tuned to get the best accuracy in each simulation, and the hyperparameter's candidate can be seen in Table 4 [14].

After the data is being preprocessed, we implement the combined LSTM-CNN, CNN-LSTM, GRU-CNN, CNN-GRU model, and standard CNN, LSTM, GRU model with the optimized hyperparameter to get their accuracy. Then, compare and analyze the results.

Table 4.   List of hyperparameter and the candidate values.

Layer	Hyperparameter	Candidate values
Word embedding	Embedding size	64; 100; 128
RNN	Unit	100; 150; 200
	regularization L2 kernel	0.01; 0.001
	regularization L2 recurrent	0.01; 0.001
CNN	Region size	200; 250; 300
	Number of filters	3; 4; 5
	regularization L2	0.01; 0.001
Fully connected	regularization L2	0.01; 0.001

3.2   Results

Results accuracy of the combined and standard models are listed in Table 5. Almost all combined models have better performance than the standard models. Only the CNN-GRU model has a lower accuracy than the CNN model in Tokopedia and Lazada datasets. The LSTM-CNN, CNN-LSTM, GRU-CNN performed better than the standard models.

Looking through the results, it seems that our intuition made earlier was correct. Almost all of our combined models are higher than the standard models. The combined model improved the model implementation because each model's benefit complements the other. While CNN worked to find the local features, RNN managed to use its memory ability to improve performance.

Furthermore, using the RNN model at the beginning of the combined model and continued with CNN provides higher accuracy than the combined CNN-RNN models for the Tokopedia and Lazada datasets. It looks like the CNN-RNN models lost some of the sequence information. Suppose the order from the output of the convolutional layer doesn't give any information. In that case, the LSTM layer won't be able to use its ability and works the same as the fully connected layer. On the contrary, the RNN-CNN models give a better result because LSTM and GRU managed to find information in the present and the past. Then, CNN work to find the local features.

Table 5. Results of the combined and standard models.

Model	Tokopedia	Shopee	Lazada
LSTM-CNN	**83.03%**	82.12%	**82.99%**
CNN-LSTM	82.82%	**82.40%**	82.85%
GRU-CNN	82.98%	81.82%	82.88%
CNN-GRU	82.13%	82.30%	82.32%
LSTM	82.71%	82.08%	82.83%
GRU	81.93%	81.72%	82.29%
CNN	82.45%	81.60%	82.69%

4   Conclusion

In this study, we have experimented to find each model's best performance and analyze whether the combined models are better than the standard models. As a result, the average accuracy produced by the combined LSTM-CNN, CNN-LSTM, GRU-CNN, CNN-GRU model on the three datasets is 82.71%, 82.69%, 82.56%, and 82.25%, respectively. Almost all of the combined models have better accuracy than the standard models. Based on our combined models, only the combined CNN-GRU model is slightly below the CNN model but has slightly better accuracy than the GRU model. Meanwhile, the other three combined models obtain slightly better accuracy than the standard models. The limitations of each of the data sets and tools developed in the research study have indicated a recommendation for further work, namely to implement the models on more data sets to be able to see a better learning on the model.

References

1. Kemp, S., Moey, S.: Digital 2019 Spotlight: E-commerce in Indonesia, https://datareportal.com/reports/digital-2019-ecommerce-in-indonesia. Accessed 01 Oct 2019
2. Liu, B.: Sentiment analysis and opinion mining. In: Synthesis Lectures on Human Language Technologies. Morgan & Claypool Publishers (2012)

3. Murphy, K.: Machine Learning: A Probabilistic Perspective. MIT Press, London (2012)
4. Devika, M.D., Sunitha, C., Ganesh, A.: Sentiment analysis: a comparative study on different approaches. Procedia Comput. Sci. **87**, 44–49 (2016)
5. Mariel, W.C.F., Mariyah, S., Pramana, S.: Sentiment analysis: a comparison of deep learning neural network algorithm with SVM and Naïve Bayes for Indonesian text (2018)
6. Kim, Y.: Convolutional neural networks for sentence classification. In: Association for Computational Linguistics (ACL), Doha (2014)
7. Hassan, A., Mahmood, A.: Deep learning for sentence classification. In: IEEE Long Island Systems, Applications and Technology Conference (LISAT), New York (2017)
8. Biswas, S., Chadda, E., Ahmad, F.: Sentiment analysis with gated recurrent units. In: Advances in Computer Science and Information Technology (ACSIT), India, vol. 2, no. 11, pp. 59–63 (2015)
9. Wang, X., Jiang, W., Luo, Z.: Combination of convolutional and recurrent neural network for sentiment analysis of short texts. In: COLING 2016 - 26th International Conference on Computational Linguistics, Osaka, pp. 2428–2437 (2016)
10. Sosa, P.M.: Twitter sentiment analysis using combined LSTM-CNN models. In: Academia Edu, pp. 1–9 (2017)
11. Hochreiter, S., Schmidhuber, J.: Long short-term memory. Neural Comput. **9**(8), 1735–1780 (1997)
12. Zhang, A., Lipton, Z., Li, M., Smola, A.: Dive into Deep Learning (2020)
13. Cho, K., Merrienboer, B., van Bahdanau, D., Bengio, Y.: On the properties of neural machine translation: encoder-decoder approaches, pp. 103–111 (2015)
14. Wijaya, M.: Analisis Kinerja Modifikasi Model Gabungan Long Short-Term Memory dan Convolutional Neural Network untuk Analisis Sentimen Berbahasa Indonesia. Universitas Indonesia, Indonesia (2020)
15. Olah, C.: Understanding LSTM Networks. http://colah.github.io/posts/2015-08-Understanding-LSTMs/. Accessed 30 Oct 2019
16. Kostadinov, S.: Understanding GRU Networks. https://towardsdatascience.com/understanding-gru-networks-2ef37df6c9be. Accessed 31 Oct 2019

Machine Learning Based Biosignals Mental Stress Detection

Adel Ali Al-Jumaily[1,2(✉)] ⓘ, Nafisa Matin[3], and Azadeh Noori Hoshyar[4]

[1] School of Computing, Mathematics and Engineering, Charles Sturt University, Wagga Wagga, Australia
aal-jumaily@csu.edu.au, aal-jumaily@fbsu.edu.sa
[2] College of Computing, Fahad Bin Sultan University, Tabuk, Saudi Arabia
[3] School of Biomedical Engineering, University of Technology Sydney, Ultimo, Australia
Nafisa.Matin@student.uts.edu.au
[4] School of Engineering, IT and Physical Sciences, Federation University Australia, Brisbane, Australia
a.noorihoshyar@federation.edu.au

Abstract. Mental Stress can be defined as a normal physiological and biological reaction to an incident or a situation that makes a person feel challenged, troubled, or helpless. While dealing with stress, some changes occur in the biological function of a person, which results in a considerable change in some bio-signals such as, Electrocardiogram (ECG), Electromyography (EMG), Electrodermal Activity (EDA), respiratory rate. This paper aims to review the effect of mental stress on mental condition and health, the changes in biosignals as an indicator of the stress response and train a model to detect stressed states using the biosignals. This paper delivers a brief review of mental stress and biosignals correlation. It represents four Support Vector Machine (SVM) models trained with ECG and EMG features from an open access database based on task related stress. After performing comparative analysis on the four types of trained SVM models with chosen features, Gaussian Kernel SVM is selected as the best SVM model to detect mental stress which can predict the mental condition of a subject for a stressed and relaxed condition having an accuracy of 93.7%. These models can be investigated further with more biosignals and applied in practice, which will be beneficial for the physician.

Keywords: Mental stress · ECG · EMG · Machine learning · Support Vector Machine · Gaussian Kernel SVM

1 Introduction

Mental stress is usually defined as the disruption in regular emotional and psychological balance. When a person feels challenges to balance assigned tasks and his capability, the pressure of expectations, or any surrounding threat or danger, his physiological condition is triggered to deal with the situation, causing mental stress. In daily life, people face two types of stress, Eustress which can be considered as positive stress and distress [1]. Task-related acute mental stress is widespread in day-to-day life. Body functions

© Springer Nature Singapore Pte Ltd. 2021
A. Mohamed et al. (Eds.): SCDS 2021, CCIS 1489, pp. 28–41, 2021.
https://doi.org/10.1007/978-981-16-7334-4_3

are significantly affected to handle the stress response. Mental Stress distracts routine life by obstructing a person's social life, affecting educational and job performance and rationality, causing a severe health problem, and affecting one's competence. Long term acute stress can cause serious mental conditions and health issues. It can trigger quite a lot of health problems, specifically cardiovascular diseases [2]. The concept of stress is closely related to human life. To deal with mental stress, identifying stress inducers, body function related to the stress, the mental health condition to the stress response, and the review of biosignals linked to mental stress has significant importance. The biological signals show dissimilar characteristics in stressed and relaxed conditions. The most affected biosignals during stress response are Electrocardiogram (ECG), Electromyography (EMG), Electrodermal Activity (EDA), and respiratory rate. Recently the research interest about the mental condition and health relation to acute mental stress is overgrowing in the biomedical field. The most typical stress detection system is questionnaire-based which is time consuming and does not have precise accuracy. As various biosignals are significantly affected and change when the body functions try to deal with the stress response, it is possible to identify stress from biosignals. Many types of research are conducting to detect mental stress by applying various techniques. Machine learning is one of the most promising approaches to detect mental stress having more precise accuracy [3]. A model can be trained with features extracted from biosignals recorded during the relaxed and stressed conditions to classify relaxed and stressed by utilizing machine learning algorithms. This trained model will predict the relaxed or stressed state for new input features extracted from biosignals recorded from a person. The person can then receive any recommendation to seek a consultation with experts for further assessment for medication or take adoptive measures to reduce mental stress.

2 Literature Review on Stress and Its Effect

Mental stress is a very common incident in day to day life. It has a significant impact on the human body, mental stress, and biological function. If acute stress is overlooked for a long time, it triggers many health issues. Normally, people who experience stressed conditions more often or daily basis may have some physical issues because of prolonged stress without even realizing.

2.1 Concept of Stress

According to the published scientific articles, stress is identified as the usual reaction of a person's mental and physical state to danger or confrontation [4]. Throughout the history of human existence, every person has faced it. Because of the presence and complications of communal, individual, and environmental situations, numerous and instantaneous communication of human beings with adjoining concerns and assortment in stress representation mainly expand incidence and comprehensiveness of stress in human societies [5]. The relation between stressor and stress reaction is not common and different for individuals and not also persistent. It is different for different cases for a specific person. Rational evaluation activities control stress, and stressor is the main reason behind this [5].

Stress can be considered as a communicative medium between a human and the surrounding atmosphere [6]. Minimal amounts of stress could be preferred, beneficial, and even healthy for better performance. Stress, in its encouraging manner, can enhance the biopsychosocial condition and accelerate performance. Moreover, constructive stress is believed to be vital for inspiration, adaptation, and response to the situation. But elevated and excessive stress levels might cause physical, emotional, and social troubles and even severe damage to humans [7].

2.2 Stressors Maintaining and Integrity

The biochemical or organic mediator, surrounding situation, exterior provocation, or an incident triggering stress to an individual is known as a stressor [8]. According to psychological studies incidences, or situations that people may think challenging, difficult, and/or frightening personal well-being are described as a stressor [9]. Incidents or substances that can cause a stress reaction may contain the following issues: Ecological stressors (excessively high or low environmental temperatures, higher pitched sound, over-brightness, congestion); everyday "stress" incidents, Changes in lifestyle and demand, Workplace stressors (elevated work requirement vs. minimal job power, constant or prolonged physical exertion and effort, persuasive efforts, excessive stances, workplace chaos [10]), Chemical stressors (tobacco, alcohol, medications, drugs), Societal stressor (social and family pressures) and Traumatic stressor: remembrance of previous painful, shocking or distressing experience which affects perception and still influences the emotional condition of a human being.

According to numerous researchers, the stress reaction can be categorized as acute or chronic, considering the period of the stressor and the individual's adaptive mental and communicative handling approaches. Chronic stress describes as a specific type of stress which affects individuals every day autonomously of the occurrence of significant stressors and can last for years. Acute stress reactions have a time-constrained incident which defined by (i) the period of the provoking surrounding incident or circumstance, (ii) self-controlling physical (e.g., parasympathetic system stimulation), behavior (e.g., an effective handling approach is triggering mind refocusing), or emotional activities (like an evaluation of the stressor or effective control of the stress experience). Chronic stress can be considered as the consequence of frequent and recurring acute stressors occurring for a prolonged period. This paper concentrates on acute mental/emotional stress, which is task correlated.

2.3 Stress Effects on Emotion and Mental Health

In psychology, the term stress refers to a broad section of an adversely filled emotional condition like anxiety, bad temper, nervousness, resentment, fear, overstimulation, dissatisfaction, and sadness. However, some researchers validate that stress may positively affect the immune reaction while facing certain circumstances [11]. Effects on physical condition and functioning while facing stress can fluctuate, differing on the individual's approach to stress [12]. Therefore, stress response should not be considered as a specific hypothesis but as a phase including several emotional expressions. Even though having the dilemma of definition, stress is considered a condition of adverse valence and conclusive arousal.

The valence of emotional states related to experiencing stressors is vastly time related. While: stressor is anticipated and soon after facing a stressor, negative feelings and reactions are probable to be provoked. However, it may be possible to rapidly experience optimistic sentiments on stressor removal. While experiencing acute stress, a person can have the following emotional reaction: Anxiety, Disturbance, Irrelevance, Guiltiness, Anger, Hallucinations, Sensitivity, Unresponsiveness, Mood Swings, Lack of concentration, Uneasiness, Feeling insecure, Feeling anxious, Feeling depressed, Feeling useless, Defensiveness, Quick change in perception, Severe imagining, Presuming the most terrible, Lack of inspiration, Poor Memory, Stiffness, Intolerance, Feeling hopeless, Feeling deprived, Shortness of breath, Avoidance, Negligence, Tobacco addiction, Over intake of carbohydrate, Cravings, Bite Nails, Speak loudly and shout sometimes, Shuffling, Lack of energy, Oversleep or insomnia, Alcohol addiction [13].

2.4 Mental Stress Impact on Health

Stress has a significant effect on our physical condition and health. When the stress response is severe and acute stress becomes constant (prolonged), it quickly impacts an outburst that changes the body's mechanism and affects the abilities. To preserve our health, the immune function is affected when a person reacts to the stressor. People who feel stressed repeatedly are prone to have a weakened immune function. There is a close relationship between cardiovascular disease and acute mental stress [14]. The consequences of prolonged stress negatively impact the cardiac system, which is both stimulatory and inhibitory [15]. Several studies claimed that stress triggers autonomic nervous system stimulation, which indirectly modifies the cardiac functionality mechanism, which induces increase heart rate, high blood pressure, and, in severe cases, heart attack [16]. Stress can cause a change in respirational function [17]. In addition, mental stress may lead to the possible adoption of activities that are not safe for the heart, like addiction to tobacco, alcohol, drugs. Some research indicates that serious mental stress can cause unexpected and sudden death [18]. The impacts of stress related to the gastrointestinal (GI) function of the body can be reviewed through two parts of the GI system. Initially, stress can change in diet choice in food intake [19]. Furthermore, from various studies, dietary habits have special impacts on stress reaction [20], indicating a consensual relation among diet and response to stress. Researchers have proved that stress modifies the food intake procedure, abdominal absorbency, mucus, saliva, and intestinal acid excretion, ion channel activity, and stomach infection [21].

2.5 Biosignals Connected to Predict Acute Mental Stress

Biosignals can be defined as time-dependent parameters of a human's biological system and categorized into two major types [22].

- Physical signals
- Physiological signals

Physical biosignals are processes of body distortion because of muscle movement. This type of signal involves pupil diameter, eyeball movements, winks, skull, body, leg, arm partial intentional posture/activities, breathing, facial representations and expression, tone of speaking. Physiological indicators are straight associated with the dynamic biological system, like cardiovascular function (Electrocardiogram [ECG], brain activity (EEG), exocrine function (sweating evaluated via electrodermal activity [EDA])], Blood Volume Pulse, and muscle movement assessed through electromyography (EMG). This paper utilized ECG, heartbeat, and EMG signal analysis to detect acute mental stress. The main feature of ECG, which is affected during stress, is known as R-peaks. The distinctive peak points in ECG are symbolized by the letters P, Q, R, S, and T. The most important feature in ECG is the R peaks. These peaks are mainly used to analyze of ECG by identifying the location of this peak across consecutive R peaks periods, which is known as RR intervals (RRI) [23].

2.6 Machine Learning Approach to Detect Mental Stress

In the field of emotion recognition machine learning approach has received a lot of attention. Numerous published works are going on in this area. The resolution of machine learning approach is to create a model which ensure to perform a specified task. The consistent algorithm is selected according to the required task to be performed. Selecting algorithm is difficult as several algorithms are present.

With machine learning algorithm by using EEG signal the highest accuracy has been received as 99.9% and 99.26% in identifying stressed and non-stressed state [24]. From social media data depression detection by Multi Kernel SVM has been proposed where SVM shows 16.54% reduced error in identifying depressed people [25]. SVM is one of the best approaches for mental state recognition as it contains a regularization factor and feature ensuring better generalization aptitude to avoid over fitting. SVM can be used to determine both classification and regression task. It approaches an assurance on the test error rate. As a result, SVM model proposed a balanced model.

3 Data Collection

To train the model for identifying the stressed and relaxed state, a website called 'Physionet' has been chosen in this study. It is an open database for biosignals. The selected database is called "Stress recognition in automobile driver" [26]. The recordings contain physiological data while completing a real-world driving task. It refers to task-related acute mental stress. ECG and EMG signals are used in this study. These signals were continuously recorded throughout the whole process, where drivers were required to follow a specific road in the Boston region. Recordings from 16 healthy drivers, is used here to train the model.

In this databank, an investigational procedure was constructed and validated to recognize the recognition of stress because driving in severe road traffic situations forms biological signals of the healthy subjects. Corresponding to the proposed protocols, volunteers were driving a car while maintaining a determined direction, and their physiologic responses were observed by evaluating numerous documented biological signals such as Electrocardiogram (ECG), Electromyogram (EMG), skin conductivity, and respiration [27]. To take part in this test, drivers were expected to get a valid driver's license and to give permission to have video and physical signals documented throughout the whole experiment time. Everyone's driving license and consent were verified thoroughly. Prior to the starting of the experiment, a map for the selected driving lane was shown to the drivers and intended guidelines for keeping the drives stable were delivered. Directions were provided to follow speed restrictions and not to turn on the radio, music, or any kind of electronics. During the experiment, a spectator was present in the car with the driver to resolve any kind of issues, problem, or query from the driver, to observe physical signal reliability, and to observe driving actions in the video recorder. The attendant for the driver is required to be seated in the back seat crosswise behind the driver to prevent obstructing the natural performance of the driver. For every volunteer driver, the route and guidelines were the same. In every case, the whole experiment took 50–70 min. Each time for each driver two 15-min rest segments were followed at the starting and at the completion of the drive. Throughout these times the driver was being seated at the car in the garage having the eyes closed. After 15 min rest period, divers completed their drive on a preset route facing regular traffics and after completing the drive they came back to the garage and completed the resting period.

To record data, every participant was required to were electrode sensors, one for electrocardiogram (ECG) on the chest and one sensor for electromyogram (EMG) recordings, placed on the left shoulder. The car used for every recording was custom-made and the brand was the Volvo S70 chain station wagon. All sensors were linked to a FlexComp which is an analog-to-digital converter. This helped to keep the volunteer optically separated from the power source. The FlexComp attachment was attached to a computer in the car. The ECG electrodes were used to record ECG and were placed on the chest with a customized lead II alignment to reduce movement artifacts and to amplify the amplitude of the R-peaks, as R peaks are mostly used features to analyses the mental stressed condition. The EMG sensor was put on the trapezius muscle skin (shoulder), as according to numerous study the trapezius muscle is a significant indicator of the response to emotional stress [28]. Many feature extraction techniques had been proposed for different applications [29], In this paper, for extracting features and train the model, the first 15 min is considered as relaxed condition (while the drivers were required to sit closing their eyes) and after 5 min of completing relaxing period from the 20th minute of the experiments (5 min after starting the drive), next 15-min (20th–35th minute) is considered as task-related stressed (acute) condition. The sampling frequency for our data is 496 Hz.

4   Methodology

After selecting the data, we started work on training the model. Specific steps were followed in this method. They are: importing data, preprocessing, and filtering, select features from the signal, extract the features, and train the model with the features. Every phase was done in MATLAB (see Fig. 1).

Fig. 1. Training model stages

4.1 Filtering and Preprocessing

To work with biosignals preprocessing and filtering are needed to eliminate noises. We have done reprocessing for both ECG and EMG signals. ECG has a low-frequency noise in recorded ECG defined as baseline wander. To eliminate the baseline wander and another noise a bandpass filter with cutoff frequency. 5 Hz–200 Hz is applied. DC offset is also removed from the recordings. As a bandpass filter Chebyshev filter type II is used. It holds no ripple at the passband and has equiripple at the stopband. The gain of this filter is:

$$G_n(\omega, \omega_0) = \frac{1}{\sqrt{(1 + \frac{1}{\varepsilon^2 T_n^2(\frac{\omega_0}{\omega})})}}$$

After that, to eliminate the powerline interference a notch filter at 50 Hz was applied. The standard formula for notch filter can be described as:

$$H(s) = \frac{s^2 + \omega_0^2}{s^2 + \omega_c s + \omega_0^2}$$

Here $\omega 0$ is for the main rejected frequency and ωc is the rejected band width.

The main noise in the EMG signal is motion artifact. The most functional and important EMG signal features are prominent between the frequency band of 50–150 Hz. To eliminate the powerline interference a notch filter at 50 Hz was applied. For filtering the EMG signal a bandpass filter with cutoff frequencies of 20 Hz to 160 Hz is applied.

4.2 Feature Extraction

For feature extraction in ECG, we used time domain analysis. We have calculated the R peaks. The mean RR interval, mean Heart Rate, Standard Deviation of RR interval, and coefficient of variance for RR interval are calculated as ECG features.

For EMG feature extraction time domain analysis has been done. The root mean square (RMS) value and the mean absolute value of the EMG signal have been calculated. The extracted features are presented in Table 1.

Table 1. Extracted features.

1	mean_R	CV_RR	SD_RR	mean_H	RMS_EI	MAVem	Condition
2	0.7367	0.4545	0.2308	81	0.0432	0.0197	Relaxed
3	0.8448	0.3041	0.2569	71	0.0154	0.0099	Relaxed
4	0.742	0.4207	0.2995	73	0.0075	0.0053	Relaxed
5	0.7417	0.3551	0.2634	80	0.0332	0.0097	Relaxed
6	0.6634	0.2642	0.1753	90	0.0372	0.0167	Relaxed
7	0.7212	0.2768	0.197	74	0.0071	0.0021	Relaxed
8	0.8058	0.3082	0.2437	70	0.0154	0.0101	Relaxed
9	0.7752	0.3146	0.2417	67	0.0611	0.0114	Relaxed
10	0.7243	0.2872	0.208	81	0.024	0.0125	Relaxed
11	0.8023	0.336	0.2696	69	0.0143	0.0089	Relaxed
12	0.7285	0.2942	0.2003	72	0.0191	0.0123	Relaxed
13	0.6598	0.2575	0.1699	81	0.0065	0.0043	Relaxed
14	0.6598	0.2575	0.1699	82	0.0066	0.0063	Relaxed
15	0.7716	0.317	0.2446	72	0.0459	0.0087	Relaxed
16	0.625	0.2809	0.1694	76	0.0116	0.0057	Relaxed
17	0.7518	0.2814	0.1815	72	0.0566	0.0316	Relaxed
18	0.5077	0.4207	0.2258	118	0.1731	0.0985	Stressed
19	0.7889	0.0717	0.0566	78	0.0157	0.0195	Stressed
20	0.72	0.3703	0.2683	83	0.0528	0.0159	Stressed
21	0.6998	0.3281	0.2296	86	0.1531	0.0685	Stressed
22	0.5999	0.1874	0.1124	100	0.098	0.0422	Stressed
23	0.7119	0.205	0.1479	84	0.0673	0.0128	Stressed
24	0.7906	0.2286	0.1842	76	0.0162	0.0102	Stressed
25	0.7685	0.2284	0.1771	78	0.0395	0.0432	Stressed
26	0.7045	0.2201	0.1551	86	0.1087	0.0317	Stressed
27	0.7829	0.2315	0.2282	78	0.0146	0.0095	Stressed
28	0.6811	0.2327	0.1695	89	0.0315	0.0141	Stressed
29	0.6001	0.21	0.126	100	0.0507	0.0139	Stressed
30	0.5696	0.1603	0.0913	106	0.0527	0.0159	Stressed
31	0.6418	0.2789	0.2069	88	0.159	0.0754	Stressed
32	0.5031	0.2599	0.1624	100	0.0764	0.0326	Stressed
33	0.6433	0.2784	0.181	93	0.1036	0.0355	Stressed

4.3 Modeling

The modelling part is done in MATLAB. For training, the model Supervised Machine Learning is applied to classify stressed and relaxed states. We got a binary class label (two class types) for our extracted features (relaxed and stressed).

One of the most current supervised machine learning methods is Support Vector Machines (SVM) model. This is defined as a learning algorithm that evaluates the information applied for classification and regression evaluation. Here the model is provided with a set of training data, each of which is graded as fitting to the predefined classifications. SVM training algorithm allocates new samples to one class or to the other class. An SVM model interprets of the sample features as places in space, which is charted for the samples of the individual classifications, that are separated by a clear difference. This gap is as broad as conceivable. New sample features are then plotted into that similar area and predicted to fit into a class by focusing on the edge of the gap at which they drop. SVM algorithm has four main benefits:

- SVM includes a regularization factor and feature, which ensure a good generalization ability to prevent overfitting.
- SVM applies the kernel trick. It effectively deals with non-linear data set.
- SVM can be applied to resolve both classification and regression assignments.

- The SVM model is balanced as it approximates a guarantee on the test error rate.

For training, the model to detect relaxed and normal conditions, the three approaches have been applied: Linear Kernel SVM, Gaussian Kernel SVM, and Sigmoid Kernel SVM.

Linear Kernel: The most uncomplicated approach is SVM is linear kernel function. It is presented with the inside result x and y along with a noncompulsory constant c.

$$k(x, y) = x^T y + c$$

Gaussian kernel: It is an implementation of radial basis function kernel (RBF).

$$k(x, y) = \exp(-\frac{\|x - y\|^2}{2\sigma^2})$$

On the Other Hand, it can also be applied utilizing

$$k(x, y) = \exp(-\gamma \|x - y\|^2)$$

The modifiable factor σ performs a key part in the execution of the kernel, and it must be thoroughly modified toward the application. If it is miscalculated, the exponential will perform nearly as linear, and the upper-dimensional prediction will begin to miss its non-linear function.

Sigmoid Kernel: The Sigmoid Kernel also is known as Multilayer Perceptron (MLP) kernel. The Neural Networks field gives the idea of the Sigmoid Kernel

$$k(x, y) = \tanh(\alpha x^T y + c)$$

Sigmoid Kernel has two modifiable factors. One is the slope α and the other one is cut off constant c. Two values (0.5 and 1) for alpha is applied in this study to train model. Cross validation is performed for evaluating the outcomes of a generalized statistical evaluation to an individual data set. It includes splitting the experiment data into corresponding subsections, while performing on one subsection and keep the other subset for validation. Cross validation relates the processed of fitness during prediction for deriving a more precise assessment of the prediction performance of the trained model. For each model cross validation is done with an 80% hold out. Hold out is applied for splitting up the dataset to a 'train' and 'test' set. To train the model the training set is used, and the test set is applied to observe the performance of trained model for unseen data. Our training set is 80% and testing set is 20%.

5 Result

We have trained four models. We have labeled relaxed as a positive class and stressed condition as a negative class as we have binary classification. For the performance evaluation, we have calculated the true positive (TP), False Positive (FP), True Negative (TN), and False Negative (FN).

True Positive: The model predicted the relaxed condition, and it is true.

False Positive: The model predicted that the subject's features as a relaxed condition, but it is not true. The model predicted the stressed condition as a relaxed condition.

True Negative: The model predicted that the subject is in a stressed condition, and it is true.

False Negative: The model predicted that the subject is in a stressed condition, and it is not true.

We have used a confusion matrix to plot the result. To evaluate the implementation of a classification model, a Confusion matrix is generally used. It is a N x N matrix, where the number of target classes is presented by N. This matrix relates the real target values with the predicted values that demonstrate the classification model's performance and error. Figures 2, 3, 4, 5 show the confusion matrix for our trained models. While the class prediction performances for all models are given in Table 2.

To compare all the models at both training and testing sessions, the accuracy rate, error rate, specificity, and Sensitivity is calculated. The most excellent sensitivity for a model is 1.0, while the worst is 0. The top specificity for a model is 1, while the worst is 0. The formula to calculate accuracy rate, error rate, specificity, sensitivity is given below:

Accuracy: $(TP + TN)/(TP + TN + FN + FP) \times 100$
Sensitivity $= TP/(TP + FN)$
Specificity $= TN/(TN + FP)$
Error Rate $= (FP + FN)/(TP + TN + FN + FP)$

The calculated accuracy rate, error rate, specificity, sensitivity for all four models are in Table 3.

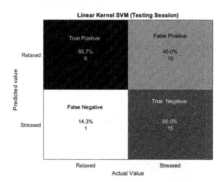

Fig. 2. Confusion Matrix for Linear Kernel SVM

From the calculation given in Table 3, we can see Linear model gives the lowest accuracy, and Gaussian Kernel Model Gives the highest accuracy. Evaluating performance (based on accuracy percentage, Error percentage, Sensitivity, and Specificity) for all four trained models, we get the best performance for the Gaussian Kernel model at the testing session. At the testing phase, the Gaussian model has an accuracy of 93.7% to predict the condition. It has an error rate of 6.3%, Sensitivity of 0.93, and specificity

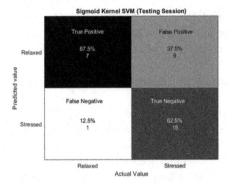

Fig. 3. Confusion Matrix for Sigmoid Kernel ($\alpha = 1$) SVM

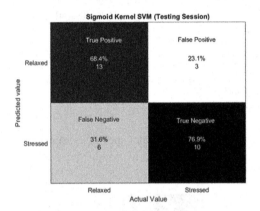

Fig. 4. Confusion Matrix for Sigmoid Kernel ($\alpha = 0.5$) SVM

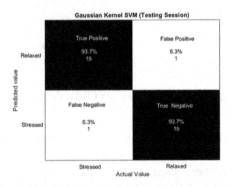

Fig. 5. Confusion Matrix for Gaussian Kernel SVM

Table 2. Class prediction percentage for all four models

SVM type	True Positive (TP)	False Positive (FP)	False Negative (FP)	True Negative (FP)
Linear Kernel	85.7%	40%	14.3%	60%
Sigmoid Kernel ($\alpha = 1$)	87.5%	37.5%	12.5%	62.5%
Sigmoid Kernel ($\alpha = 0.5$)	82.4%	13.3%	17.6%	86.7%
Gaussian Kernel	93.7%	6.3%	6.3%	93.7%

Table 3. Performance of four trained model

SVM type	Accuracy	Error rate	Sensitivity	Specificity
Linear Kernel	72.85%	27.15%	0.857	0.60
Sigmoid Kernel ($\alpha = 1$)	75%	25%	0.875	0.625
Sigmoid Kernel ($\alpha = 0.5$)	84.55%	15.45%	0.824	0.867
Gaussian Kernel	93.7%	0.63%	0.93%	0.93

of 0.93. This model can be used for further prediction. The confusion matrix of the Gaussian kernel model has given here.

Most research works and published papers regarding stress recognition mainly utilize the task sessions replicated in the laboratory atmosphere [30]. Sometimes it cannot completely replicate real life tasks, pressure, and tension. Also, people can react differently in the laboratory setup and real life situations. This paper presents a model where utilized data is recorded while driving on busy roads, which is a real-life task-related stress. The Gaussian Kernel Model gave the highest accuracy of 93.7% to predict task related stress and non-stressed state.

6   Conclusion

A Machine learning based acute (task related) mental stress detection system from bio-signals has been developed and analyzed in this research paper. The investigation of various parameters related to mental stress recognition has been undertaken. The general concept of mental stress, stress inducers, and the impact of the stress on mental and physical health has been reviewed. The relation between acute stress and biosignals has also been studied and reviewed with available literature. Based on the literature review, two types of biosignals – EMG and ECG- have been applied to train models that can predict the stressed and relaxed condition. Using an open access database, four types of models have been trained. Among all the four models, the Linear Kernel SVM model provides the lowest accuracy of 72.85%, and the Gaussian Kernel SVM model demonstrates the best performance having an accuracy of 93.7%. This work can be further extended for real time data of various biosignals under different mental health conditions to improve accuracy.

References

1. Fevre, M.L., Kolt, G.S., Matheny, J.: Eustress, distress and their interpretation in primary and secondary occupational stress management interventions: which way first? J. Manag. Psychol. **21**(6), 547–565 (2006)
2. Das, S., O'Keefe, J.H.: Behavioral cardiology: recognizing and addressing the profound impact of psychosocial stress on cardiovascular health. Curr. Atheroscler. Rep. **8**, 111 (2006)
3. Khalil, R.M., Al-Jumaily, A.: Machine learning based prediction of depression among type 2 diabetic patients. In: 2017 12th International Conference on Intelligent Systems and Knowledge Engineering (ISKE), Nanjing, pp. 1–5 (2017). https://doi.org/10.1109/ISKE.2017.825 8766
4. Cannon, W.B.: The wisdom of the body (1992)
5. Edwards, J.A., Webster, S., Van Laar, D., Easton, S.: Psychometric analysis of the UK Health and Safety Executive's Management Standards work-related stress Indicator Tool. Work Stress **22**(2), 96–107 (2008)
6. Lazarus, R.S.: Emotion and adaptation (1991)
7. Tucker, J.S., Sinclair, R.R., Mohr, C.D., Adler, A.B., Thomas, J.L., Salvi, A.D.: A temporal investigation of the direct, interactive, and reverse relations between demand and control and affective strain. Work Stress **22**(2), 81–95 (2008)
8. Sato, T., et al.: Restraint stress alters the duodenal expression of genes important for lipid metabolism in rat. Toxicology **227**(3), 248–261 (2006)
9. Deckers, L.: Motivation Biological, Psychological, and Environmental, pp. 208–212. Routledge, New York (2018)
10. Roster, C.A., Ferrari, J.R.: Does work stress lead to office clutter, and how? Mediating influences of emotional exhaustion and indecision. Environ. Behav. (2019)
11. Dhabhar, F.S.: Immune function, stress-induced enhancement. Encyclopedia Stress **2**, 455–461 (2007)
12. Crum, A.J., Salovey, P., Achor, S.: Rethinking stress: the role of mindsets in determining the stress response. J. Pers. Soc. Psychol. **104**(4), 716–733 (2013)
13. Greenberg, N., Carr, J.A., Summers, C.H.: Causes and consequences of stress. Integr. Comp. Biol. **42**(3), 508–516 (2002)
14. Rozanski, A., Blumenthal, J.A., Kaplan, J.: Impact of psychological factors on the pathogenesis of cardiovascular disease and implications for therapy. Circulation **99**, 2192–2217 (1999)
15. Engler, M.B., Engler, M.M.: Assessment of the cardiovascular effects of stress. J. Cardiovasc. Nurs. **10**, 51–63 (1995)
16. Vrijkotte, T.G.M., van Doornen, L.J.P., de Geus, E.J.C.: Effects of work stress on ambulatory blood pressure, heart rate, and heart rate variability. Hypertension **35**(4), 880–886 (2000)
17. Kreibig, S.: Autonomic nervous system activity in emotion: a review. Biol. Psychol. **84**(3), 394–421 (2010)
18. Pignalberi, C., Ricci, R., Santini, M.: Psychological stress and sudden death. Ital. Heart J. (Suppl.) **3**, 1011–1021 (2002)
19. Bagheri Nikoo, G., et al.: Effects of systemic and intra-accumbal memantine administration on the impacts of plantar electrical shock in male NMRI mice. Physiol Pharmacol. **18**(1), 61–71 (2014)

20. Yaribeygi, H., et al.: The impact of stress on body function: a review. EXCLI J. **16**, 1057–1072 (2017)
21. Ghanbari, Z., Khosravi, M., Hoseini Namvar, F., Zarrin Ehteram, B., Sarahian, N., Sahraei, H.: Effect of intermittent feeding on metabolic symptoms of chronic stress in female NMRI mice. Iran South Med. J. **18**(5), 982–999 (2015)
22. Everly, G.S., Lating, J.M.: The anatomy and physiology of the human stress response. In: Everly, G.S., Lating, J.M. (eds.) A Clinical Guide to the Treatment of the Human Stress Response, pp. 17–51. Springer, New York (2013). https://doi.org/10.1007/978-1-4614-553 8-7_2
23. Kaniusas, E.: Fundamentals of biosignals. In: Kaniusas, E. (ed.) Biomedical Signals and Sensors I, pp. 1–26. Springer, Heidelberg (2012). https://doi.org/10.1007/978-3-642-248 43-6_1
24. Attallah, O.: An effective mental stress state detection and evaluation system using minimum number of frontal brain electrodes. Diagnostics (2020)
25. Peng, Z., Hu, Q., Dang, J.: Multi-kernel SVM based depression recognition using social media data. Int. J. Mach. Learn. Cybern. **10**(1), 43–57 (2017). https://doi.org/10.1007/s13042-017-0697-1
26. https://www.physionet.org/content/drivedb/1.0.0/. Accessed 08 Sept 2021
27. Healey, J.A., Picard, R.W.: Detecting stress during real-world driving tasks using physiological sensors. IEEE Trans. Intell. Transp. Syst. **6**(2), 156–166 (2005)
28. Dumitru, V.M., Cozman, D.: The relationship between stress and personality factors. Hum. Vet. Med. **4**, 34–39 (2012)
29. Khushaba, R.N., Al-Ani, A., Al-Jumaily, A.: Feature subset selection using differential evolution. In: Köppen, M., Kasabov, N., Coghill, G. (eds.) ICONIP 2008. LNCS, vol. 5506, pp. 103–110. Springer, Heidelberg (2009). https://doi.org/10.1007/978-3-642-02490-0_13
30. Němcová, A., et al.: Multimodal features for detection of driver stress and fatigue: review. IEEE Trans. Intell. Transp. Syst. **22**(6), 3214–3233 (2021)

Sentences Prediction Based on Automatic Lip-Reading Detection with Deep Learning Convolutional Neural Networks Using Video-Based Features

Khalid Mahboob$^{(\boxtimes)}$ ⓘ, Hafsa Nizami, Fayyaz Ali, and Farrukh Alvi

Department of Software Engineering, Sir Syed University of Engineering and Technology, Karachi, Pakistan
kmahboob@ssuet.edu.pk

Abstract. Lip-reading is the process of deciphering text from a speaker's visual interpretation of facial, lip, and mouth movements without using audio. The challenge is traditionally divided into two stages: creating or learning visual characteristics and prediction. End-to-end techniques for deep lip-reading have been popular in recent years. Existing work on end-to-end models, on the other hand, only does word classification rather than sentence-level sequence prediction. Longer words improve human lip-reading ability, suggesting the relevance of characteristics that capture the temporal context in an inconsistent communication channel. In this study, an end-to-end model based on deep learning convolutional neural network shave been employed to develop an automated lip-reading system that uses a re-current network spatiotemporal convolutions, and the connectionist temporal classification loss to translate a variable-length series of video frames to text. The accuracy of the trained lip-reading process in predicting sentences was evaluated using video-based features.

Keywords: Lip-reading · Convolutional neural networks · Model

1 Introduction

With their remarkable potential to extract patterns and detect trends from difficult or imprecise data, convolutional neural networks may be used to extract patterns and detect too complex trends for humans or other computer approaches to discover. A trained convolutional neural network can be considered an expert in the category of data it is assigned to examine [1]. LIP-READING, one of the simplest techniques to identify speech, uses a convolutional neural network. It is a relatively new approach for voice recognition that is frequently used. A lip-reading method can be described that takes into account both the shape of the lips and the intensity of the mouth area [2].

© Springer Nature Singapore Pte Ltd. 2021
A. Mohamed et al. (Eds.): SCDS 2021, CCIS 1489, pp. 42–53, 2021.
https://doi.org/10.1007/978-981-16-7334-4_4

This is also a technique for identifying the contents of speeches using just visual data rather than audible data. It is intended to be applied in situations where speech recognition technology is difficult to use, such as in high-noise conditions to gather audio data in a public area where producing a voice is difficult, and to assist people with hearing and speech impairments in communicating [3]. Despite the fact that lip-reading has been explored for decades, there are several obstacles in lip-reading tasks, including the identification target: singular sound, separated word, constant word, and phrase, accessible modalities: audible or only-visual, face orientation, and language [4].

As discussed earlier, lip-reading is the process of extracting visual speech characteristics from a person's lips. The inner and outer lip contours carry the greatest visual speech information, but it has also been discovered that information regarding the appearance of teeth and tongue gives crucial speech signals. The disparity in pronunciation and relative accent of words and phrases is due to the variety of languages spoken throughout the world. Developing a software program that interprets spoken words automatically and properly based simply on the speaker's visual lip movement becomes extremely difficult [5].

Lip-reading methods have improved significantly as a result of the advent of deep learning. The 3D-CNN (usually a 3D convolutional layer tracked by a deep 2D Convolutional Network) has recently gained popularity as a preferred front-end option. CNN, which comprises alternating convolutional and pooling layers, has been an effective model for extracting visual information for image identification and classification applications. The inner product of the linear filter and the receptive field is computed by the convolutional layers, which are then followed by a non-linear activation function (e.g. ReLu, sigmoid, tanh, etc.) [6].

The CNN's output features are input into two Bidirectional Gated Recurrent Network (GRU) layers, which are then followed by a linear transformation and a softmax over the corpus through each time step (which is a character-based interpretation in this situation). A Connectionist Temporal Classification (CTC) network with a softmax output layer with many labels in the vocabulary with one blank character is used to train this end-to-end model. The CTC determines the probability of all possible string combinations [6, 7].

The purpose of this study is to propose a system that will help disabled people that cannot hear properly so that they can get the lip-synch through CNN that will generate text for them so that they can have their conversation properly. The system features a deep learning model for lip-reading detection that will consider the video as an input and generate text as the description. The main objective of this study is to predict sentence-level sequences based on automatic lip-reading detection using a convolutional neural network with video-based features [14]. Several model hyperparameters, such as picture size, filter size, number of filters, activation function, types, and layer layout, number of layers, were optimized to find the best architecture for CNN [15]. For imaging data classification, CNN is recognized as more effective and efficient [16].

The paper is organized as follows: Sect. 2 offered a comprehensive literature review. The pre-processing and methodology are stated in Sect. 3. Section 4 offered brief results and discussion on an exploration followed by the last section i.e., Sect. 5 with a conclusion.

2 Literature Review

The visual recognition system proposed in this study reads the gestures made by the lips of the user and tries to guess the word that is being said by the user. There is a scarcity of such tools which could be used in the generation of the content using visual lip gestures which could be helpful for people with speech and auditory disabilities. It is a difficult task to develop a system that could read the lip expression of the user and translate it in a language that the disabled person best understands.

Vaishali A. Kherdekar et al. [1] have presented the experiment for speech recognition of mathematical phrases which is useful to people with disabilities. The CNN model is used to increase the recognition accuracy. The researchers have chosen 17 mathematical terms that are often used in mathematical expressions. Because of its speed, the Rectified Linear Unit Activation Function is utilized to train CNN. For Adam and Adagrad optimizers, this study examines the model for MFCC and Delta MFCC features. The results demonstrate that for both Adam and Adagrad optimizers, Delta MFCC provides an accuracy of 83.33 percent. It shows that Delta MFCC is superior to MFCC in terms of outcomes. Adagrad with Delta MFCC trains the model earlier than Adam, according to the results.

Densely Connected Temporal Convolutional Network (DC-TCN) was introduced by Pingchuan Mal et al. [2] to recognize people's isolated words. To capture more strong temporal characteristics, they added dense connections to the network. To improve the model's classification power, the technique employs the Squeeze and-Excitation block, a lightweight attention mechanism.

Tasuya Shirakata et al. [3] used a technique for identifying what individuals are saying using just visual data rather than auditory data. It is useful for handicapped persons and patients. It refers to the location where the noise problem arises. In the lip reading approach, it incorporates facial expression elements such as expression-based feature and action-based unit feature. Experiments were carried out in the OuluVS, CUAVE, and CENSREC-1-AV public databases.

Karan Shrestha [5] used lip-reading techniques to detect the visual interpretation based on the movement of the face, mouth, and lips without the need for audible data. To predict real words, CNN architectures are implemented. To train the robust lip-reading system, the entire dataset is utilized in the LRW dataset. In addition, the lightweight architecture may be advanced.

Denis Ivanko et al. [7] analyzed the impact of several audiovisual fusion methods on word recognition accuracy rates in English and Russian (on GRID and HAVRUS corpora, respectively). GMMCHMM, DNN-HMM, and end-to-end techniques were examined as tree audiovisual modalities integration strategies. The conventional GMM-CHMM method produced the best recognition results. The GRID dataset has demonstrated that NN-based approaches are superior to the old GMM-CHMM approach to audiovisual speech recognition in all experiments.

M.Rahman et al. [8] suggest an automated speech recognition system built on Support Vector Machine (SVM) along with the help of dynamic time warping (DTW) for Bengali speaking individuals. The data was collected from 40 Bangla-speaking individuals for five different Bengali words with minimal noise environment and acoustics with high amplitudes. DTW was used after determining the feature vectors. A reliable model was proposed that was tested with 12 speakers with the recognition of 86.08%. The only apparent setback for their proposed system was that this system was specifically developed for Bengali speech recognition and cannot be generalized.

Garg et al. [9] proposed a combination of the CNN and LSTM deep learning methods and applied it to the lip gestures data collection problem. CNN in this scenario was used for feature extraction (reading lip gestures) and LSTM for classification.

Because CNN contains characteristics such as dimensionality reduction and parameter sharing, it outperforms other machine learning algorithms. The number of parameters in CNN is decreased as a result of parameter sharing, and therefore the computations are also reduced. Further, when CNN's feature extraction grows deeper, it delivers improved image identification (encompasses additional layers). That is why CNN is selected for automatic lip-reading in this study.

3   Pre-processing and Methodology

The system we developed based on CNN, takes a video as an input and predicts the speaker's lip-synch which is different from the study in [10] as they proposed the lip images feature. Because the model is trained on the grid dataset, it will predict data from the grid dataset. The model has three spatiotemporal convolutions, channel-wise dropout, and spatial max-pooling layers, as well as two Bi-GRUs layers, all of which include rectified linear unit (ReLU) activation functions and a softmax activation function for sequence classification. The overall flow of the system processing is shown in Fig. 1.

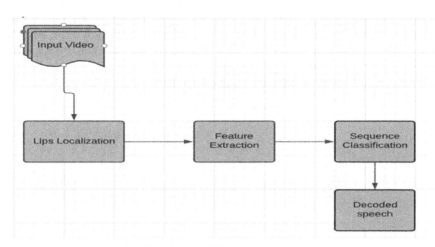

Fig. 1. Overall flow of the process

The desktop PC on which the model runs is the hardware component; the operating system is the software component; the operating system we chose is Linux (Ubuntu). And the video should be in mpg format, according to the model. The user may produce a video on any device, convert it to mpg format, and then execute a program on Linux. The system is not user-friendly since the user must run a command to make a prediction, and the model does not operate in real-time.

The data is in the form of a video file with the extension.mpg, which will be utilized as an input. The video will be processed by the model's learned weights, and the model will predict the speaker's phrase based on the grid data set corpus. Pre-processing is a crucial stage in any deep learning or machine learning system. The mouth has been cropped to a size of 100×50 pixels every frame as part of the preprocessing. To get a zero mean and unit variance, the RGB channels normalized throughout the whole training set. To minimize overfitting, we have added basic transformations to the dataset.

Keras comes pre-loaded with many algorithms designed particularly for deep learning. Different layers exist in models. We have used an input layer that accepts input in the form of img_frames, img_width, img_height, and img_channel if the image data format is Theano will take precedence. For padding zeros in three dimensions, we have used the ZeroPadding3D layer. Because the data is in video format, we have used Conv3D layers and the rectified linear unit as the activation function, which makes a positive input equal and a negative input zero, and is the usual activation function for deep learning applications. The Batch Normalization layer has been used to normalize the input [11].

We have utilized Spatial Dropout3D layers to keep the model from being too tight. We have also developed MaxPooling3D layers, which will employ three-dimensional max pooling to discover the most features. As an input to the wrapper layer, we applied Time Distributed wrapper layers with flattening layers. We have also used BiDirectional wrapper layers with GRU layers as wrapper inputs. BiDirectional wrapper layers help GRU layers train quicker and more efficiently. For sequence-wise classification, we employed a dense layer with a softmax activation function.

The output data is a video with sentences predicted in it, with the predicted sentence written on the console and as subtitles. The grid corpus data set is implemented for training the model, which consists of 34 speakers telling 1000 words each. NumPy, Keras, and Tensor Flow are some of the libraries, used for face detection. Python is the programming language that is used here, and all of the libraries are written in Python. The entire system flow architecture is illustrated in Fig. 2.

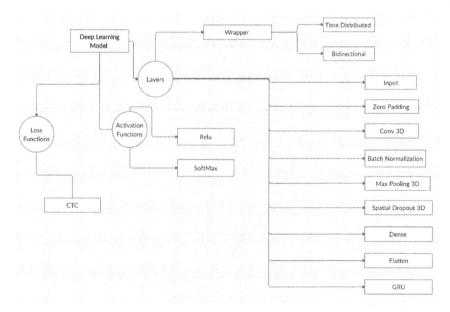

Fig. 2. System flow architecture

4   Results and Discussion

As a dataset, the GRID corpus was employed. It includes videos of each of the 34 speakers speaking 1000 sentences (17 male, 16 female). The sentences are structured in the following way: command + color + preposition + letter + digit + adverb. Bin, lay, place, and the set is among the commands, and the colors are blue, green, red, and white. At, by, in, and with are the five prepositions used, and all Latin letters except W are used. The digits range from 0 to 9, and the adjectives are: again, now, please, soon. 'bin white at f zero again' is an example statement from the dataset. Transcription of the words said, as well as information on when each word is pronounced, is included with each video [12]. The architecture of the proposed convolutional neural network (CNN) is shown in Fig. 3.

5D vector input

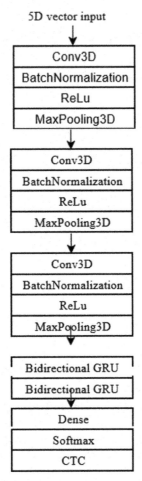

Fig. 3. The architecture of the proposed convolutional neural network (CNN)

Each video in the dataset is divided into 75 frames to prepare it for training. Each frame is then evaluated with the Face Recognition API for Python, which is based on dlib's deep learning-based face recognition. It calculates the (x, y) coordinates of 67 facial landmarks, such as the eyes, nose, mouth, and chin, as illustrated in Fig. 4. The speaker's mouth coordinates are recorded, while the rest are deleted. The mouth is then normalized to the smallest (x, y) coordinates feasible [13].

Fig. 4. The 67 Facial landmarks identified with dlib's facial recognition

Because letters may be missing or due to other problems, the CTC layer's output frequently requires processing. The output was run through a spelling algorithm that searched up words in the dictionary and picked the one with the shortest Levenshtein distance, or the term with the most occurrences in the training labels if there were many possibilities. With kenLM and ARPA-format, a 3-g model was created and implemented in the model assessment procedure. Following the spelling process, the CTC output was routed to the LM. Speech recognition accuracy can be improved by using a mix of CTC and an N-gram LM. Different speakers lip-synching a sentence without audio is depicted in Fig. 5.

The ARPA model was created using a text file containing all GRID corpus sentences. The probability of a word was provided in log10, and a term that was not in the lexicon was assigned an absolute penalty of -100 probability because the model did not know the correct probability of that word. The sequence of sentences with the accuracy achieved is shown in Table 1 and Fig. 6. It is important to note that from the experiment results, Letter {A − Z/a − z} has achieved the highest accuracy i.e. 79% whereas Digit {0–9} has achieved the least accuracy i.e. 44.5%.

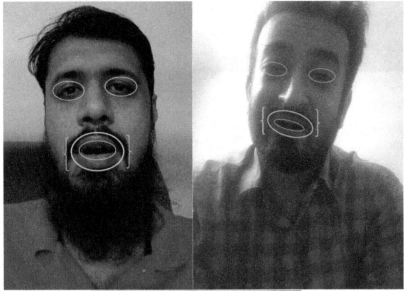

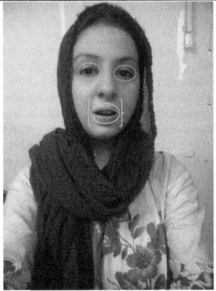

Fig. 5. Three speakers lip-synching a sentence

Table 1. Structure of the sentences with their Accuracies

Sentences	Accuracy
Command	74.00%
Color	64.00%
Preposition	66.15%
Letter	79.00%
Digit	44.50%
Adverb	76.00%

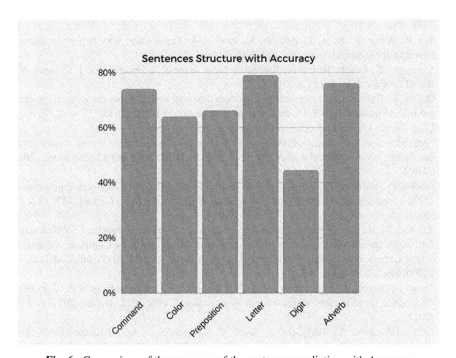

Fig. 6. Comparison of the sequence of the sentences prediction with Accuracy

5   Conclusion

The major goal of this study was to look into the possibility of employing facial landmarks based on deep learning convolutional neural networks to train a model for lip-reading with video-based features. The facial landmark representation proposed in a study captured sufficient information about the speaker. However, it appears that this depiction does not include all of the essential characteristics for lip-reading that may be identified in videos. Except for the lips, the present technique for extracting the speaker's facial features is unable to recognize any aspects of the mouth. The capacity to track teeth and tongue, based on the results and discussion, would most certainly improve the model's ability to discriminate between visages and therefore improve accuracy.

Because various resolutions result in varying distances between the coordinates, normalizing the distance between the coordinates would reduce the disparities between videos and therefore improve the model's generalization ability. Additional specialized landmark extractions perhaps capture more critical characteristics, such as teeth and tongue, to increase the accuracy of the suggested model.

References

1. Kherdekar, V.A., Id, E., Naik, S.A.: Convolution neural network model for recognition of speech for words used in mathematical expression. Turk. J. Comput. Math. Educ. **12**(6), 4034–4042 (2021)
2. Ma, P., Wang, Y., Shen, J., Petridis, S., Pantic, M.: Lip-reading with densely connected temporal convolutional networks, pp. 2857–2866 (2021)
3. Shirakata, T., Saitoh, T.: Lip reading using facial expression features. Int. J. Comput. Vis. Signal Process. **1**(1), 9–15 (2020)
4. Ozcan, T., Basturk, A.: Lip reading using convolutional neural networks with and without pre-trained models. Balk J. Electr. Comput. Eng. (July), 195–201 (2019)
5. Shrestha, K.: Lip reading using neural network and deep learning
6. Fernandez-Lopez, A., Sukno, F.M.: Survey on automatic lip-reading in the era of deep learning. Image Vis. Comput. [Internet] **78**, 53–72 (2018). https://doi.org/10.1016/j.imavis.2018.07.002
7. Ivanko, D., Ryumin, D., Karpov, A.: An experimental analysis of different approaches to audio–visual speech recognition and lip-reading. Smart Innov. Syst. Technol. **187**, 197–209 (2021)
8. Rahman, M.M., Roy Dipta, D., Hasan, M.M.: Dynamic time warping assisted SVM classifier for bangla speech recognition. In: 2018 International Conference on Computer, Communication, Chemical, Material and Electronic Engineering (IC4ME2) (2018). https://doi.org/10.1109/ic4me2.2018.8465640
9. Garg, A., Noyola, J.: Lip reading using CNN and LSTM. In: Proceedings of the 30th IEEE Conference on Computer Vision Pattern Recognition, CVPR 2017, January 2017, p. 3450 (2017)
10. Li, Y., Takashima, Y., Takiguchi, T., Ariki, Y.: Lip reading using a dynamic feature of lip images and convolutional neural networks. In: 2016 IEEE/ACIS 15th International Conference on Computer and Information Science (ICIS) (2016). https://doi.org/10.1109/icis.2016.7550888
11. Rahmani, M.H., Almasganj, F.: Lip-reading via a DNN-HMM hybrid system using combination of the image-based and model-based features. In: 3rd International Conference on Pattern Recognition and Image Analysis, IPRIA 2017, April 2017, pp. 195–199 (2017)
12. Adeel, A., Gogate, M., Hussain, A., Whitmer, W.M.: Lip-reading driven deep learning approach for speech enhancement. IEEE Trans. Emerg. Top Comput. Intell. **5**(3), 481–490 (2019)
13. Vakhshiteh, F., Almasganj, F.: Lip-reading via deep neural network using appearance-based visual features. In: 2017 24th National and 2nd International Iranian Conference on Biomedical Engineering (ICBME) (2017). https://doi.org/10.1109/icbme.2017.8430230
14. Lu, Y., Li, H.: Automatic lip-reading system based on deep convolutional neural network and attention-based long short-term memory. Appl. Sci. **9**(8) (2019)

15. Jameel, S.M., Hashmani, M.A., Rehman, M., Budiman, A.: Adaptive CNN ensemble for complex multispectral image analysis. Complexity (2020). https://doi.org/10.1155/2020/836 1989

16. Jameel, S.M., Hashmani, M.A., Alhussain, H., Rehman, M., Budiman, A.: An optimized deep convolutional neural network architecture for concept drifted image classification. In: Bi, Y., Bhatia, R., Kapoor, S. (eds.) IntelliSys 2019. AISC, vol. 1037, pp. 932–942. Springer, Cham (2020). https://doi.org/10.1007/978-3-030-29516-5_70

Unsupervised Learning Approach for Evaluating the Impact of COVID-19 on Economic Growth in Indonesia

Marieta Monica[✉], Nadiah Ulfa Ayuningtiyas, Harun Al Azies, Muhammad Riefky, Hidayatul Khusna, and Santi Puteri Rahayu

Institut Teknologi Sepuluh Nopember, Kota Surabaya, Indonesia
{marietamonica.206003,nadiah.206003,harunazies.206003,
muhammadriefky.206003}@mhs.its.ac.id, hidayatul@its.ac.id,
santi_pr@statistika.its.ac.id

Abstract. The spread of COVID-19 that occurred in several parts of Indonesia resulted in the economy getting worse. Almost all business fields in Indonesia experienced a contraction. Each province has a different impact from one another so that the policies taken cannot be generalized. Therefore, this study was conducted to group provinces based on the value of Gross Regional Domestic Product (GRDP) in 2019 and 2020 using unsupervised learning. The year 2019 represents the economic conditions before COVID-19 and 2020 represents the conditions during the COVID-19 pandemic. The unsupervised learning method used in this research is the K-Means, K-Medoids, SOM, as well as Hybrid SOM-K-Means methods. From the grouping results obtained, then a comparison of the results of the four methods will be carried out to obtain the best method based on the Silhouette Index. The results show that based on 2019 data, the grouping of provinces using the K-Medoids, SOM and hybrid SOM-K-Means methods is three clusters. While the results of grouping using the K-Means method are as many as two clusters. On the other hand, based on 2020 data, both the K-mean, K-Medoids, SOM, and hybrid SOM-K-Means methods show the same results, namely the grouping is carried out in two clusters. The best method based on 2019 GRDP data is K-Means with two groups. Meanwhile, in 2020, the best method obtained is the K-mean, K-Medoids, and SOM methods with two groups. In addition, all economic growth indicators have contracted or decreased due to the impact of the COVID-19 pandemic.

Keywords: Unsupervised learning · COVID-19 · Gross Regional Domestic Product (GRDP)

1 Introduction

Since the announcement of the first case of COVID-19 in March 2020, Indonesia has still not been able to overcome the spread of corona virus, especially in several regions in Indonesia. Until now, the addition of COVID-19 cases is still happening. The first

A. Mohamed et al. (Eds.): SCDS 2021, CCIS 1489, pp. 54–70, 2021.
https://doi.org/10.1007/978-981-16-7334-4_5

impact caused by the COVID-19 virus is the fragility of the Indonesian health system. According to Todaro and Smith [1], health is the basic capital in economic development in a country. Thus, this pandemic has resulted in the economy in Indonesia getting worse [2]. Gross Regional Domestic Product (GRDP) is one of the important indicators used to determine the economic condition of a region, both based on constant prices and current prices [1]. GRDP itself is an indicator of economic growth. The existence of a relationship between GRDP and COVID-19 in the research carried out by Indayani and Hartono [3]. The study said that the economic growth in Indonesia decreased during the COVID-19 pandemic [3]. The GRDP grouping in this study was conducted to find out which provinces have similar GRDP due to the COVID-19 pandemic. This can be used by local governments to make policies related to the economic growth of the region.

Considering that Indonesia's economic condition is in danger of worsening, it is important to make plans to improve the economy as a benchmark for Indonesia's performance. Currently, the administrative region of Indonesia consists of 5 major islands (Sumatra, Java, Kalimantan, Sulawesi, Papua) and four archipelagos (Riau, Bangka Belitung, Nusa Tenggara, Maluku) which are divided into 34 provinces. Each province has a different level of economic growth. The resilience of good domestic economic growth is supported by the existence of potential sectors, optimization of inter-regional trade and increased product added value. This study was conducted to classify provinces based on the value of Gross Regional Domestic Product (GRDP) in 2019 and 2020 using unsupervised learning as an expectation that they can be input for planning economic development strategies in Indonesia. The unsupervised learning method used in this research is the K-Means, K-Medoids, SOM, and Hybrid SOM-K-Means methods. From the grouping results obtained, then a comparison of the results of the four methods will be carried out to obtain the best method based on the Silhouette Index (SI).

This study presents a novelty in the method used to group together the dominant dimensions of each district/city in Indonesia. Based on previous studies, no clustering studies were found by comparing four approaches to clustering regions to see the impact of the pandemic on economic growth. Therefore, this study will be the first study to explore the clustering of provinces in Indonesia to see the changes in economic growth due to the COVID-19 pandemic using four cluster methods at a time. The advantage obtained from the results of this study is that it can know the grouping between the provinces according to the value of the regional gross domestic product (GDP) in 2019 and 2020.

2   Unsupervised Learning

Unsupervised Learning is a type of machine learning algorithm that is used to conclude datasets consisting of input data labeled responses [4]. The most commonly used unsupervised learning method is cluster analysis. Cluster analysis aims to allocate a group of individuals in independent groups so that individuals in the same group are similar each other, while individuals in different groups have different characteristics. In this study, the algorithms used, involve K-Means, K-Medoids, SOM and Hybrid SOM-K-Means, will be discussed in more detail in the next section. Furthermore, biplot analysis is utilized to understand which variable that significantly contributes to each cluster.

2.1 Biplot Analysis

Biplot analysis was introduced by Gabriel in 1971. Biplot analysis is a form of multiple variable analysis that can provide a graphic description of the closeness between objects, diversity or correlation of variables and the relationship between objects and variables [5]. Biplot analysis can directly display the most dominant variable from a group of objects [6]. The important things that can be obtained from the biplot display are the closeness between the observed objects, the diversity of variables, the correlation between variables, and the value of variables on an object [7, 8].

Biplot analysis is based on Singular Value Decomposition (SVD). SVD aims to describe an X matrix of size n x p, consisting of a multiple variable that is corrected to its mean, into 3 matrices as follows:

$$_n\mathbf{X}_p = {}_n\mathbf{U}_r \, {}_r\mathbf{L}_r \, {}_r\mathbf{A}_p^T; r \leq \{n, p\} \tag{1}$$

U and A is a matrix with orthonormal columns ($\mathbf{U}^T\mathbf{U} = \mathbf{A}^T\mathbf{A} = \mathbf{I}_r$) and L is a diagonal matrix containing the roots of the eigenvalues of $\mathbf{X}^T\mathbf{X}$, that are $\sqrt{\lambda_1} \geq \sqrt{\lambda_2} \geq \ldots \geq \sqrt{\lambda_r}$. The columns of matrices $\mathbf{A}$ and $\mathbf{U}$ are eigenvectors of $\mathbf{X}^T\mathbf{X}$ which are obtained from $u_i = \frac{1}{\sqrt{\lambda_1}} a_i$. According to Jollife [9], Eq. (1) can be written as:

$$\mathbf{X} = \mathbf{U}\mathbf{L}^\alpha \mathbf{L}^{\alpha-1} \mathbf{A}^T; r \leq \{n, p\} \tag{2}$$

Gabriel [10] suggested the approximate size of the X matrix with Biplot in the form:

$$\rho^2 = \frac{\lambda_1 + \lambda_2}{\sum \lambda_i} \tag{3}$$

If the value of ρ^2 is getting closer to 1, it means that the biplot provides a better presentation of the data regarding the information contained in the actual data.

2.2 K-Means Clustering

K-Means is a technique that is quite simple and fast in the object clustering process. It can group large amounts of data. The followings are the algorithm of the K-Means method [11].

1. Randomly select the initial k centroids in the data.
2. Determine the distance of each observation to the center of the cluster.
3. Group each word based on its proximity to the centroid (smallest distance) using Euclidean Distance. The Euclidean distance formula is shown in Eq. (4).

$$d_{ik} = \sqrt{\sum_{j=1}^{m} (x_{ji} - y_{jk})^2} \tag{4}$$

where,

d_{ik}: Euclidean distance from the i-th observation to the k-th cluster center.

x_{ji}: The j-th variable on the i-th observation.

y_{jk}: The j-th variable at the k-th cluster center.

4. Recalculate the new cluster center (centroid).
5. Repeat steps 2 to 4 until the members in each cluster do not change.

If the number of the k-th cluster is unknown, we can use the Variance Ratio Criterion (VRC) to determine the optimum number of k-clusters [12]. The formula for the Variance Ratio Criterion can be seen in Eqs. (5).

$$VRC = \frac{\sum_{k=1}^{K} \sum_{l=1,l \neq k}^{K} \sum_{j=1}^{m} \left(y_{jk} - y_{jl}\right)^2 / (K - 1)}{\sum_{k=1}^{K} \sum_{i=1}^{l} \sum_{j=1}^{m} \left(x_{ji} - y_{jk}\right)^2 / (N - K)} \tag{5}$$

where K and N are the total number of objects and the number of clusters in partition, respectively.

2.3 K-Medoids Clustering

The k-medoid method was developed by Kaufman and Rousseeuw in 1987 [13]. The K-Medoid algorithm is often referred to as the Medoid Partitioning Algorithm (PAM). The K-Medoid Method has similarities to the K-Means Method, both methods include the Partitioning Method. The partitioning method is a method of grouping data into several clusters without any hierarchical structure between them. This algorithm uses objects in a set of objects to represent a cluster. The selected object is called the medoid.

The K-Means algorithm and the K-Medoids algorithm have differences in determining the center of the cluster. The K-Means algorithm is determined from the average value in each cluster, while K-Medoids uses data objects as representatives (medoids) [14]. The K-Medoids algorithm is used to overcome the weakness of the K-Means algorithm which is very sensitive to outliers because these objects are very far/characteristically located far away from most of other data. So, if they are included in a cluster, this type of data can distort the mean value, the average value of the cluster [15].

2.4 Self-Organizing Map (SOM)

Self-Organizing Map (SOM) is one of the artificial neural network models that uses unsupervised learning methods with the aim of grouping data. The most important feature of SOM is that it summarizes data by comparing clusters and each cluster is projected onto a map node [16]. SOM consists of two layers, namely the input layer and the output layer. The formation of SOM involves three characteristic processes, namely [17]:

a. Competition: Neurons (output nodes) in the SOM compete to represent the best input sample as measured using the discriminant function. The determination of the winner of the competition uses the best-matching unit (BMU) function. The BMU function that is often used is the Euclidean distance.
b. Cooperation: The winning node determines the spatial location of the cooperating node environment. These nodes, sharing common features, enable each other to learn something from the same input.

c. Adaptation: The weight vectors of the winner and its neighboring units in the map are adjusted in favor of higher values of their discriminant functions. Through this learning process, the relevant nodes become more similar to the input sample.

Kohonen's SOM basic algorithm can be summarized as follows [18].

1. Initialization. Choose the dimension and size of the output space.
2. Sampling. Randomly select an input vector $x(t)$ from the training data set.
3. Similarity Matching. Compute the Euclidean distances between the input vector $(x(t))$ and each output node's weight vector $(w_i(t))$. Find the best matching node c(t) at iteration t by applying the minimum distance criterion like Eq. (6).

$$c(t) = \arg \min_t\{\|x(t) - w_i(t)\|\}; i = 1, 2, \ldots, n \tag{6}$$

4. Weight Updating. Adjust the weights of the winning node and its neighborhood according to their distances to the winning node by using the Eq. (7). For the winning node the neighborhood function h_{ci} (t) will be equal to 1.

$$w_i(t + 1) = w_i(t) + \alpha(t)h_{ci}(t)[x(t) - w_i(t)] \tag{7}$$

5. Parameter Adjustment. Set $t = t + 1$. Adjust the neighborhood size and the learning rate.
6. Continuation. Keep returning to Step 2 until the change of the weights is less than a prespecified threshold value or the maximum number T of iterations is reached. Otherwise stop.

2.5 Silhouette Coefficient

Silhouette Coefficient is a method used to evaluate the results of clustering by checking how well the resulting clusters [19]. The following are the steps for calculating the silhouette coefficient.

1. Calculate the average distance from the i-th observation with all observations in the same cluster $(a(i))$ using Eq. (8).

$$a(i) = \frac{\sum_{j \in C_i, i \neq j} dist(i, j)}{|C_i| - 1} \tag{8}$$

where,
C_i: the number of observations in the *cluster* C_i
j: another observation in *cluster* Ci
dist(i, j): Euclidean distance between i-th observation dan j-th observation.
2. Calculate the average distance from the i-th observation with all observations in different clusters $(b(i))$ and the smallest value is obtained using Eq. (9).

$$b(i) = \min_{C_l:1 \leq l \leq k, l \neq 1}\left\{\frac{1}{|C_l|}\sum_{j \in C_l} dist(i, l)\right\} \tag{9}$$

where,
l: another observation in the different *cluster*

3. Calculate the *silhouette coefficient* using Eq. (10).

$$s(i) = \frac{b(i) - a(i)}{\max\{a(i), b(i)\}} \tag{10}$$

The silhouette coefficient value ($s(i)$) is between -1 to 1. When the silhouette coefficient value is around 1, the grouping is far from other clusters [20].

2.6 Gross Regional Domestic Product (GRDP)

According to the Central Bureau of Statistics, GRDP is the gross added value of all goods and services created or produced in the domestic territory of a country arising from various economic activities in a certain period regardless of whether the production factors are owned by residents or non-residents [21]. GRDP at Current Prices or known as nominal GRDP is compiled based on prices prevailing in the calculation period and aims to see the structure of the economy. The aggregate income presented in GRDP at Current Prices is assessed on the basis of the prices prevailing in each year, both at the time of assessing production and intermediate costs as well as during the assessment of GRDP. GRDP at Current Prices can show the ability of economic resources produced by a region. The greater value of GRDP means greater the ability of large economic resources, and vice versa [22].

Meanwhile, GRDP at Constant Prices is compiled based on prices in the base year. GRDP at Constant Prices aims to measure economic growth and see the development of aggregate income from year to year. Meanwhile, all income aggregates in question are assessed on the basis of fixed prices so that the development of income aggregates from year to year is only influenced by real production developments, not due to price increases or inflation [22].

3 Data and Methodology

In this sub-chapter, the data and methodology to assess the impact of COVID-19 on economic growth in Indonesia will be discussed. Table 1 presents the indicator variables of economic growth in Indonesia. The data analysis using some unsupervised learning methods utilized the variables presented in Table 1. Each variable consists of 68 observations, representing 34 provinces in Indonesia for a period of 2019 and 2020.

The variables presented in Table 1 have different data units. Therefore, the data analysis using some unsupervised learning methods is performed based on data transformation or standardization. The following are the research steps:

1. Explore the indicator variables of economic growth using statistical description.
2. Conduct biplot analysis in 34 provinces in each year (2019 and 2020).
3. Group the provinces in each year using K-Medoids Clustering, K-Means Clustering, Self-Organizing Map (SOM), and Hybrid SOM - K-Means Clustering.
4. Compare and find the best-unsupervised learning method using the largest Silhouette coefficient.
5. Draw a conclusion.

Table 1. Research variable*

Variable	Unit
GRDP at Constant Prices (X1)	Billion Rupiah
GRDP at Constant Prices per Capita (X2)	Thousand Rupiah
GRDP Growth Rate at Constant Prices (X3)	%
GRDP Growth Rate at Constant Prices per Capita (X4)	%
Distribution of GRDP at Current Prices (X5)	%

*Data Source: Central Bureau of Statistics, 2021

4   Result and Discussion

4.1 Descriptive Statistic

Figure 1 presents the boxplots of the indicator variables of economic growth in Indonesia. In general, the data distribution for the five variables shows that there are outliers. The data distribution for variables X1, X2, and X5 shows that these variables have many outliers in the form of upper outliers, both in 2019 (before the COVID-19 pandemic) and 2020 (COVID-19 pandemic). These outliers indicate that these variables in some regions are still very important compared to other regions.

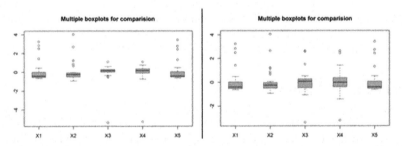

Fig. 1. Boxplot of research variables in 2019 (before the COVID-19 pandemic) and 2020 (during the COVID-19 pandemic)

Based on Fig. 1, it can also be seen that the diversity of each variable is relatively the same between the year before the pandemic and during the pandemic, except the variable GRDP growth rate (X3) in 2019 that was slightly narrower than the other variables. In this case, the outliers are always included in the next analysis because the method used is a robust clustering method against outliers or the presence of outliers will not affect the results.

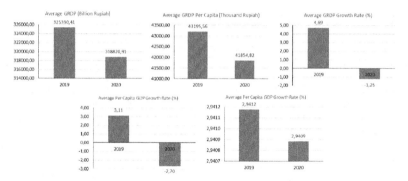

Fig. 2. The average growth of research variables

Figure 2 shows the average comparison of each variable between 2019 before the COVID-19 pandemic and 2020 when the COVID-19 pandemic occurred. Based on Fig. 2, all research variables have decreased, this can be indicated by the GRDP sector in Indonesia being affected by COVID-19.

4.2 Biplot Analysis

The results of the biplot analysis for all provinces in Indonesia during 2019 and 2020 are presented in Fig. 3.

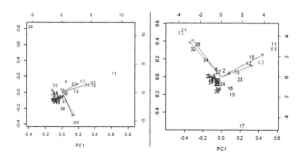

Fig. 3. Biplot analysis in 2019 (left) and 2020 (right)

Based on the outputs presented in Fig. 3, important information is obtained from the biplot analysis, namely:

a. **Proximity between Objects**

This information is used as a guide to find out which provinces have similar characteristics with other provinces. Provinces that are located close together are said to have fairly close characteristics in common (Table 2).

Table 2. Comparison of proximity characteristics of provinces in 2019 and 2020

2019
Group 1 : Papua **Group 2** : DKI Jakarta, West Java, Central Java and East Java. **Group 3** : Aceh, North Sumatra, West Sumatra, Riau, Jambi, South Sumatra, Bengkulu, Lampung, Kepulauan Bangka Belitung, Riau Kepulauan, DI Yogyakarta, Banten, Bali, NTB, NTT, West Kalimantan, Central Kalimantan, South Kalimantan, East Kalimantan, Kalimantan North, North Sulawesi, Central Sulawesi, South Sulawesi, Southeast Sulawesi, Gorontalo, West Sulawesi, Maluku, North Maluku and West Papua.
2020
Group 1 : DKI Jakarta, West Java, Central Java and East Java. **Group 2** : Aceh, North Sumatra, West Sumatra, Riau, Jambi, South Sumatra, Bengkulu, Lampung, Kepulauan Bangka Belitung, Riau Kepulauan, DI Yogyakarta, Banten, Bali, NTB, NTT, West Kalimantan, Central Kalimantan, South Kalimantan, East Kalimantan, Kalimantan North, North Sulawesi, Central Sulawesi, South Sulawesi, Southeast Sulawesi, Gorontalo, West Sulawesi, Maluku, North Maluku, West Papua and Papua.

b. **Interpretation of Variable Values in an Object**

This information is used to determine the characteristics in each province. Provinces that are located in the direction of the variable vector indicate the high value of the variable in the province or it can be interpreted that the variable value in the region is above the average (Table 3).

Table 3. Comparison of variable values in provinces in 2019 and 2020

2019	2020
1. The Provinces of DKI Jakarta, West Java, Central Java, East Java, North Sumatra, Banten, Riau, Kepulauan Riau and East Kalimantan have values of $X1$, $X2$ and $X5$ above the average 2. Provinces of North Sumatra, West Sumatra, South Sumatra, Bengkulu, Lampung, DI Yogyakarta, Banten, Bali, NTT, West Kalimantan, Central Kalimantan, East Kalimantan, North Kalimantan, North Sulawesi, Central Sulawesi, South Sulawesi, Southeast Sulawesi, Gorontalo, West Sulawesi, Maluku and North Maluku have value of $X3$ and $X4$ above the average	1. The Provinces of DKI Jakarta, West Java, Central Java, East Java, North Sumatra, Riau, and East Kalimantan have values of $X1$, $X2$ and $X5$ above the average 2. Provinces of North Sumatra, Jambi, South Sumatra, Bengkulu, Central Sulawesi, South Sulawesi, North Maluku and Papua have value of $X3$ and $X4$ above the average

c. **Variable Diversity**

This information is used to see the diversity of variables in each province. It can be strategically estimated aspects that must be improved in order to improve economic conditions in Indonesia (Table 4).

Table 4. Comparison of variable diversity values in 2019 and 2020

2019	2020
The longest variable vector is the X3 and X4. It means that X3 and X4 has the greatest diversity. Meanwhile, the shortest variable vector (least variance) is X2	The longest variable vector is the X3 and X4. It means that X3 and X4 has the greatest diversity. Meanwhile, the shortest variable vector (least variance) is X2

d. **Correlation Between Variables**

Correlation or relationship between variables can be seen from the biplot image. Two variables that have a positive correlation value will be described as two lines with the same direction or forming a narrow angle (Table 5).

Table 5. Comparison of correlation between variables in 2019 and 2020

No	2019	2020
1	The variables of GRDP at Constant Prices (X1), GRDP at Constant Prices per Capita (X2) and Distribution of GRDP at Current Prices (X5) influence each other and are positively correlated. This is determined from the magnitude of the very small angle. So if GRDP increases, then GRDP per capita and distribution of GRDP will also increase	The variables of GRDP at Constant Prices (X1), GRDP at Constant Prices per Capita (X2) and Distribution of GRDP at Current Prices (X5) influence each other and are positively correlated. So if GRDP increases, then GRDP per capita and distribution of GRDP will also increase
2	The variable rate of GRDP Growth Rate at Constant Prices (X3) and the rate of GRDP Growth Rate at Constant Prices per Capita (X4) is positively correlated, which means that if the rate of GRDP increases, the rate of GRDP per capita will also increase. This is determined by the size of the small angle	The variable rate of GRDP Growth Rate at Constant Prices (X3) and the rate of GRDP Growth Rate at Constant Prices per Capita (X4) is positively correlated, which means that if the rate of GRDP increases, the rate of GRDP per capita will also increase

4.3 K-Medoids Clustering

K-Medoids clustering is a non-hierarchical clustering method that partitions data into one or more clusters/groups. This cluster algorithm uses objects in a set of objects to represent a cluster. The selected object is later called the medoid [13]. Based on the validation results, it was found that the optimal number of clusters was 3 clusters (k = 3) for 2019 and 2 clusters (k = 2) for 2020. Here are the results of clusters using the method K-Medoids for 2019 and 2020.

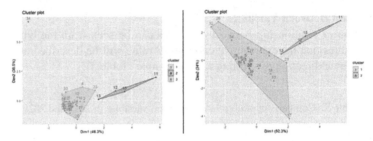

Fig. 4. Cluster plot of K-medoids clustering in 2019 (before the COVID-19 pandemic) and 2020 (COVID-19 pandemic)

Figures 4 use the K-Medoids algorithm with Euclidean distance. According to the two images, the results of the grouping of K-Medoids in 2019 (before the COVID-19 pandemic) the province of Papua became the only member of clusters 3 and four provinces were members of cluster 2, namely DKI Jakarta, West Java, Central Java, and East Java, the remaining 29 provinces are members of cluster 1. The results of clustering in 2020 (COVID-19 pandemic) show that four provinces are members of cluster 2, namely DKI Jakarta, the province of Java, West Java, Central and East Java, and 30 provinces are members of cluster 1.

4.4 K-Means Clustering

Figure 5 uses the K-Means algorithm with Euclidean distance. From the two images showing the same results, the results of grouping of K-Means in 2019 (before the COVID-19 pandemic) and in 2020 (COVID-19 pandemic) present the same composition, namely four provinces are members cluster 1, namely DKI Jakarta, West Java, Central Java, and East Java and there are 30 provinces which are members of cluster 2.

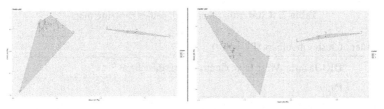

Fig. 5. Cluster plot of K-Means in 2019 (before the COVID-19 pandemic) and 2020 (COVID-19 pandemic)

4.5 Self-Organizing Map (SOM)

Table 6. Silhouette coefficient SOM

Data	Silhouette coefficient	Number of clusters
GRDP 2019	0.645	3
GRDP 2020	0.5491	2

The next grouping uses the Self-organizing Map (SOM) method. The selection of the number of clusters uses the silhouette coefficient method with the help of R software and the silhouette coefficient value is obtained as Table 6. The determination of the number of clusters will also be used to determine the size of the grid on the map. In Fig. 6 it can be seen that for 2019, the first cluster consists of 29 provinces, the second cluster consists of 1 province, and the third cluster consists of 4 provinces. Meanwhile, in 2020, the first cluster consists of 5 provinces and the second cluster consists of 29 provinces.

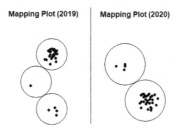

Fig. 6. Mapping plot of GRDP in 2019 and 2020 by province

The details of the cluster members (Province) in each cluster are presented in Table 7.

Table 7. Cluster members with self-organizing map

Year	Cluster	Cluster members (Province)
2019	1	DKI Jakarta, West Java, Central Java, East Java
	2	Papua
	3	Aceh, North Sumatra, West Sumatra, Riau, Jambi, South Sumatra, Bengkulu, Lampung, Bangka Belitung Islands, Riau Islands, DI Yogyakarta, Banten, Bali, West Nusa Tenggara, East Nusa Tenggara, West Kalimantan, Central Kalimantan, South Kalimantan, East Kalimantan, North Kalimantan, North Sulawesi, Central Sulawesi, South Sulawesi, Southeast Sulawesi, Gorontalo, West Sulawesi, Maluku, North Maluku, West Papua
2020	1	DKI Jakarta, West Java, Central Java, East Java, East Kalimantan
	2	Aceh, North Sumatra, West Sumatra, Riau, Jambi, South Sumatra, Bengkulu, Lampung, Bangka Belitung Islands, Riau Islands, DI Yogyakarta, Banten, Bali, West Nusa Tenggara, East Nusa Tenggara, West Kalimantan, Central Kalimantan, South Kalimantan, East Kalimantan, North Kalimantan, North Sulawesi, Central Sulawesi, South Sulawesi, Southeast Sulawesi, Gorontalo, West Sulawesi, Maluku, North Maluku, West Papua, Papua

In addition, using SOM can be analyzed related to the dominant variables in each cluster. These variables describe the characteristics of each cluster. The way to read the codes plot on the SOM is to look from the bottom to the right if there is a node beside it, then continue up, and so on. In Fig. 7 for 2019, the third cluster is dominated by the variable GRDP Growth Rate at Constant Prices and per Capita. While the second cluster is dominated by the variable GRDP at Constant Prices per Capita. The first cluster has the same similarity for the five variables. In 2020, the second cluster is dominated by the variable GRDP at Constant Prices, GRDP at Constant Prices per Capita, and Distribution of GRDP at Current Prices. This is different from the case in the first cluster which is dominated by the variable GRDP Growth Rate at Constant Prices and per capita.

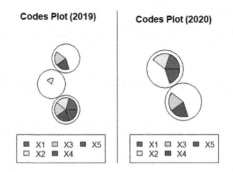

Fig. 7. Codes plot of GRDP in 2019 and 2020 by province

4.6 Hybrid SOM - *K-Means* Clustering

Clustering is also carried out using the Hybrid SOM-K-Means method. This aims to see whether a more complex model will give the best grouping results. In this method, the determination of the grid is set at 5x5 because it is sufficient to represent the data under study. Grid size that is too large will affect the unrepresentation of the data in the grouping. The determination of the number of clusters for the K-Means method has been carried out in the previous sub-chapter and obtained 3 clusters for 2019 and 2 clusters for 2020. This information is used to build groups using Hybrid SOM-K-Means.

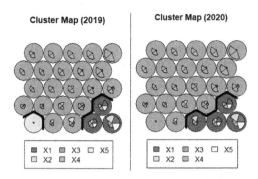

Fig. 8. Cluster map of GRDP in 2019 and 2020 by province

Figure 8 is the result of grouping with this method. Grouping in 2020 resulted in groups and members of the same group using the SOM method. While in 2019 there were differences in the first cluster, Central Java Province entered the third cluster. Judging from the dominant variable, it is different from the SOM method. The first cluster is marked with red nodes on the 2019 Cluster Map where the characteristics of the cluster are dominated by variables X1, X2, and X5. The second cluster is marked with a light green node where the characteristics of the cluster are dominated by the X2 variable. While the third cluster is dominated by orange nodes where the characteristics of the cluster are dominated by X3 and X4 variables. The first cluster is marked with a red node on the 2020 Cluster Map where the characteristics of the cluster are dominated by variables X1, X2, and X5. While the second cluster is marked with orange nodes where the characteristics of the cluster are dominated by X3 and X4 variables.

4.7 Comparison Unsupervised Learning Method

The comparison of the clustering methods in Table 8 can be concluded that the most optimal groups for the 2019 GRDP are 2 clusters using the K-Means method, likewise for GRDP 2020, it's just that the SOM, K-Medoids and K-Means methods provide the same criteria for clustering goodness. Cluster members in 2020 from the SOM method are different from the K-Medoids and K-Means methods. Therefore, considering that the two methods have the same members, for the 2020 cluster, it is recommended to use K-Medoid/K-Means. Table 9 is a descriptive statistical table of economic growth

Table 8. Comparison unsupervised learning method with silhouette coefficient

Method	GRDP 2019		GRDP 2020	
	Silhouette Coefficient	Number of Clusters	Silhouette Coefficient	Number of Clusters
SOM	0.6450	3	0.5491	2
SOM-K-Means	0.5494	3	0.4424	2
K-Medoids	0.6450	3	0.5491	2
K-Means	0.7359	2	0.5491	2

* Columns marked in yellow are selected clusters

indicators for each cluster in 2020. In this table, it can be seen that in cluster one, the focus of the evaluation of the impact of COVID-19 on economic growth is the first (X1), second (X2), and fifth indicators. (X5). This is because the three indicators have a lower average value than the second cluster. On the other hand, the focus of the evaluation is on cluster two, namely the third (X3) and fourth (X4) indicators.

Table 9. Descriptive statistics (average) economic growth indicators for each cluster in 2020

Variable	Cluster 1	Cluster 2
GRDP at Constant Prices (X1)	167,194.40	1,456,019.75
GRDP at Constant Prices per Capita (X2)	38,583.10	66,392.75
GRDP Growth Rate at Constant Prices (X3)	-1.09	-2.46
GRDP Growth Rate at Constant Prices per Capita (X4)	-2.62	-3.23
Distribution of GRDP at Current Prices (X5)	1.53	13.47

* Columns marked with yellow have a smaller mean

Based on the results of the clustering, it is shown that the territory of Indonesia is divided into 2 clusters, namely the non-Java area cluster (Cluster 1) and the cluster consisting of parts of the Java Island (DKI Jakarta, East Java, West Java and Central Java) (Cluster 2). This shows that the economy in Indonesia is still centered on the island of Java. This statement is in accordance with the results of a survey conducted by the OECD Economic Surveys which states that the economy on the island of Java generates 55% of Indonesia's GRDP [23]. This statement is also in line with research conducted by Rahma et al. (2019) on the Development of a Composite Measure of Regional Sustainable Development in Indonesia which states that the provinces on the island of Java tend to be more advanced in terms of economic development than other regions [24]. This is because most of the infrastructure in the country, such as roads, ports, and electricity networks, is concentrated on the island of Java.

5   Conclusion and Future Research

Evaluation of the impact of COVID-19 on economic growth in Indonesia using the unsupervised learning method gives the following results. All indicators of economic growth contracted or decreased due to the impact of the COVID-19 pandemic. This is evidenced by historical data on the average economic growth in 34 provinces. Based on biplot analysis and unsupervised learning, 34 provinces in Indonesia are divided into two groups. The first cluster is dominated by non-Java regions with a focus on evaluating the impact of COVID-19 on economic growth, namely GRDP, GRDP per capita, and distribution of GRDP at Current Prices. While the second cluster is dominated by the Java region with a focus on evaluating the impact of COVID-19 on economic growth, namely the GRDP growth rate and the per capita GRDP growth rate.

For future research can use cluster analysis which includes all improvements in the methodology that support researchers in optimizing important decisions such as algorithm choice, number of clusters, algorithm parameters, etc. For a better analysis, it can be continued with the relevant analysis to be able to know the influential variables.

References

1. Todaro, M.P., Smith, S.C.: Economic Development, 11th edn. Pearson Educated Limited, Harlow (2011)
2. Hartati: Penggunaan Metode ARIMA dalam Meramal Pergerakan Inflasi. Jurnal Matematika Sains dan Teknologi, pp. 1–10 (2017)
3. Indayani, S., Hartono, B.: Analisis Pengangguran dan Pertumbuhan Ekonomi Sebagai Akibat Pandemi COVID-19. Jurnal Perspektif **18**(2), 201–208 (2020)
4. Nurhayati, B., Iswara, R.P.: Pengembangan Algoritma Unsupervised Learning Technique Pada Big Data Analysis di Media Sosial Sebagai Media Promosi Online Bagi Masyarakat (2019)
5. Widowati, W., Muzdalifah, L.: Perbandingan Analisis Biplot Klasik dan Robust Biplot Pada Pemetaan Perguruan Tinggi Swasta di Jawa Timur. Jurnal Riset dan Aplikasi Matematika **1**(1), 17–26 (2017)
6. Ariawan, I.A., Iputu, E.N., Niluh, P.S.: Komparasi Analisis Gerombol (Cluster) dan Biplot dalam Pengelompokkan. E-Jurnal Matematika **4**(2), 17–22 (2013)
7. Sartono, B., Affendi, F.M., Syafitri, U.D., Sumertajaya, I.M., Angraeni, Y.: Analisis Peubah Ganda. Fakultas Matematika dan Ilmu Pengetahuan Alam IPB, Bogor (2003)
8. Mattjik, A.A., Sumertajaya, I.M.: Sidik Peubah Ganda. IPB Press, Bogor (2011)
9. Jollife, I.T.: Principal Component Analysis, 2nd edn. Springer, New York (2010)
10. Gabriel, K.R.: The biplot graphic display of matrices with application to principal component. Biometrika **58**(3), 453 (1997)
11. Thomas, S., Harode, U.: A comparative study on k-means and hierarchical clustering. Int. J. Electron. Electron. Comput. Syst. **4**, 5–11 (2015)
12. Calinski, T., Harabasz, J.: A dendritc method for cluster analysis. Commun. Stat. **3**(1), 1–27 (2007)
13. Kaufman, L., Rousseeuw, P.J.: Chapter 3: Clustering Large Applications (Program CLARA), pp 126–163. Wiley, Hoboken (2008)
14. Kaur, N.K., Kaur, U., Singh, D.D.: K-Medoid clustering algorithm-a review. Int. J. Comput. Appl. Technol. (IJCAT) **1**(1), 2349–1841 (2014)
15. Han, J., Kamber, M.: Data Mining: Concept and Techniques. Morgan Kauffman Publisher, Waltham (2006)
16. Badran, F., Yacoub, M., Thiria, S.: Self-organizing maps and unsupervised classification. In: Badran, F., Yacoub, M., Thiria, S. (eds.) Neural Networks, pp. 379–442. Springer, Heidelberg (2005). https://doi.org/10.1007/3-540-28847-3_7

17. Haykin, S.: Neural Networks: A Comprehensive Foundation, 2nd edn. Prentice-Hall, Englewood Cliffs (1999)
18. Asan, U., Ercan, S.: An introduction to self-organizing maps. In: Computational Intelligence Systems in Industrial Engineering, pp. 295–315. Atlantis Press, Paris (2012)
19. Han, J., Kamber, M., Pei, J.: Data Mining Concepts and Techniques. Morgan Kaufmann Publisher (2012)
20. Kogan, J.: Introduction to Clustering Large and High-Dimensional Data. Cambrige University Press, New York (2007)
21. Central Bureau of Statistics Yogyakarta: Produk Domestik Regional Bruto Kota Yogyakarta Menurut Lapangan Usaha 2012–2016. BPS Yogyakarta, Yogyakarta (2017)
22. Department of Communication, Information, Statistics and Encoding of Paser Regency: Analisis Produk Domestik Regional Bruto Kabupaten Paser Menurut Lapangan Usaha. Tana Paser: Department of Communication, Information, Statistics and Encoding of Paser Regency (2019)
23. OECD: OECD Economic Surveys: Indonesia 2018, OECD Publishing, Paris (2018). https://doi.org/10.1787/eco_surveys-idn-2018-en
24. Rahma, H., Fauzi, A., Juanda, B., Widjojanto, B.: Development of a composite measure of regional sustainable development in Indonesia. Sustainability **11**(20), 5861 (2019). https://doi.org/10.3390/su11205861

Rainfall Prediction in Flood Prone Area Using Deep Learning Approach

Siti Zuhairah Ramlan[1] and Sayang Mohd Deni[2(✉)]

[1] Faculty of Computer and Mathematical Sciences, UiTM,
40450 Shah Alam, Selangor, Malaysia
[2] Center for Statistics and Decision Science Studies, Institute for Big Data Analytics and
Artificial Intelligence (IBDAAI), Faculty of Computer and Mathematical Sciences, UiTM,
40450 Shah Alam, Selangor, Malaysia
sayang@tmsk.uitm.edu.my

Abstract. Flood is a catastrophic event that contributes to the impact on socio-economic of a developing country. Flood prone area is also known as the location at risk as heavy rainfall attribute to flood events. This circumstance leads to the effective flood mitigation phase. One of the critical problems facing by responsible government agencies is to minimize future uncertainties. This study explores four deep learning methods with univariate rainfall temporal data in gauging station near to flood prone area. Four models tested are Multi-layer Perceptron MLP, Long Short Term Memory LSTM, Stacked-LSTM, and hybrid model Convolutional Neural Network CNN-LSTM. The aim of this paper is to compare and determine the best method for rainfall prediction of next day event. Model comparison is conducted by comparing the correlation coefficient, Root Mean Square error (RMSE) and Mean Absolute Error (MAE). Based on the selected locations, the results showed at Kuantan station generally underfitting, meanwhile the Kuala Krai station does not showed discrepancies between training and testing dataset. It could be concluded that by adding the complexity to the model, will not significantly improved the model prediction. The LSTM model with 16 memory blocks was outperformed in both locations. The potential of deep learning methods should be considered for rainfall amount prediction due to less complexity and much easier to be applied into dataset for model fitting purposes. It may assist to accommodate strategic precautions for flood mitigation phases in flood prone areas.

Keywords: Floods mitigation · Long Short Term Memory (LSTM) · Rainfall prediction · Univariate time series

1 Introduction

Due to its unique tropical location, Malaysia faces heavy rainfall between November to March every year. Owing to its climate, Malaysia is prone to a series of unfortunate flooding events due to massive precipitation. Flooding in Malaysia is a common seasonal phenomenon. Approximately 29,800 km^2 or 9% of the total land area is in flood

© Springer Nature Singapore Pte Ltd. 2021
A. Mohamed et al. (Eds.): SCDS 2021, CCIS 1489, pp. 71–88, 2021.
https://doi.org/10.1007/978-981-16-7334-4_6

prone areas and effects about 4.82 million residents, which is around 22% of Malaysia's total population [1]. In Malaysia, the east coast is identified as flood prone areas since it suffers from a series of flood events every year. In year 2013, around 14,044 people evacuated in Kuantan due to major damages occurred in terms of electricity, road structures, buildings and belonging. In the year 2013, the government suffered from the significant financial cost of repairing flood damages [2]. Many agencies donated millions of Ringgit throughout the rescue and aid mission [3]. This situation worrying the public as it has a negative impact affecting the community's quality of life [1].

The best way to deal with these issues is to build flood management systems for critical situation awareness and decision-making processes [4]. As stated by [1] if disaster prevention and preparedness were successfully managed, it will help in reducing the burden of flood management department. The government and the respective authorities will prepare the action plan to be implemented during severe flood event including evacuation for flood victim. Managing flood risks required effective flood mitigation that involves in-depth planning and preparation before the actual incident happens. This minimizes the damages caused and leads to an improved quality of life. Natural disasters may contribute to a high level of stress and other psychological problems to the flood victims [5].

Previous research acknowledges that accurate rainfall forecasting remains a complex task and lead to one of the challenges in almost over the years not only in Malaysia but throughout the world. This problem was difficult to handle in the rainfall prediction process as it is dependent on the nature of rainfall data that typically is non-linear and random. Frequency, intensity, and amount are the main attribute of rainfall time series. Several methods expressed future events as a linear function of the past data. It requires identifying the type of relationship among variables compared to the neural network which is data driven approach model. This model depends on availability of the data to be learned without prior knowledge of distribution and hypothesis of the relationship and it is the non-linear function [6].

Machine learning always required structured data, whereas feature engineering is needed for the algorithm to learn from labelled data. Furthermore, the accuracy could not be improved by adding more data into the algorithm. Although deep learning is subset of machine learning, the main different relies on layers of neural network. Deep learning is essentially a large neural network, which contains input, output and hidden layers in the network. The success of deep learning is connected to two major factors. First, the deep learning performance continues to grow as it offered more data for training. This differs from other machine learning techniques where it becomes a state of little change or no change after adding more data. Second, deep learning models learn good features that represent the data without having the great domain knowledge for feature extractions.

Rainfall data is often collected with missing value, making it impossible to capture important information. Incomplete data can be caused by damaged or malfunctioning measuring instruments which cause measurement errors and geographical data gaps or changes to instrumentation over time, such as a change in the location measurement, a change in data collectors, the measurement inconsistency, or extreme topical changes in a climate zone [7].

There have been many studies to investigate rainfall prediction, most early studies rainfall forecast by using box Jenkin's methodology and regression model. Both techniques are linear model. [8] implements the Auto-regressive Integrated Moving Average (ARIMA) model coupled with the Ensemble Empirical Mode Decomposition (EEMD). The result stated that EEMD-ARIMA improves ARIMA for annual runoff time series forecasting.

Alternatives to linear model, [9] make a prediction by identifying the seasonal rainfall peak periods by using the Fourier series. Daily rainfall data collected from 12 rain gauge stations at Kelantan. The obtain result is used to describe the seasonal variation, estimate the wet period, the dates of rainfall peaks and the probability of amounts falling that exceed certain thresholds.

This has also been investigated in studies [10] proposed a hybrid model to downscale monthly precipitation in Minab basin, Iran. The model based on a feed forward neural network (FFNN) optimize by Particle Swarm Optimization (PSO) algorithm. The robust and reliability of the model evaluate with an Artificial Neural Network (ANN) model trained by Levenberg-Marquardt (LM) algorithm. The result shows that ANN-PSO performs better than ANN-LM.

[11] build LSTM network for monthly rainfall prediction in Camau, Vietnam. It covers 39 years. This network will be compared with ANN and Seasonal Artificial Neural Network (SANN). This performance evaluated by comparing the correlation coefficient, RMSE and MAE. The validation results shows LSTM model has captured exceptionally good accuracy in monthly rainfall data. The authors concludes that LSTM network can be a promising model in hydrological and climatic applications for estimating more accurate precipitation predictions.

While in Malaysia, the authors [12] bring information that monitoring severe weather events has been increasing due to many disaster events that have occurred in recent years and predicting the trend of precipitation is a difficult task. A study implements ANN compared with ARIMA for historical rainfall data from 116-gauge stations for 50 years which is 1965 to 2015. Before modelling, missing data are treated by using the Expectation-Maximization (EM) algorithm. The performance suggests that ANN model outperforms ARIMA model.

A comparative study between recurrent neural network (RNN) and LSTM network conducted by [13] for monthly rainfall forecasting from the year 1871–2016 in India using long sequential raw data for time series analysis. The author recommends LSTM as a potential alternative to assist in assessing climatic phenomena such as rainfall forecasting. The author suggests extending the application of LSTM model in other climatic extremes such as a flood.

To the best of our knowledge, the application of LSTM network in daily rainfall amount prediction received less attention among of Malaysian researcher and in the field of study throughout Malaysian region. The outcomes of the study is expected to assist in providing the input for early warnings system for potential occurrence of flood and drought events. It is of interest to know whether the advanced deep learning technique helps in making prediction on the flood prone areas. The main issues of the study is to explore on the possible application of deep learning approach which could be used for decision making in flood mitigation and rainfall prediction analysis. This led to research question and objective to find the best univariate time series model in flood prone area. The objective of the study is to assess the performance of time series models in predicting rainfall amount using deep learning approach and to identify the best univariate time series using a deep learning approach for rainfall prediction. This research will be focused on the application of the model in daily rainfall amount at two rain gauge stations including Kuantan and Kuala Krai which are located at Pahang and Kelantan states of Peninsular Malaysia. Four deep learning approaches were selected such as multi- layer perceptron (MLP), LSTM, Stacked-LSTM and convolutional neural network (CNN-LSTM).

2 Methodology

2.1 Study Area and Data Preparation

The data used in this study consists of daily rainfall amounts (mm) which were obtained from Malaysia Meteorology Department. The total sample size of 15706 from January 1, 1975, to December 31, 2017 were collected from Kuantan rain gauge station Meanwhile, the dataset of 14600 data points for the period from January 1, 1975, to December 31, 2014 were obtained from the Department of irrigation and Drainage (DID) for Kuala Krai rain gauge station.

Due to the shape of the dataset which is highly skewed to the right, data normalization technique will be implemented before inputting the series to the training module. Data normalization rescales the values into a range of 0 to 1 as expressed in Eq. (1) to avoid effect of scale in the deep learning architecture [14].

$$\text{Xchanged} = \frac{X - \text{Xmin}}{\text{Xmax} - \text{Xmin}} \tag{1}$$

Where X is input data of time series variables. Xmin is the minimum amount of X, while Xmax is the maximum amount of X.

To fit the supervised learning model, the univariate timeseries data must be transformed into a shift approach named the sliding window. This approach frames a data set of the time series into independent and dependent variables manner. However, the 2-dimension structure of the supervised learning model will be converted to a 3-dimensional structure to fit LSTM model, or 4-dimension structure to fit hybrid CNN-LSTM.

The samples of the series will be restructured to attribute value and convert to supervised learning [15] during the process. The series can be arranged into the attribute value by using previous time steps or lagged as input variables and use the next time step as the output variables. This process could be achieved by creating a loop to arrange the univariate data series. The looping will be used to prepare the dataset in the form of dimensional array desired for model training. Since the data will be used to predict the event for the next day, then a lag of 365 days will be applied in the data series. The data series then will be divided into x array for 365 lags, y array with the value of 1.

2.2 Model Development

The data is divided into two sets. The first set is used to estimate the parameters of the model, this part is called training. The second set is known as the testing set, and it contains unknown data from the previously defined model, which is used to estimate the prediction using the parameters. The test set enables the model to estimate parameters on data that were not available to the model when the parameters were first calculated. This allows model performance to be evaluated, for which the actual outcome compares it to the predicted outcome. From Table 1 stated the length of the dataset used for each portion.

Table 1. Length of training and testing dataset

Station	Kuantan	Kuala Krai
Training Set	10965	10731
Testing Set	4700	4599

In this stage, a modelling technique or algorithm will be selected to be used with the prepared dataset. Training the first prepared dataset to create the best-fitted model. Next, testing on the generated model will be conducted to validate the quality and validity of the model. One or more algorithms can be built on the prepared dataset. Finally, the models could be assessed carefully to ensure that the created models met the research objectives. In Fig. 1, left hand side display the flow of training procedure and right-hand side display the testing procedure.

In the training procedure, the first step is to normalize the dataset with min-max normalization which aims to simplify the calculations continue with segment the univariate time series data input into supervised learning using a sliding window of 365 days. The next step is to train the prepared dataset with MLP, LSTM, Stacked- LSTM and CNN-LSTM. The training process use 70% segment data ratio.

The weight and loss are calculated for prediction approximation and coefficient detail after learning data procedure were conducted. The testing procedure then began with the same algorithm iteratively generating the prediction details. A backpropagation algorithm is used to accelerate convergence towards the desired minimum error value. Once it had reached the desired value or maximum epoch, it was terminated. The hyperparameter setting is fixed for the epoch is 500 and batch size of 200 for each of test. We also set the model with 'early stopping', which means that the training may not complete the entire 500 epoch iteration because it monitors the validation loss while not improving at the same time to prevent the model from becoming overfitting.

Artificial Neural Network (ANN). ANN design was inspired by brain tissue composed by cells called neurons. Neuron exchange signals with one another forming a dense and complex network.

It consists of input layer, takes signal (values) and passes them on the next layer.

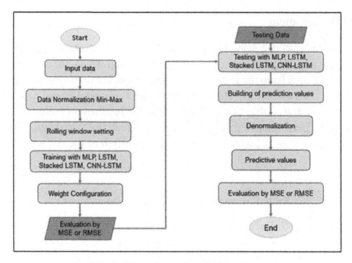

Fig. 1. Flowchart of modelling phase

Then, hidden layer has neurons which apply different transformations to the input data. After that, output layer or last layer in the network and receives input from the last hidden layer. In this network, weights (parameters) represent the strength of the connection between units. A weight brings down the importance of the input value and lastly activation functions are used to introduce non-linearity to neural networks.

To get a better view of neural network, we can take linear regression model where x is input data or temporal vector, w as weight or parameter and b is biased. Linear regression when only one independent variable included while multiple regression deals with more than one independent variable.

$$z = x_1 * w_1 + x_2 * w_2 + \cdots + x_n * w_n + b * 1 \tag{2}$$

$$\hat{y} = \sigma(z) = \frac{1}{1 + e^{-z}} \tag{3}$$

From multiple regression, when the activation function sigmoid function is introduced, it changes to logistic regression. Logistic regression can be viewed as a small neural network where $\hat{y} = \sigma(z)$. The equation expressed in Eq. 3 and hyperbolic tangent expressed in Eq. 4.

$$\tanh(z) = \frac{e^z - e^{-z}}{e^z + e^{-z}} \tag{4}$$

When compute the model with using 4 neuron unit, the total parameter can be achieved when we calculate as the equation below:

$$\text{No of Parameters} = (i * h + h * o) * (h + o) \tag{5}$$

where i is an input that equals 365, h is neuron unit in a hidden layer which is 4 and o is our output that equal to 1. The model summary shows a total of parameters is equal to 129.

Backpropagation. Result $\hat{y}$ from forward propagation, then loss could be calculated. Gradient descent taking the derivative of the loss function with respect to the weights in the model. Gradient descent carries out minimization process by first calculating the gradient of the loss function and then updating the weights in the network accordingly. It will iterate until it converges. Backpropagation is a tool that gradient descent uses to calculate the gradient of the loss function.

Long Short Term Memory (LSTM). The LSTM network was introduced by [16]. This is the special model of the RNN and allows for a model to make use the passes value on the input value to the output value, this behaviour makes this model suita-ble for the sequence model. The model was able to retain state from one iteration to the next by using output as input for the next step. Traditional RNN can hardly re-member sequences with a length of over 10 [17]. It will suffer for a longer time de-pendency as the vanishing gradient problem.

The RNN's problem is solved by LSTM. The architecture's specialty is to avoid long-term dependency issues, and multiple gates provide flexible control over infor-mation flow. [18] describe LSTM cell consists of various gates and their functions are as follows:

- Input gate consists of the input entered.
- Cell state runs through the entire network and can add or remove information with the help of gates to filter the input.
- Forget gate decides the fraction of the information to be allowed.
- Output gate consists of the output generated by LSTM.
- Sigmoid activation function layer generates numbers between zero and one, describing how much of each component should be let through.
- Hyperbolic tangent activation function layer generates a new vector, which will be added to the state (Fig. 2).

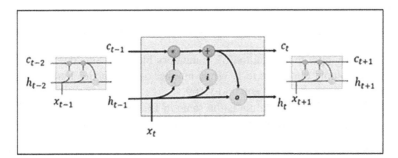

Fig. 2. Internal Operation of LSTM

LSTM special forms of recurrent network have multiple gates to offer flexible control to allow information through in each memory cell. Each of the operation gates contains an activation function. Set of operation in memory cell express in the equation below.

Forget gate: where the value (0, 1) result from the sigmoid activation function

$$f_t = \sigma(W_f x_t + U_f h_{t-1} + b_f) \tag{6}$$

Input gate: where the value (0, 1) result from the sigmoid activation function

$$i_t = \sigma(W_i x_t + U_i h_{t-1} + b_i) \tag{7}$$

Candidate value that could be added to the state, where the value in the range $(-1, 1)$ results from a hyperbolic tangent

$$\tilde{c}_t = tanh(W_t x_t + U_{\tilde{c}} h_{t-1} + b_{\tilde{c}}) \tag{8}$$

From Eq. 6 to 8, new cell state updated by Eq. 9.

$$c_t = f_t \odot c_{t-1} + i_t \odot \tilde{c}_t \tag{9}$$

Output gate: where the value (0,1) result from the sigmoid activation function

$$o_t = \sigma(W_o x_t + U_o h_{t-1} + b_o) \tag{10}$$

New hidden state:

$$h_t = tanh(c_t) \odot o_t \tag{11}$$

where $\odot$ denotes element-wise multiplication, W_t, W_f, W_i, W_o and U_t, U_f, U_i, U_o are adjustable weights, x_t is an input or timesteps, h_{t-1} earlier hidden state, are adjustable bias. The new c_t and h_t will repeat the same operation to the next memory cell with the new x_t value included in the network. Multiple LSTMs can be stacked to form more complex structures.

When compute LSTM network using 16 memory blocks unit, the total parameters can be achieved when we calculate as the Eq. 12

$$\text{No of Parameters} = g * [h(h + i) + h] * (h + o) \tag{12}$$

where i, is an input that equal to 365, h memory blocks in a hidden layer which is 16 and o is our output that equal to 1. While number of FFNNs in a unit where LSTM has 4. The total of parameters is equal to 1,169.

For stacked LSTM, the equation is the same in 13, however, there is two layer the calculation will be conducted in each layer.

$$\text{No of Parameters in 1}^{\text{st}} \text{ layer} = g * [h(h + i) + h] \tag{13}$$

The first layer i is equal to 1, h memory blocks used is equal to 13, and g number of FFNN in a unit where LSTM has 4. Whereas in the second layer i is equal to 13 as this is the output from the first layer. h memory blocks used is equal to 4 and o is our output that equal to 1, the total parameter is equal to 1,073.

$$\text{No of Parameters in 2}^{\text{nd}} \text{ layer} = g * [h(h + i) + h] * (h + o) \tag{14}$$

Hybrid CNN – LSTM. CNN-LSTM is a combination of a convolutional neural network and a long short-term neural network. The CNN network success in pattern recognition and image classification. Hence, implement CNN for feature extraction before input to the LSTM layer. The convolutional neural networks CNNs are a special kind of deep neural network. It differs from standard neural network architecture where the statistical connection between input and output is built through ordered connected layers of neurons.

CNNs involves two special matrix operators: a convolutional layer, and a pooling layer. Units in convolutional layers are only connected to specific local patches through a set of learnt filters [19]. Convolutional networks use convolution operation instead of general matrix multiplication in at least one of their layers [18]. A convolutional neural network involves four steps. 1) Convolutional; 2) Max pooling; 3) Flattening; 4) Full connection.

The input data will be filtered through a feature detector. It strides where the movement from top left to right from input matrices accordance to feature detector. The element-wise multiplication 3 by 3 feature detector matrices and produces value in a feature map. This operation reduces the dimension of input data.

Next pooling to ensure that network has a property called spatial in variance where the network must have some flexibility to find a certain feature. Several types of pooling can be used such as max pooling or mean pooling. For example, max pooling selects a maximum value from the feature map (Fig. 3).

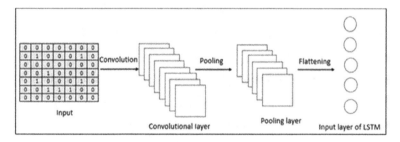

Fig. 3. CNN Step for Feature Extraction

Next, flattening is applied to pooled feature map matrices into a column. Mainly, just take the numbers row by row in the pooled feature map and arrange it into a single one-dimensional (1D) vector. The reason is to prepare the single input timestep to the long short-term memory layer for further processing. 1D CNN is mainly used for sequence data, 2D CNN frequently used for image and text processing while 3D CNN is used for medical image and video recognition. Thus, 1D CNN is adopted in this study [20]. The full connection is adding the whole artificial neural network or in our case LSTM network to CNN.

For CNN, the number of parameters calculated as per Eq. 15,

$$\text{No of Parameters} = \left[i * (f * f) * o\right] + o \tag{15}$$

where i number of input maps (or channels). f filter size (length) and lastly, o number of output maps (or channels. this is also defined by how many filters are used).

2.3 Model Evaluation

An evaluation for test samples was used to determine the model accuracy and how well it fitted the data. Cost function measured by the difference between known observations and predicted observations. The objective of many algorithms is to minimize this cost function using a gradient descent algorithm or other similar optimization techniques. Therefore, the best model can be selected [21].

The error term is defined as

$$e_t = y_t - \hat{y}_t \tag{16}$$

where y_t is known observations and $\hat{y}_t$ is predictor observation.

The root mean squares error defined as the square root of the average sum of squares of the error term:

$$RMSE = \sqrt{\frac{\sum e_t^2}{n}} \tag{17}$$

where n is a total of a sample dataset.

The mean absolute error defined as the average absolute value of error term

$$MAE = \frac{\sum |e_t|}{n} \tag{18}$$

where n is a total of a sample dataset.

The correlation coefficient, to measure the similarity between predicted and actual. Correlation is defined as the statistical association between two variables, it exists between two variables when one of them is related to the other in some way.

$$r = \frac{\sum (E - \overline{E})(O - \overline{O})}{\sqrt{\sum (E - \overline{E})^2 \sum (O - \overline{O})^2}} \tag{19}$$

where E is estimate value and $\overline{E}$ average of estimate value, while O observed value and $\overline{o}$ average of the observed value.

[22] stated the goodness-of-fit of a hydrologic or hydroclimatic model should be assessed with correlation measures and additional evaluation measures such as RMSE or MAE.

3   Result and Discussion

3.1   Preliminary Studies

Table 2 show the summary of the dataset. The minimum value for each location is zero which signifies that no rain, the maximum value of 527.5 mm and 388.5 mm are observed

Table 2. Analysis of daily rainfall amount

Station	Min	Max	Median	Average	Std	Count	Kurtosis	Skewness
Kuantan	0	527.5	0.1	8.0765	21.7255	15695	78.6658	6.8409
Kuala Krai	0	388.5	0.0	6.6545	17.7823	14600	64.8564	6.4025

at Kuantan and Kuala Krai station respectively. The standard deviations are observed quite large which 21.7255 mm in Kuantan and 17.7823 in Kuala Krai, which may lead complexity in the estimation procedure.

Rainfall amount pattern are random with a seasonal of extreme value. The histogram plot for each location in Kuantan and Kuala Krai as in Fig. 4. The histogram for each location shows that distribution highly skewed.

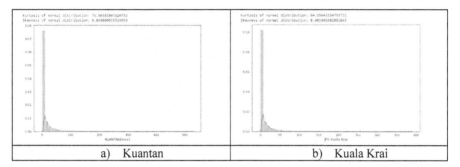

a) Kuantan   b) Kuala Krai

Fig. 4. Histogram of rainfall amount

3.2 Performance Evaluation

Each location dataset entered to LSTM network, Stacked-LSTM, CNN-LSTM, and MLP network are train with five different memory units which is (4, 8, 16, 32, 64). After several tuning with the unit selected, we will choose the best unit based on their loss values and how similar between predicted and actual value for each of the algorithms. This comparison has been created to demonstrate the performance between the four models. The experimental evaluation summary for both stations is given in Table 3.

Referring to Table 3, LSTM with 16 memory blocks will benefit the rainfall amount of data prediction in Kuantan. The MAE is the lowest in training 10.5377 and testing dataset 10.5412. Although the correlation coefficient is slightly lower than MLP with 4-unit neuron but a different range of loss value between the training and testing is not too obvious.

However, the MLP with 4-unit neurons showed the lowest RMSE which is 16.2217 in training, but stacked LSTM showed the lowest RMSE in testing 16.2662 at Kuala Krai station. In contrast, stacked LSTM exhibited lowest MAE in training and testing LSTM showed the highest correlation in training which is 0.3921 as well as the difference of

Table 3. Summary of evaluation for both gauging station

Dataset		Training			Testing			Training Time (s)
Statistical Performance		r	RMSE	MAE	r	RMSE	MAE	
Kuantan	MLP 4	0.3734	19.4459	11.0638	0.3055	21.1472	11.3571	17
	LSTM 16	0.3374	19.6764	10.5377	0.3261	20.8865	10.5412	363
	Stacked LSTM (4,3)	0.3383	19.6759	10.6592	0.326	20.8946	10.635	962
	CNN LSTM (4,3)	0.33	19.7334	10.6448	0.3136	20.9809	10.6448	324
Kuala Krai	MLP 4	0.3989	16.2217	8.4793	0.3634	16.7181	8.8019	77
	LSTM 16	0.3921	16.2998	8.609	0.4216	16.2836	8.6361	3,223
	Stacked LSTM (8, 5)	0.3848	16.3287	8.4163	0.4259	16.2662	8.4874	3,427
	CNN LSTM (4,3)	0.3281	16.7101	8.4222	0.3497	16.8331	8.5084	192

loss between training and the testing dataset is not too obvious. This signifies that the model would fit in training and testing dataset without severe underfitting or overfitting. Thus, the LSTM model is chosen.

3.3 Model Comparison and Discussion

To further investigate, the loss function graph is examined. The training procedure attempt to achieve the lowest loss value possible by examined MSE (loss) against epochs training.

In Kuantan station, Fig. 5 (a) MLP with the 4-unit neuron, the loss curves are going smoothly down, indicate that the model improves as it is training. However, after it converges, the training's loss value is lower than validation. This signifies underfit, from the graph after a certain epoch the loss value becomes stagnant. Good fit example after running the model creates a line plot showing the train and validation loss meeting at the intersection point.

For Fig. 5 (b) LSTM with 16 memory unit, the loss curves decrease as epoch increase. Then again, the training loss is lower than validation loss, this model underfit. This situation also occurs in Fig. 5 (c) Stacked-LSTM (4,3) where the first layer with 4 memory unit and second layer with 3 memory unit used. Figure 5 (d) CNN- LSTM. Generally, all models have a huge difference value between in training and testing loss.

During this work, we discovered that simple model benefits in Kuantan station prediction. Adding complexity in the model is not improving the model. Our findings are somewhat surprising since [11] stated the accuracy of LSTM model significantly depends on the number of the memory block. The increasing memory block may capture the input data better as the complexity of LSTM architecture of input gate, output gate and forget gate to filter the input and contribute model to learn better.

This denotes that if the range value of data from our data distribution is highly skewed predicted model tends to be underfit. Moreover, it is hard to detect the extreme value and at the same time to minimize generalization error. It may point out that it is hard to train when we have highly imbalanced data or skewed data [20].

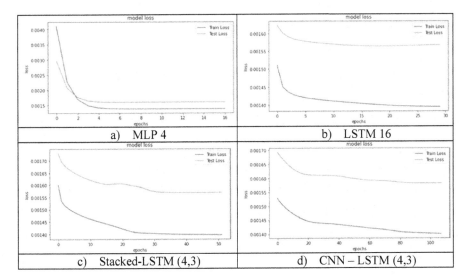

Fig. 5. Loss function in Kuantan Station

Overall, for Kuantan station, the model experience severe underfit, for model LSTM has slightly different between training and testing datasets. Hence, LSTM with 16 memory blocks is chosen as in Fig. 5 (b).

In the testing dataset, the blue and red lines represent the actual value against the predicted value. Figure 6 predicts rainfall amount for the next 200 days. The model generally captured the rain amount, but the higher peak is not able to be detected by the predicted model generated.

Following that, we may refer to Fig. 7 for Kuala Krai station. For Fig. 7 (a) MLP with 4 neuron units in Kuala Krai station, the loss function rapidly decreases as epoch increases. The training loss is slightly lower than the testing loss. The loss stagnant after a certain point means that loss does not improve after several epochs. This model shows a slight underfit.

Figure 7 (b) LSTM with 16 memory blocks has rapid change at the beginning of epoch. In the end, there is a spike where the point intersection between the training and testing meets. At this point, the training will come to end.

Figure 7 (c) Stacked-LSTM loss curves slowly decreases after certain epoch and has a spike and the end. The model is underfitting since no signal of the training and testing

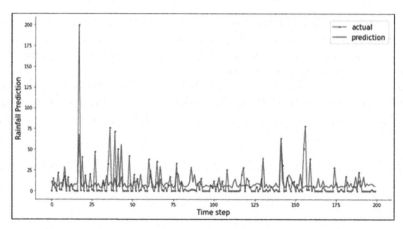

Fig. 6. Actual vs predicted for Kuantan using LSTM

meet at the intersection point. Figure 7 (d) CNN-LSTM unit in Kuala Krai station, the loss function rapidly decreases after certain epoch training. The training loss is slightly lower than the testing loss. The loss stagnant after a certain point means that loss does not improve after several epochs. This model shows a slight underfit.

365 days sliding windows allows the model to learn feature or series within that period to predict next day rainfall amount. Since the rainfall event in Malaysia is yearly recurring events therefore the choice number of windows was subjected to their characteristics as stated by [23]. It may help the model learn useful insight from the data.

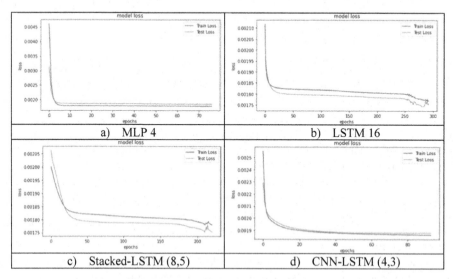

Fig. 7. Loss function of Kuala Krai Station

Overall, for Kuala Krai station the model converges quite fast, only for model stacked LSTM slightly slow. LSTM with memory block 16 is chosen as in Fig. 7 (b).

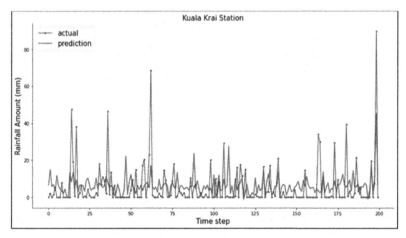

Fig. 8. The Actual and predicted rainfall amount (mm) using LSTM at Kuala Krai station

In the testing dataset, the blue and red lines represent the actual value against the predicted value. This graph shows the next 200 days of prediction. The predicted model can capture the rain amount, but the sharp fluctuation which may be due to outlier is not able to be detected (Fig. 8).

In general, since both locations having heavy-tailed distributions, this model tends to fall either overfitting or underfitting if trained with imbalance univariate time series and occurrence of extreme events lead to heavy tail distribution according to a large value of the standard deviation which increases the difficulty of prediction [23].

4   Conclusion and Recommendation

4.1   Conclusion

The aim of this study is to find the prediction of rainfall amount on the following day. The sequential data will be fed into a deep learning model after the data processing step. The result would be evaluated to identify the best univariate model. Data within flood prone areas are examined. Four deep learning methods are discussed are MLP, LSTM, Stacked-LSTM and CNN-LSTM to predict the rainfall amount and each of the methods is tune with five different sets of a parameter value which is 4, 8, 16, 32 and 64.

After examining, we discovered that adding the complexity of the parameter into the deep learning technique did not significantly improve the performance of the model. Therefore, the simpler model would perform better with the univariate rainfall amount prediction model for both location in comparison of model evaluation and performance of model either tend to underfitting or overfitting.

However, in summary, this work only offers a limited aspect of the univariate data. The hyperparameter setting with a sliding window size of 365, with an epoch of 500 and a batch size of 200, this lag will predict the amount of rain for the next day.

The choice of window size enables the model to obtain a useful pattern of data when model training. When we have a huge difference in the range of data point it will cause a model hard to train.

Overall, we should consider deep learning approaches are promising to cope with the univariate time series data in addition to non-stationary and seasonal feature as rainfall amount prediction. Besides, deep learning has flexibility in hyperparameter tuning it makes an easier adjustment during deployment to adapt in future changes.

4.2 Recommendation

In this study, the attribute of rain is random and by analysing heavy rain amount data that lead to a flood are extreme value. From the result in the previous section, we selected LSTM with 16 memory blocks in Kuantan and LSTM with 16 memory blocks chosen for Kuala Krai station. The simpler model leads to the shortest training time of the model. One of the most important findings relates to the deep learning model, that the variety of parameter and hyperparameters setting that offered easier adjustment and flexibility to the preferred station location.

The potential of this approach has not proven, it may not be appropriate to extrapolate this finding to the other flood prone area. It is difficult to assess this fully because of the random nature of the data. Our hope to examine with other type of data scaling technique or data normalization, using different sliding windows size implemented to data series and different optimizer used. Hence, in future, we required more relevant features that contribute to rainfall nature and nearby gauging station data to the flood prone area. Several interesting aspects may be explored further by implementing other hybrid methods to deep learning techniques to tackle certain limitations of the model and improve the efficiency of the imbalance data.

Acknowledgement. The authors wish to thank Malaysian Meteorological Department for the data and sponsorship from Faculty of Computer and Mathematical Science, Universiti Teknologi Mara (UiTM). The authors are also indebted to the staff of Drainage and Irrigation Department for providing the daily rainfall data for this study. They also acknowledge their sincere appreciation to the reviewers for their valuable suggestion and remarks to improve the manuscript.

References

1. Yusoff, I., Ramli, A., Alkasirah, N., Nasir, N.: Exploring the managing of flood disaster: a Malaysian perspective. Malaysian J. Soc. Space **14**(3), 24–36 (2018). https://doi.org/10.17576/geo-2018-1403-03
2. Zaidi, S.M., Akbari, A., Ishak, W.M.F.: A critical review of floods history in kuantan river basin: challenges and potential solutions. Int. J. Civ. Eng. Geo-Environ. **5** (2014)
3. Salleh, A.H., Ahamad, M.S.S.: Flood risk map (case study in Kelantan). IOP Conf. Ser.: Earth Environ. Sci. **244**, 012019 (2019). https://doi.org/10.1088/1755-1315/244/1/012019

4. Hu, C., Wu, Q., Li, H., Jian, S., Li, N., Lou, Z.: Deep learning with a long short-term memory networks approach for rainfall-runoff simulation. Water (Switzerland) **10**(11), 1–16 (2018). https://doi.org/10.3390/w10111543
5. Mohd Taib, Z., Jaharuddin, N.S., Mansor, Z.: A review of flood disaster and disaster management in Malaysia. Int. J. Account. Bus. Manag. **4**(2), 98–106 (2016)
6. Toth, E., Brath, A., Montanari, A.: Comparison of short-term rainfall prediction models for real-time flood forecasting. J. Hydrol. **239**(1–4), 132–147 (2000). https://doi.org/10.1016/S0022-1694(00)00344-9
7. Sattari, M.T., Rezazadeh-Joudi, A., Kusiak, A.: Assessment of different methods for estimation of missing data in precipitation studies. Hydrol. Res. **48**(4), 1032–1044 (2017). https://doi.org/10.2166/nh.2016.364
8. Wang, W.-C., Chau, K.-W., Xu, D.-M., Chen, X.-Y.: Improving forecasting accuracy of annual runoff time series using ARIMA based on EEMD decomposition. Water Resour. Manag. **29**(8), 2655–2675 (2015). https://doi.org/10.1007/s11269-015-0962-6
9. Mah Hashim, N., Mohd Deni, S., Shariff, S., Tahir, W., Jani, J.: Identification of seasonal rainfall peaks at kelantan using Fourier Series. In: Tahir, W., Abu Bakar, S.H., Wahid, M.A., Mohd Nasir, S.R., Lee, W.K. (eds.) ISFRAM 2015, pp. 169–179. Springer, Singapore (2016). https://doi.org/10.1007/978-981-10-0500-8_14
10. Alizamir, M., Moghadam, M.A., et al.: A hybrid artificial neural network and particle swarm optimization algorithm for statistical downscaling of precipitation in arid region, January 2017. http://journals.modares.ac.ir/browse.php?a_code=A-24-13886-1&slc_lang=en&sid=24
11. Anh, D.T., Bui, M.D., Rutschmann, P.: Long short term memory for monthly rainfall prediction in Camau, Vietnam, September 2018
12. Sulaiman, J., Wahab, S.: Heavy rainfall forecasting model using artificial neural network for flood prone Area. In: Kim, K.J., Kim, H., Baek, N. (eds.) ICITS 2017. LNEE, vol. 449, pp. 68–76. Springer, Singapore (2018). https://doi.org/10.1007/978-981-10-6451-7_9
13. Kumar, D., Singh, A., Samui, P., Jha, R.K.: Forecasting monthly precipitation using sequential modelling. Hydrol. Sci. J. **64**(6), 690–700 (2019). https://doi.org/10.1080/02626667.2019.1595624
14. Hernández, E., Sanchez-Anguix, V., Julian, V., Palanca, J., Duque, Néstor.: Rainfall Prediction: a deep learning approach. In: Martínez-Álvarez, F., Troncoso, A., Quintián, H., Corchado, E. (eds.) HAIS 2016. LNCS (LNAI), vol. 9648, pp. 151–162. Springer, Cham (2016). https://doi.org/10.1007/978-3-319-32034-2_13
15. Parmezan, A.R.S., Souza, V.M.A., Batista, G.E.A.P.A.: Evaluation of statistical and machine learning models for time series prediction: identifying the state-of-the-art and the best conditions for the use of each model. Inf. Sci. (Ny) **484**, 302–337 (2019). https://doi.org/10.1016/j.ins.2019.01.076
16. Hochreiter, S., Schmidhuber, J.: Long short-term memory. Neural Comput. **9**(8), 1735–1780 (1997)
17. Kratzert, F., Klotz, D., Brenner, C., Schulz, K., Herrnegger, M.: Rainfall – runoff modelling using Long Short-Term Memory (LSTM) networks, pp. 6005–6022 (2018)
18. Selvin, S., Vinayakumar, R., Gopalakrishnan, E.A., Menon, V.K., Soman, K.P.: Stock price prediction using LSTM, RNN and CNN-sliding window model. In: 2017 International Conference on Advances in Computing Communication Informatics, ICACCI 2017, vol. 2017-Janua, pp. 1643–1647 (2017). https://doi.org/10.1109/ICACCI.2017.8126078
19. Miao, Q., Pan, B., Wang, H., Hsu, K., Sorooshian, S.: Improving monsoon precipitation prediction using combined convolutional and long short term memory neural network. Water **11**(5), 977 (2019). https://doi.org/10.3390/w11050977

20. Li, T., Hua, M., Wu, X.: A hybrid CNN-LSTM model for forecasting particulate matter (PM2.5). IEEE Access **8**, 26933–26940 (2020). https://doi.org/10.1109/ACCESS.2020.297 1348
21. Boukary, N.A.: A comparison of time series forecasting learning algorithm on the task of predicting event timing (2016)
22. Legates, D.R., McCabe, G.J.: Evaluating the use of 'goodness-of-fit' measures in hydrologic and hydroclimatic model validation. Water Resour. Res. **35**(1), 233–241 (1999). https://doi.org/10.1029/1998WR900018
23. Hossain, N., Hossain, S.R., Azad, F.S.: Univariate time series prediction of reactive power using deep learning techniques. In: 2019 International Conference on Robotics Signal Processing Technology, pp. 186–191 (2019)

Auto-DL: A Platform to Generate Deep Learning Models

Aditya Srivastava$^{(\boxtimes)}$ (D), Tanvi Shinde, Raj Joshi, Sameer Ahmed Ansari,
and Nupur Giri

Vivekanand Education Society's Institute of Technology, Mumbai, India
{2017.aditya.srivastava,2017.tanvi.shinde,2017.raj.joshi,
2017.sameer.ansari,nupur.giri}@ves.ac.in

Abstract. Deep Learning (DL) model building is a tedious and taxing process. The number of prerequisites is high and a lot of time is invested. Hence, there is a scope of Automation. Code to build DL models follows a standard structure, broadly classified into four categories (Imports, Data Input, Model Creation, and Evaluation). The work in this paper proposes to automate this core structure and build a Graphical User Interface (GUI) based tool/platform called "Auto-DL" which, on defining the task and training data, generates code in python for the specified deep learning model. The paper then discusses the platform capabilities and evaluates it and the generated code against various quantitative and qualitative parameters.

Keywords: Deep Learning (DL) · Code generator · Neural Networks (NN) · Keras · PaaS

1 Introduction

Advanced AI adopters are seeing the most acute shortages of talent, cited by 23% in a recent Deloitte survey [6]. The lowest competency rates were observed in Deep learning techniques with Generative Adversarial Networks (Neural Network) (7%); Recurrent Neural Networks (15%) and Convolutional Neural Networks (26%) [7]. This calls for some radical change in the process of building Deep Learning models. Currently, they require a good amount of coding skills, managing libraries, choosing from different programming languages, and expertise in multiple domains, including and not limited to mathematics, deep learning, and statistics. This inhibits the enthusiasm of many and gives an impression of the process being esoteric.

The work in this paper proposes an alternative to the traditional deep learning practice by providing an easy-to-use tool that can also be used as an aid to develop Deep Learning Models, that doesn't make the users dependent on it for their deep learning needs. The work carried out could have the following impact:

1. Students, researchers, and professionals can get started with Deep Learning, without worrying about the choice of programming language, library management, dependency installation, etc., and only focus on Deep Learning.

© Springer Nature Singapore Pte Ltd. 2021
A. Mohamed et al. (Eds.): SCDS 2021, CCIS 1489, pp. 89–103, 2021.
https://doi.org/10.1007/978-981-16-7334-4_7

2. People with less to no coding background along with the ones with different coding backgrounds can come together to research/collaborate on a deep learning project.
3. For businesses and enterprises using deep learning, fast and hassle-free development of DL models is ensured, without worrying about the code implementation details.

2 Related Work

Tools like AutoHTML [2] create HTML pages with CSS based on a drawing created by the user. This is achieved by creating an intermediate representation of various blocks in the webpage. That is where the concept of creating such an intermediate representation for Deep learning models was established. PMML (Predictive Model Markup Language) [1] represents Machine learning (ML) models in XML format. But, PMML was not built for Deep Learning as it did not support many deep learning layers, so our team created DLMML (Deep Learning Model Markup Language) to represent Deep learning models in an intermediate JSON representation.

A paper [3] by Adithya Balaji and Alexander Allen uses open-source datasets and benchmarks some AutoML solutions. TPOT, H2O's AutoML and Auto-sklearn, were tested. Their analysis and benchmarks would provide guidance to Auto-DL, regarding the shortcomings of current Auto-ML solutions, and thus help us prioritize our feature development.

All open-source tools that the authors of the article [10] came across were not totally "no-code", to start the AutoML process they require users to write at least a few lines of code and need an active development environment. Auto-DL targets the generation of Deep Learning models without writing a single line of code.

The focus of Google AutoML [5] is to automate the workflow of machine learning by eliminating intermediate steps involved, like data exploration, model selection, etc. But in this process, understanding the model becomes abstruse. Hence, our focus while building Auto-DL has been transparency with respect to the architecture of the model generated as well as the provision of code for the generated solution.

Some existing products related to AutoML are IBM Watson AutoAI [8] - A Platform for automating the process of AI and Flash-ML [9] that provides UI for Pytorch models. Also, H2O [10] automates data science processes and Datarobot [11] provides solutions for Enterprise AI. MAKEML [12] is a Platform for Object Detection and Segmentation models whereas Rapid Miner [13] and TPOT [14] are tools for data mining and AutoML respectively.

In the testing & result analysis sections, Table 3 compares some features of Auto-DL against these products and the traditional way of Deep Learning. Many AutoML solutions rely on trial-and-error mechanisms and select a model from a set of existing ones, which makes tweaking the model a difficult process. Auto-DL allows customization at the layer level. Users can also use prebuilt models for transfer learning. Auto-DL plans on providing real-time collaboration while building deep learning models. Auto-DL also provides deployment solutions, but the users are not dependent on the platform for using the model and have complete authority over it.

3 Methodology Applied

The pith of Auto-DL is based on the observation that a lot of code that's written for creating deep learning models is repetitive and could be automated. Users today re-write or copy-paste this stub every time they create deep learning models. Import statements, data preprocessing, model training, and testing are major constituents of the stub. These parts of the code are automated in Auto-DL. The core, i.e. the model-building part, is semi-automated and the input is taken from the user via the frontend drag and drop GUI. For designing a model, the user specifies the data type of the training and testing data. Based on that, preprocessing parameters are captured, E.g. Image Augmentation, target size, and batch size. On dropping a layer, Auto-DL informs the user about the compatible layers that could follow this layer, as well as warns the user if he selects an incompatible layer. This is performed by checking the shapes of the selected layer. This would prevent the model training from failing at a later stage. The editable hyperparameters are Optimizer, Loss, Epochs, Learning Rate, Verbosity, and Metrics. Inputting these parameters is tantamount to filling an online form, and the desired model code is generated via a button.

Figure 1 (below) demonstrates how the Auto-DL platform works on a high level. Using the front end (GUI)[1] the user's first step is Authentication[2]. After logging in, the user can create or edit projects by specifying the details mentioned in Table 1 (below).

Table 1. Requirements for a project

Task	Whether the problem statement is Classification, Regression, etc
Name	Identification for the project
Data directory	The path of the dataset
Language	The language for the generated code. Currently, our platform supports Python
Library	Framework for generated solution Currently our platform supports Keras
Output File Name	Name for the generated file

This data is sent to the Auto-DL Backend[3] which is then forwarded to DLMML[4] (Deep Learning Model Markup Language), which parses the IR to generate its code in the specified language and framework. Training of the model can be commenced from the platform, or the code can be downloaded and run on the user's machine (preferred).

The data used for training is present in the local File System, and it is encouraged that the code is shipped to the data and not vice versa. The trained model and various explainability graphs, like training progress, can also be stored on the local machine. This provision assures that confidential data need not be uploaded to a cloud.

4 Auto-DL Platform

The Auto-DL platform can be classified into 3 components, the Frontend (A), the Backend (B), and the Database (C).

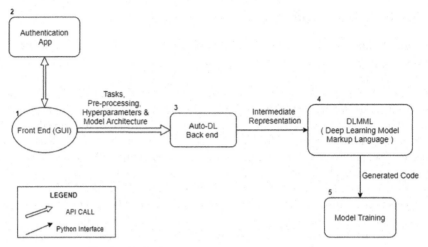

Fig. 1. Block diagram showing Auto-DL components

4.1 React Application (Frontend)

The client-side application is made with React. It offers an interface to input various parameters and hyperparameters. It also provides example values for reference. Furthermore, it encompasses the Drag and Drop feature for designing models. Various projects can be managed from the home page. Generated results like code and training results are accessed from here. The built model and the generated code are available to be downloaded. Figure 2 and Fig. 3 (below) are the home page and the project creation modal. Whereas, Fig. 4 and Fig. 5 show the model building page and download code UI of the application.

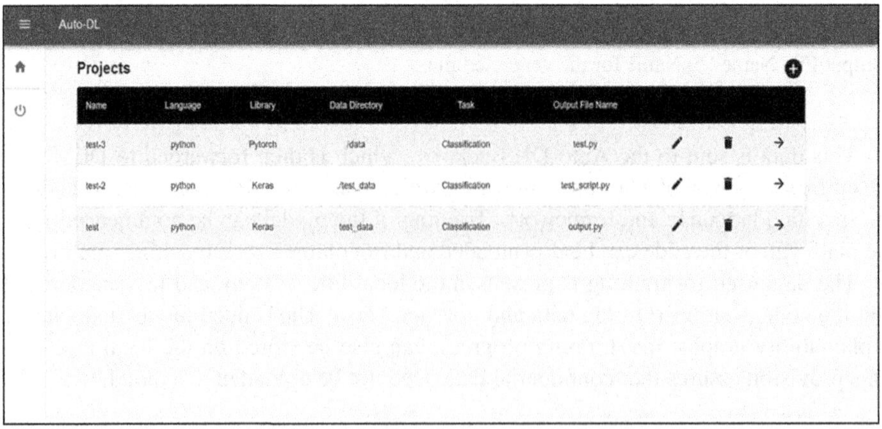

Fig. 2. Auto-DL platform home page

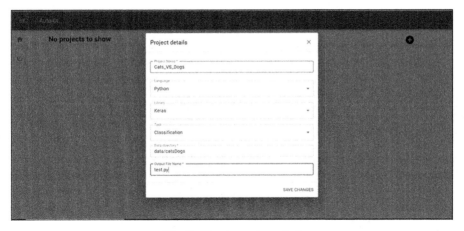

Fig. 3. Step 1: create project

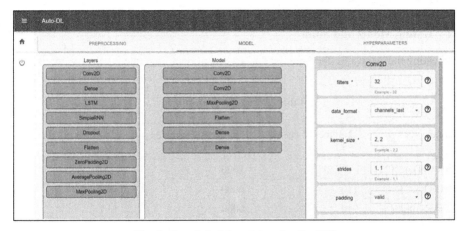

Fig. 4. Step 2: build models using the GUI

Fig. 5. Step 3: Download and Run the Generated Code

4.2 Django Application (Backend)

The Backend or the server-side application is written in Django Rest Framework. Its tasks are handling Authentication Requests with the database, providing API endpoints for the Frontend and managing project-related data and metadata, parsing of IR, converting it to code and managing training of deep learning models.

4.3 MongoDB Instance (Database)

This NoSQL database comprises User and Session data. It also contains data to support IR creation. The React app and the Django app are dockerized and placed in separate containers, whereas the MongoDB instance is on AWS cloud which is a Replica Set with 3 nodes.

5 Testing and Result Analysis

5.1 Testing Model and Suite

For testing the Auto-DL platform the datasets mentioned in Table 2 below were used.

Table 2. Datasets for testing

Name	Number of classes	Number of Samples per class	Avg No of layers Required
Cats vs Dogs [21]	2	1000	6
MNIST [22]	10	6000	9
CIFAR-10[23]	10	6000	15

5.2 About Datasets

1. Cats vs Dogs: The dataset consists of 2000 images of cats and dogs. The task is to label an image as a cat or dog i.e. a classification task.
2. The MNIST dataset consists of images of handwritten digits from 0–9, the task is to classify an image of a digit into one of the 10 classes.
3. The CIFAR-10 is a popular dataset used to train computer vision models. It contains 10 classes which are aeroplanes, cars, birds, cats, deer, dogs, frogs, horses, ships, and trucks.

5.3 Platform Usability Report

Some assumptions and definitions:

Parse Time *is the time taken for Auto-DL's Generator to parse the Intermediate Representation and generate code in python using the Deep learning framework (Keras/PyTorch) specified by the user.*

Build Time *is defined as the time taken to build a deep learning model on the Auto-DL platform such that the model is parsed and compiled correctly without any errors or warnings displayed on the webpage.*

The following chart (Fig. 5) shows the average Parse Time in Keras (red) and PyTorch (blue) for models with varying depth, having 200 to 1000 layers. For each depth (number of layers) a thousand random models were generated, and their Parse Time was recorded.

From Fig. 6, we can infer the following:

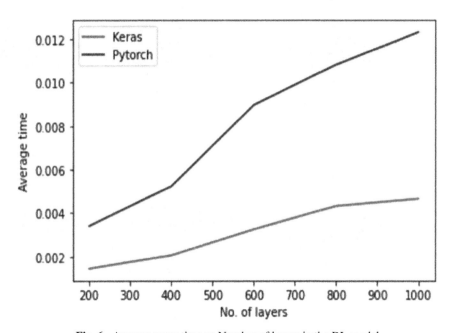

Fig. 6. Average parse time vs Number of layers in the DL model

- Parsing Keras models takes less time than PyTorch models.
- Even for a 1000 layer deep neural network, the parsing was roughly 1/10th of a second for PyTorch and 1/250th of a second for Keras.

Saving time is one of the main goals of Auto-DL and Figs. 7, 8, and 9 show how Auto-DL helped various users to generate a deep learning model of their choice in minutes. For this research purpose, a set of 10 users were given the task to generate models using Auto-DL, for datasets mentioned in Table 2.

All the 3 models were generated thrice by each user and the Build Times were recorded. Figure 7, below shows the build time for the Cats vs Dogs Classification model. Whereas Figs. 8 and 9 represent build times for MNIST and CIFAR-10 classification respectively.

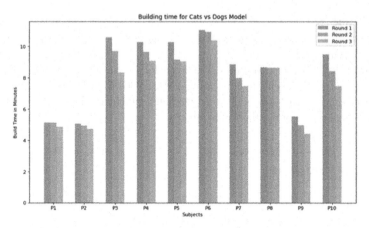

Fig. 7. Build Time for Cats vs Dogs

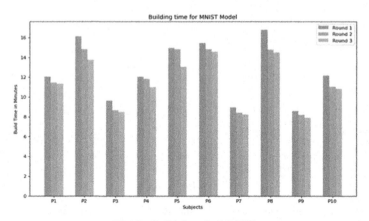

Fig. 8. Build time for MNIST

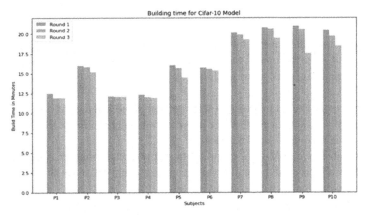

Fig. 9. Build time for CIFAR-10

From the above figures we can infer that irrespective of the model, the users take less time once they get acquainted with the platform.

The above Fig. 10 shows the average time taken by users for building the three different models on the Auto-DL platform. The upper bound for simple model building can be safely set to 20 min, which is still very fast compared to the hours that traditional deep learning takes.

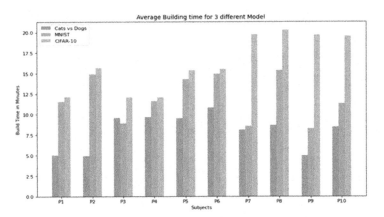

Fig. 10. Average building time

5.4   Comparative Study with Other Products

Table 3 demonstrates features of Auto-DL as compared to some other products mentioned in part II as well as the traditional way of Deep Learning, that is by writing code manually. The comparison is as based on the following parameters -

1. User Interface: The UI of various tools was compared and the easy accessibility for developing solutions using the user interface was judged. Auto-DL's user interface aims to maximize the User Experience.
2. Customizable model: Models generated by various platforms were modified and the ease and availability of the feature were captured. Auto-DL allows its users to easily modify the generated deep learning models.
3. Code: The tools were tested against their ability to generate code.
4. Ease to use: The solutions were checked for user-friendly and assisting systems like info buttons, example values, tutorials, etc.
5. Data locality: This test asks the question, does the data stay on-premises (checkmark) or it needs to be uploaded to the cloud (cross mark) for processing? Confidential data is sometimes exposed to security risks on the cloud. Not to forget the time required to upload GBs of data to the cloud.
6. Transfer learning: Support for Transfer Learning as a feature was tested. Only Auto-DL and IBM AutoAI from the cohort provide this feature.
7. Model Deployment: Trained model deployment feature was tested in various platforms.

Table 3. Comparative study Auto-DL vs Other AutoML platforms

	Auto-DL	AutoAI	Flash-ML	H2O	Datarobot	MAKEML	Rapid miner	TPOT	TraditionalDL
UI	✓	✓	✓	✓	✓	✓	✓	✗	✗
Customizable model	✓	✗	✓	✗	✗	✗	✗	✓	✓
Code Generation	✓	–	✓	✗	✗	✗	✗	✓	–
Ease to use	✓	✓	✓	✓	✓	✓	✓	✗	✗
Data Locality	✓	✗	–	✗	✗	✗	–	✓	–
Transfer learning	✓	✓	✗	✗	✗	✗	–	✗	✓
Deployment	✓	✓	–	✓	✓	✗	–	✗	–

5.5   Generated Code Evaluation

The generated code was tested against the following Software Engineering Maintainability parameters:

Code Quality. As Auto-DL generates code in python, the generated python code's quality was tested using Pylint [22]. The following figure (Fig. 11) shows the generated source code quality for 1000 different random models generated using Auto-DL. The source codes quality remains within the range of 9.40–9.75 which qualifies the threshold for production-level code as mentioned in the "Pylint Static Code Analysis | GitHub Action to fail below a score threshold" article by Analytics Vidhya [23].

Halstead Metrics [18]: These metrics are calculated statistically and provide a means to judge the code based on parameters like volume, difficulty, effort, etc. The following table (Table 4) shows values of these metrics applied on the source code generated by Auto-DL for the datasets mentioned in Table 2. The metrics were calculated using Radon [25].

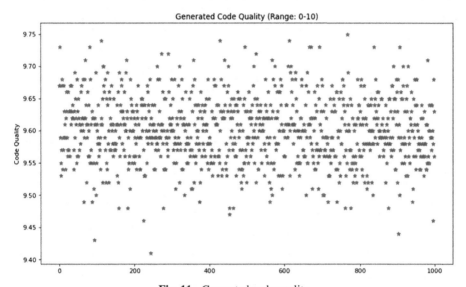

Fig. 11. Generated code quality

Even though, from Table 2 it can be observed that the number of deep learning layers required for building a deep learning model for different datasets varies; Table 4 shows that Halstead metrics remain constant. This behavior shows that an increase in the depth of the deep learning model is not detrimental to the generated code's quality and complexity. The number of bugs delivered in each case is negligible (0.0095).

5.6 Stress Testing the Auto-DL Platform

The platform was deployed on AWS t2-micro instance [20] (1 GB RAM, 1 vCPUs). The deployed architecture was stress-tested using:

i.   Cloud-based platform → Loader [26].
ii. Local testing tool → Siege [19].

Table 4. Halstead metrics

Metrics	Cats vs Dogs	MNIST	CIFAR-10
Number of Distinct Operators ($\eta 1$)	3	3	3
Number of Distinct Operands ($\eta 2$)	6	6	6
Number of Operators (N1)	3	3	3
Number of Operands (N2)	6	6	6
Vocabulary (η)	9	9	9
Length (N)	9	9	9
Calculated Length (N^)	20.2646	20.2646	20.2646
Volume (V)	28.5293	28.5293	28.5293
Difficulty (D)	1.5	1.5	1.5
Effort (E)	42.7939	42.7939	42.7939
Time (T)	2.3774	2.3774	2.3774
Bugs (B)	0.0095	0.0095	0.0095

Testing using Loader

The hosted platform was tested using the online tool with a varying load over a period of 1 min. Table 5 shows the results of these tests.

Table 5. Stress test results -- loader

Nodes	Avg. response time (ms)	Error rate (%)
1000	188	0
2000	562	1.1
5000	3162	22.1
10000	5147	40.8

Where,

1. Nodes: Concurrent clients/users
2. Avg. Response time: Average roundtrip time in milliseconds to get a response from the server throughout the test
3. Error Rate: Percentage of timed out requests

The server was still available (up and running) even if some requests timed out. On manually testing the server right when the test ended, showed that the server was running.

Testing using Siege

Two stress tests were conducted with different configurations as mentioned below:

1. Stress Test-1: The platform was attacked by a cluster of 100 nodes concurrently over a time duration of 60 s. The server survived the attack (Availability: 100%).
2. Stress Test-2: An experiment similar to Stress Test-1 was repeated by a cluster of 1000 nodes.

Similar experiments with 2000, 5000 & 10,000 nodes were planned but could not be executed due to hardware restrictions.

Table 6 shows the results of the stress tests [24]. These tests were conducted using the NPM package siege [19]. A clear inference can be drawn from the results that the platform, even when deployed on minimalistic hardware shows 100% availability (uptime) and there were no failed transactions. This buttresses the claim that the platform is durable.

Table 6. Stress test results

Metric	Test 1	Test 2	Unit
Transactions	571	1989	Hits
Availability	100	100	%
Elapsed Time	59.86	59.88	secs
Data Transferred	132.13	35.73	Mb
Response Time	8.56	11.81	secs
Transaction Rate	9.54	33.22	/sec
Throughput	2.21	0.6	Mb/sec
Concurrency	81.64	392.21	–
Successful Transactions	571	1989	–
Failed Transactions	0	0	–
Longest Transaction	53.07	44.19	–
Shortest Transaction	0.12	0.26	–

6   Conclusion

Auto-DL makes building Deep Learning models easier by automating most of its parts. With the use of Auto-DL, programming expertise as a prerequisite would be eliminated.

With the "next layer suggestion" and just-in-time error checking mechanisms, Auto-DL can serve as a good learning/teaching tool especially for people just starting with AI/DL. People with less to no programming background can get hands-on experience with building DL models. Experienced DL users can also utilize the platform to make DL models in a matter of minutes. These engagements will help to increase the acceptance and use of AI/DL in day-to-day life.

7 Future Scope

Auto-DL as a platform and as an idea is far from completion. Time and manpower were a bottleneck and the following features for the platform are pipelined or are a work in progress. Presently, Auto-DL supports image input type. The addition of other types of datasets like videos, CSV, text, etc. would make it diverse. For scalability, queuing requests, and maintaining workflows are planned. Capturing the progress of training would enable the platform to implement training analysis and make the model more explainable. Adding real-time collaboration would possibly make Auto-DL the first Deep Learning PaaS to provide the feature. Parameters like learning rate, optimizer, etc. affect the accuracy of the generated model and can be automated in the future to optimize without much human intervention. With the use of our Intermediate Representation of Deep Learning models, it is possible to achieve portability between different frameworks like Keras and PyTorch.

References

1. Predictive Model Markup Language (PMML): A XML based standard for predictive model generation, Wikipedia, 10 Oct 2005. https://en.wikipedia.org/wiki/Predictive_Model_Markup_Language. Accessed Jan 2021
2. AutoHTML: Parse Images to generate HTML Pages, Github, 06 Oct 2019. https://github.com/ADI10HERO/AutoHTML. Accessed Jan 2021
3. Balaji, A., Allen, A.: Benchmarking automatic machine learning frameworks arXiv:1808.06492 [cs.LG] (2018)
4. AutoML Software/Tools in 2021: In-depth Guide. https://research.aimultiple.com/auto-ml-software. Accessed Apr 2021
5. Google Cloud AutoML. https://cloud.google.com/automl. Accessed Apr 2021
6. The AI Talent Shortage Isn't Over Yet. https://deloitte.wsj.com/cmo/2020/10/16/the-ai-talent-shortage-isnt-over-yet/. Accessed Apr 2021
7. A majority of data scientists lack competency in advanced machine learning areas and techniques. https://businessoverbroadway.com/2018/02/18/a-majority-of-data-scientists-lack-competency-in-advanced-machine-learning-areas-and-techniques. Accessed Apr 2021
8. AutoAI with IBM Watson Studio. https://www.ibm.com/in-en/cloud/watson-studio/autoai. Accessed Apr 2021
9. Flash-ML - drag & drop machine learning models w/ PyTorch flexibility. https://www.producthunt.com/posts/flashml. Accessed May 2021
10. H2O Driverless AI. https://www.h2o.ai/products/h2o-driverless-ai/. Accessed May 2021
11. DataRobot. https://www.datarobot.com/. Accessed May 2021

12. MakeML - create object detection and segmentation ML models without code. https://mak eml.app/. Accessed May 2021
13. RapidMiner. https://rapidminer.com. Accessed May 2021
14. TPOT. http://automl.info/tpot. Accessed May 2021
15. Measuring Software Maintainability. https://quandarypeak.com/2015/02/measuring-sof tware-maintainability/. Accessed May 2021
16. Pylint: python code static checker. https://pypi.org/project/pylint/. Accessed May 2021
17. Pylint Static Code Analysis—Github Action to fail below a score threshold. https://medium. com/analytics-vidhya/pylint-static-code-analysis-github-action-to-fail-below-a-score-thresh old-58a124aafaa0. Accessed May 2021
18. Halstead Metrics. https://radon.readthedocs.io/en/latest/intro.html#halstead-metrics. Accessed May 2021
19. siege.js. https://www.npmjs.com/package/siege. Accessed May 2021
20. Amazon EC2 Instance Types. https://aws.amazon.com/ec2/instance-types/. Accessed May 2021
21. Cats vs Dogs Dataset. https://drive.google.com/file/d/16R6_yzSUHQ3BZezOBzouXBYJO_VcYW8F/view. Accessed July 2021
22. THE MNIST DATABASE of handwritten digits. Yann LeCun, Courant Institute, NYU Corinna Cortes, Google Labs, New York Christopher J.C. Burges, Microsoft Research, Redmond. http://yann.lecun.com/exdb/mnist/
23. (CIFAR-10) Learning Multiple Layers of Features from Tiny Images, Alex Krizhevsky (2009). https://www.cs.toronto.edu/~kriz/learning-features-2009-TR.pdf
24. Understanding Siege's Results. https://www.digitalocean.com/community/tutorials/how-to-benchmark-a-website-with-firefox-siege-and-sproxy-on-ubuntu-16-04. Accessed May 2021
25. Radon. https://radon.readthedocs.io/en/latest/. Accessed May 2021

A Smart Predictive Maintenance Scheme for Classifying Diagnostic and Prognostic Statuses

Revi Asprila Palembiya[1], Muhammad Nanda Setiawan[1], Elnora Oktaviyani Gultom[1], Adila Sekarratri Dwi Prayitno[1], Nani Kurniati[2], and Mohammad Iqbal[1](✉)

[1] Department of Mathematics, Institut Teknologi Sepuluh Nopember, Sukolilo, Surabaya 60111, Indonesia
iqbal@matematika.its.ac.id

[2] Department of Industrial and System Engineering, Institut Teknologi Sepuluh Nopember, Sukolilo, Surabaya 60111, Indonesia
nanikur@ie.its.ac.id

Abstract. This study attempts to propose a smart predictive maintenance method to classify manufacturing machines' diagnostic and prognostic statuses. The main goal of this study is to reduce the manual predictive maintenance budgets of manufactures in Indonesia. In the proposed method, we perform feature maps to obtain the binary states of sensor data, which is further clustered into the machine's error states (diagnostic status) and the machine' useful life states (prognostic status). Moreover, the proposed method comprises the two states predictions of machines based on *Deep Long Short Term Memory*. The proposed method is demonstrated on the Rawmill and Kiln machines of a cement factory in Indonesia for evaluation performances. Without labelling manually, we investigated the annotation of both states, which are similar to the ground truth. In addition, the proposed method can achieved high accuracy and outperformed to another baseline method.

Keywords: Smart predictive maintenance · Cement factory · Diagnostic state prediction · Prognostic state prediction · Deep learning

1 Introduction

Manufacturing sectors play an essential role to the global economic growth. According to the Minister of Economy of the Republic of Indonesia, Indonesia can be mentioned as an industrial country by viewing the economic sides. Moreover, the proportion of added value to Indonesia's GDP on manufacturing industry sector surprisingly hit over 20.61% in 2020[1] with the production value as the most influence variable for the National economic growth.

Speaking on the optimal industry production, a machine maintenance process should be considered as one of the significant step over the whole manufacturing systems[2]. By

---

[1] https://www.bps.go.id/indikator/indikator/view_data/0000/data/1214/sdgs_9/1.

[2] https://mobile.aditama-finance.com/berita/detail/320/Pentingnya-Perawatan-Mesin-Industri-yang-teratur-dan-terencana.

© Springer Nature Singapore Pte Ltd. 2021
A. Mohamed et al. (Eds.): SCDS 2021, CCIS 1489, pp. 104–117, 2021.
https://doi.org/10.1007/978-981-16-7334-4_8

means that, we minimize the total number of broken machine and suppress all possibility harm to the machine. Further, we expect to maintain the production remain in prime conditions. To do so, a Predictive Maintenance (PdM) is applied by predicting the specific crashes machine and then deciding the maintenance step [1]. An PdM has three different aspects: Remaining Useful Life (RUL), Conditional-based Maintenance (CbM), and Prognostic Health Management (PHM). In PHM, we detect the deterioration machine components, identify the fault-machine types, forecast the failure machine, and control the breakdown machine process [2]. When PHM provides RUL information (known as *prognostic distance*), we further implement to decide the best timeline of the maintenance step.

However, in real application, we mostly find the PdM process working in manual fashion. Obviously, manual procedure is very costly and also become untrustworthy prediction due to human limitation. To overcome the issue, smart predictive maintenance allows to do PdM automatically by integrating Internet-of-Things and machine learning. In a brief, we read the sensor data of the machine and then predict the failure chance based on the data using machine learning method. Subsequently, we find two main challenges: (i) determine a threshold of the sensor data whether machine failure, and (ii) predict the machine failure state. Susanto, *et al.* [3] investigated the first challenge based on a statistical-based approach focusing on cement factory, especially for Rawmill engine. Unfortunately, they did not provide an information on how successfully the implementation of the generated threshold. Moreover, we argue that a statistical-based method can handle a huge amount of sensor data. In this study, we also concentrate on cement factory with additional engine: Kiln engine. Extensively, we propose a smart predictive maintenance scheme to classify both diagnostic and prognostic machine states of the two engines based on deep learning method.

In the proposed method, we answer the first challenge by bridging domain expert information and feature map representation. We next apply clustering method[3] to annotate the diagnostic and prognostic states of the two engines. For the second challenge, this study classifies the diagnostic engine state (or the error condition) and the prognostic engine state (or the failure engine condition). Hence, the goal of this study is to reduce the budget PdM on the cement factory in Indonesia by moving manual to automatic ones. Meanwhile, some researches discussed on machine learning for smart PdM with different domain applications or scenarios. In general, T. Zonta, *et al.* [5] mentioned Long Short Term Memory (LSTM) suitable for PdM as most of its data containing timeseries. Mei Yuan, *et al.* [6] proved LSTM by showing better prognostic prediction for aero engine. Suai Zeng *et al.* [7] and Olgun Aydin [8] continued the LSTM framework of the previous research. Similar to the previous domain application, T. S. Kim [9] improved Convolutional Neural Network (CNN) on prognostic prediction and showed more accurate than vanilla CNN and LSTM on their case. Despite of their improvement model results, we still found that the vanilla CNN fail to beat LSTM. Hence, we believe to proceed with deep LSTM or D-LSTM to predict both diagnostic and prognostic states of Rawmill and Kiln engines of cement factory in Indonesia.

We summarize the contributions of this study, as follows:

---

[3] In this study, we employed Kmeans since it is more efficient than K-Nearest Neighbor [4].

- We define feature map based on binary representation to assist the labeling process of diagnostic and prognostic engine states;
- We propose a smart predictive maintenance method to predict both diagnostic and prognostic engine states by implementing deep long short term memory;
- We demonstrated the proposed method on two engines of cement factory in Indonesia: Rawmill and Kiln engines.

The rest organizations of this paper are described as follows: Sect. 2 tells some related studies on smart predictive maintenance using machine learning. Section 3 explains all the details of the proposed method. Section 4 discusses the simulation results of the proposed method on Rawmill and Kiln machines. Section 5 summaries the previous discussions.

2   Related Studies

As the fundamental part to build a smart predictive maintenance framework is the integration between sensor data and machine learning techniques, we discuss some related studies on PdM using machine learning in this section. A. Theissler *et al.* reviewed on PdM studies by stating that machine learning methods provided promising results given a large dataset but they did not explain the evaluation. Starting with traditional machine learning, Z. M. Çinar, *et al.* [10] brought discussions on how the implementation of traditional machine learning, such as: Linear Regression (LR), Support Vector Machine (SVM), Multi Layer Perceptron (MLP) on PdM without any simulation evidences. S. Cho, *et al.* [11] constructed a hybrid method based on Expectation-Maximization (EM) for PdM yet they did not provide accuracy results. Susanto, *et al.* [3] formulated a statistical query to determine a threshold which is used to judge whether the Rawmill engine is damaged.

Moving to advance machine learning method, Z. Li, *et al.* [12] proposed an unsupervised approach called SAE-LSTM to solve the fault classification (or diagnostic states) of Kiln engine. Dario Bruneo, *et al.* [13] tuned the hyperparameters on LSTM for RUL estimation (or prognostic forecasting) and also compared with SVM and shallow deep neural network. Haiyue Wu, *et al.* [14] used LSTM for both diagnostic state classification and prognostic forecasting for multi-machine power systems. In the aero-engine field, M. Yuan, *et al.* [6] applied LSTM to classify the diagnostic state and estimate the prognostic state. Furthermore, S. Zheng, *et al.* [7] delivered a comparative study of prognostic forecasting on three aero-engine problems using several machine learning methods with LSTM worked accurate in their analysis. Based on the aforementioned, this study will develop a smart predictive maintenance using deep LSTM by examining diagnostic and prognostic states classification of Rawmill and Kiln engines of cement factory in Indonesia. In the next section, we will describe the proposed method.

3   Method

We attempt to design a method for a smart predictive maintenance by classifying of error and useful life states of the engine from their embedded sensor data based on deep learning. In this application, error and useful life states are referred as diagnostic and prognostic statuses, respectively. Furthermore, we mainly focuses on two engines of cement factory, namely Rawmill and Klin. The designed method has three main stages: (i) machine state annotation, (ii) machine state model construction, and (iii) future machine state prediction. Overall, the designed method can be viewed on Fig. 1.

In the first stage, the embedded sensor data reconstruct into three binary modes to represent its active states, adopting from the feature map representation on [15]. We continue to apply the representation for annotating the diagnostic and prognostic statuses of Rawmill and Kiln engines. Noteworthy that the feature map representation solely implemented for the labeling purposes. Afterward, we build two models of diagnostic and prognostic statuses for each engine using D-LSTM. At the last stage, the future of both engines state is predicted according to the constructed model. Before we further explain the proposed method's details, we will describe some necessary notations in the next section.

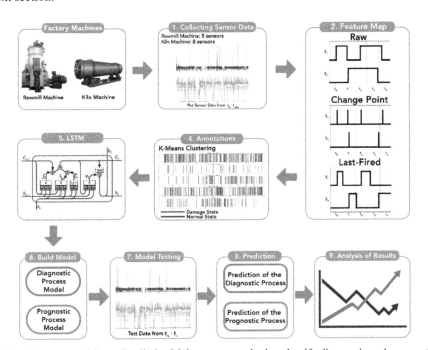

Fig. 1. The proposed Smart Predictive Maintenance method to classify diagnostic and prognostic statuses of Rawmill and Kiln engines of cement factory in Indonesia.

3.1 Notations and Problems

Following the nature application of this study, we have at most eight embedded sensors[4]. In general, we recorded the embedded sensor data into a set $S = \{s_1, s_2, \ldots, s_{|S|}\}$ with

---

[4] Kiln has full eight embedded sensors yet Rawmill only has five installed sensors: three types of rpm, one type of vibration and one type of pressure sensors.

$|S|$ is the cardinality of set S or the total number of recorded data and $s \in \mathbb{R}^n, n$ is the number of sensors. In this case, each sequence $s_i = \langle r_i^1, r_i^2, r_i^3, r_i^4, n_i^1, l_i^1, l_i^2, c_i^1, v_i^1 \rangle$ is multivariate timeseries data since it contains eight sensors information: r^1, r^2, r^3, r^4 are four types of rpm sensors (%); n is type of pressure sensor (m/bar); l^1, l^2 are two types of mass flowmeter sensors (t/h); c is type of temperature sensor $(°C)$; and v is a vibration sensor type (mm/s). In a specific case, we may have a univariate time series of vibration sensor data $\mathbf{v} = \langle v_1, v_2, \ldots v_T \rangle$ with $T = |S|$.

We assume holding information about two sets of diagnostic engine state: $D = \{d_1, d_2 \ldots\}$ and prognostic engine state: $P = \{p_1, p_2, \ldots\}$. Moreover, this study only focuses on two binary case problems with $D = \{0, 1\}$ and $P = \{0, 1\}$, where '0' refers to normal engine condition and '1' stands for error or broken engine states. We then define the two problems of this study. The **first problem** is to find a function $\mathbf{m}_d$ that maps each sequence s $\in S$ corresponding to a certain diagnostic state $d \in D$, i.e., $\mathbf{m}_d : S \rightarrow \{0, 1\}$. The **second problem** is to obtain a function $\mathbf{m}_p$ that also maps each sequence s yet with a certain prognostic state p $\in P$, i.e., $\mathbf{m}_p : S \rightarrow \{0, 1\}$. Further, this study employs a black-box model, i.e., deep long short term memory, to construct both functions or classifiers of $\mathbf{m}_d$ and $\mathbf{m}_p$. Before that, we need to describe on how to obtain the diagnostic and prognostic states of each sensor in set S.

3.2 Machine State Annotation

Originally, we wait for operator machines (human) to provide the diagnostic and prognostic statuses when generating both models $\mathbf{m}_d$ and $\mathbf{m}_p$, which is very costly. Alternatively, this study incorporates between the basic knowledge from the human experts and clustering technique to annotate the sensor data automatically (as one of the main step on smart predictive maintenance) relating to diagnostic and prognostic states information. However, the basic knowledge from the expert exclusively specifies each threshold of given sensors for diagnostic state whether the engine still 'work' properly, which is depicted on Table 1.

Table 1. Rawmill Machine regulation [2]

Equipments	Sensor types	Threshold
Motor belt conveyor (MBC)	RPM (r^1)	<33\%
Sealing air fan (SAF)	Pressure (p^1)	<9 millibar
Motor bucket elevator (MBE)	RPM (r^2)	<33\%
Rotary Feeder (RF)	RPM (r^3)	<33\%
Vibration part *mill* as a whole (VIB)	Vibration (v)	>5 mm/s

As the diagnostic threshold information is presented in binary value, we plan to discover the binary representation of each sensor by adopting the idea of the triggered sensor motion on [15]. Each sensor value is mapped into three binary representations:

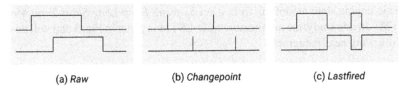

(a) *Raw*　　　　　　(b) *Changepoint*　　　　　　(c) *Lastfired*

Fig. 2. An illustration of the three binary representations of sensor data.

raw, change point, and *last fired.* An illustration of the three binary representation is shown in Fig. 2.

In a nutshell, *raw* representation of each sensor fully follows from Table 1. *Change point* representation of each sensor give value 1 whenever its *raw* representation is changing. *Last fired* representation of each sensor will be changed $(1 \rightarrow 0 \ or \ 0 \rightarrow 1)$ if any *raw* representation of other sensors is changing. We then cluster each binary representation of s yielding its diagnostic status (in line 6 of Algorithm 1). Without loss-generality, we assume that the prognostic engine state is highly related to its diagnostic state (a dependent feature). For an instance, an engine will be diagnosed in improper state when it works continuously in a very long time. Hence, we may simply judge that its useful life state at very low level. Based on that assumption, we consider to put the diagnostic engine state as one of features (in line 7) when annotating the prognostic engine state using a clustering technique (in line 9). Furthermore, to hold more evidence towards temporal information, this study performs average temporal Δt. As the results, we have a set $S' = \left\{ s'_j | 1 \leq j \leq \frac{|S|}{\Delta t} \right\}$ with $s = \left\langle s'_1, s'_2, \ldots \right\rangle$ and $s' = \frac{1}{\Delta t} \sum_{t=1}^{|s|} s_{t:t+\Delta t}$. The procedure of machine state annotation is described in Algorithm 1.

Algorithm 1 Machine State Annotation

    Input: $\mathcal{S}, \Delta t$
    Output: a set $\mathcal{S}_d$ and a set $\mathcal{S}_p$
1: **procedure** MSA($\mathcal{S}, \Delta t$)
2:     $\mathcal{S}' = \varnothing$;
3:     **for** $\forall$ sensor types $st \in \mathcal{S}$ **do**
4:         $\mathcal{S}' = \mathcal{S}' \bigcup$ feature_map(st);   ▷ binary representation
5:     **end for**
6:     $\mathcal{D} = $ Clustering($\mathcal{S}'$);   ▷ labeling diagnostic status
7:     $\mathcal{S}'_d = \{(s', d) \mid \forall s' \in \mathcal{S}', \forall d \in \mathcal{D}\}$;   ▷ add label d to s'
8:     $\mathcal{S}_d = \{(s, d) \mid \forall s \in \mathcal{S}, \forall d \in \mathcal{D}\}$;
9:     $\mathcal{P} = $ Clustering($\mathcal{S}'_d$);   ▷ labeling prognostic status
10:    $\mathcal{S}_p = \{(s, p) \mid \forall s \in \mathcal{S}, \forall p \in \mathcal{P}\}$;
11: **end procedure**

3.3 Machine State Model Construction

From Algorithm 1, we stored two labeled sequences of diagnostic and prognostic engine states S_d and S_p. We next construct diagnostic engine state model $\mathbf{m}_d$ from a set S_d and prognostic engine state model $\mathbf{m}_p$ from a set S_p using *Deep Long Short Term Memory* (D-LSTM). In this study, the architecture of D-LSTM is given in Fig. 3 and will describe in the following next paragraphs.

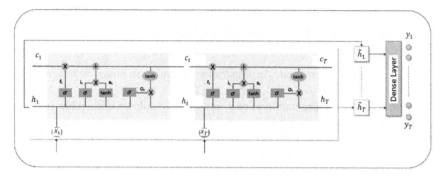

Fig. 3. The architecture of deep long short term memory.

For simplification, the following discussion will be used for both engine states. In general, we recall input sequences S and the output $Y = (y_1, y_2, \ldots, y_j)$, where v can be either diagnostic d or prognostic p. engine states. Given hidden sequences $\mathbf{h} = (h_1, h_2, \ldots, h_T)$, we compute each hidden sequence h iteratively with $t = 1, 2, \ldots, T$, as follows:

$$h_t = \phi^h(w_s \cdot s_t + w_h \cdot h_{t-1} + b_h); \tag{1}$$

with w_s, w_h and b_h are two weights and one bias parameters on hidden layers, and ϕ is a function to composite four gates: an input gate v, and output gate o, a forget gate f and a memory cell gate v. Here, we write the four gates sequentially, as follows:

- $i_t = \sigma(w_i \cdot s_t + w_{hi} \cdot h_{\{t-1\}} + w_{ci} \cdot c_{t-1} + b_i)$; *(input gate)*
- $f_t = \sigma(w_f \cdot s_t + w_{hf} \cdot h_{\{t-1\}} + w_{hc} \cdot c_{t-1} + b_f)$; *(forget gate)*
- $c_t = f_t \cdot c_{t-1} + tanh(w_i \cdot s_t + w_{hc} \cdot h_{t-1} + b_c)$; *(memory cell gate)*
- $o_t = \sigma(w_o \cdot s_t + w_{ho} \cdot h_{t-1} + w_{co} \cdot c_{t-1} + b_o)$; *(output gate)*

with $w_{(\cdot)}$, $b_{(\cdot)}$ are weight and bias corresponding to the current and next gates inside the hidden layer, and $\sigma(\cdot)$, $tanh(\cdot)$ are the activation functions. Next, we connect the information from the output gate and the memory cell gate by following a formula below.

$$\tilde{h}_t = o_t \cdot \tanh(c_t); \tag{2}$$

At the final stage D-LSTM, we compute the posterior probability between $\tilde{h}_t$ and each label or class y, as follows:

$$y_j = \arg \max_{\forall y_j \in Y} p(y_j | \tilde{h}); \tag{3}$$

$$\approx softmax(w_y \cdot \tilde{h} + b_y);\qquad(4)$$

Based on the above-mentioned, we ready to predict both diagnostic and prognostic states given new sensor data of Rawmill and Kiln engines in the next section.

3.4 Future Machine State Prediction

Let assume we have two models for diagnostic status $\mathbf{m}_d$ and prognostic status $\mathbf{m}_p$ from D-LSTM. In this step, we receive a new sensor data either from Rawmill or Kiln engines to be predicted its diagnostic and prognostic states. Let say, we collect a set of new recorded sensor data $T = \{\mathbf{t}_1, \mathbf{t}_2, \dots\}$. We then predict diagnostic status of each new sensor data according to $\mathbf{m}_d$ and also its prognostic status from $\mathbf{m}_p$. Overall, the proposed method is given in Algorithm 2.

Algorithm 2 Diagnostic and Prognostic Statuses Prediction

    Input: $\mathcal{S}, \Delta t, \mathcal{T}$
    Output: a set of predicted diagnostic status $\mathcal{T}_d$
                 and a set of predicted prognostic status $\mathcal{T}_p$
1: **procedure** DNP($\mathcal{S}, \Delta t, \mathcal{T}$)
2:     $\mathcal{S}_d, \mathcal{S}_p = \text{MSA}(\mathcal{S}, \Delta t)$;
3:     $\mathbf{m}_d = \text{D-LSTM}(\mathcal{S}_d)$;   ▷ construct diagnostic model
4:     $\mathbf{m}_p = \text{D-LSTM}(\mathcal{S}_p)$;   ▷ construct prognostic model
5:     **for** $k = 1$ to $|\mathcal{T}|$ **do**
6:         $\mathcal{T}_d = \mathcal{T}_d \cup \langle \mathbf{t}_k, \text{predict}(\mathbf{t}_k, \mathbf{m}_d)\rangle$;
7:         $\mathcal{T}_p = \mathcal{T}_p \cup \langle \mathbf{t}_k, \text{predict}(\mathbf{t}_k, \mathbf{m}_p)\rangle$;;
8:     **end for**
9: **end procedure**

---

4 Discussions

4.1 Data Information and Experiment Setups

In this study, we investigated the proposed method on two sets of sensor data of Rawmill and Kiln engines of cement factory[5]. Let S^R and S^K be two sets of embedded sensor sequences from Rawmill and Kiln engines, respectively. A set S^K consists of five sensor types: three rotational speed sensors, two pressure sensors and a vibration sensor. A set S^R consists of eight sensor types with additional two temperature sensors, and another pressure sensor. Each sensor on both engines were recorded per 10 s from the 2015 year until 2020, thus $|S^K| = |S^R| = 16,044,480$.

Before we are going further, we can see through the sensor data of Rawmill engine. We easily catch from Fig. 4 that several sensors failed to send any information yet lots of anomaly data on the 2018–2019 years can be detected easily (the purple line dots).

---

[5] The detail of data information can be found in [2, 3].

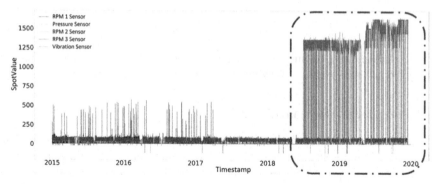

Fig. 4. Sensor data information on Rawmill engine since 2015–2019.

Hence, this study decides to play with the 2015 year data. Subsequently, $\left|S^R\right| = \left|S^K\right| = 3, 153, 600$. In this study, we set the hyperparameters for D-LSTM: Adam optimizer, learning rate $= 0.0001$, #-epoch $= 50$, batch size $= 128$, #-hidden nodes: 20, and average time $\Delta t = 1, 3, 6$ or at the time interval $t = 10$ s, 30 s, 60 s. In additions, we split the data into 80% of $|S|$ as the training set to construct the engine state model and 20% of $|S|$ for testing set to evaluate the model. Let say $z = 0.8 \times |S|$, we have $T = \{s_{z+1}, ..., s_{|S|}\}$.

4.2 State Annotation Analysis

From here, we assume to already hold nine feature representations from the combination of the three binary representations with the three average times. Next, the nine feature representations are clustered using k-Means into 'normal' and 'error' states. On the other hands, we referred the ground truth of normal engine state from [2]. We had implementing other clustering methods, such as: DB-Scan and hierarchical clustering. Unfortunately, both clustering methods need large memory and cannot be applied for this dataset. To know the similarity between the clustered and the ground-truth labels, we define the following equation:

$$\text{sim}(\%) = \frac{\# - \text{labaled state correctly}}{|S|} \times 100; \tag{5}$$

Surprisingly, the clustered diagnostic states of both engines exhibited close enough to the ground truth (see on Table 2). However, on Rawmill engine, we noticed that some binary states on changepoint representation and at an average time $\Delta t = 3$ tend to far away from the raw ones (or an original binary regulation feature). On the other sides, this study defines additional threshold for temperature sensor on Kiln engine to complete the above regulations (on the Table 1) by averaging the values. Subsequently, the similarity state on Kiln engine showed differently compare to the Rawmill engine. Hence, we continue to see the average similarity to suggest the scenario when generating the models. In this case, we can say to choose the raw and last-fired with the time interval time $t = 10$ s.

Table 2. The similarity percentage results between clustered and ground truth of diagnostic statuses of Rawmill and Kiln engines.

Scenario		Sim on rawmill (%)	Sim on Kiln (%)	Average sim
Raw	$\Delta t = 1$	**96,34**	**81,5**	**88,92**
	$\Delta t = 3$	3,72	81,35	42,535
	$\Delta t = 6$	90,87	81,35	86,11
Changepoint	$\Delta t = 1$	0,14	98,01	49,075
	$\Delta t = 3$	0,3	98,48	49,39
	$\Delta t = 6$	0,42	98,84	49,63
Lastfired	$\Delta t = 1$	**96,34**	**81,5**	**88,92**
	$\Delta t = 3$	3,72	81,35	42,535
	$\Delta t = 6$	90,87	81,35	86,11

4.3 Diagnostic and Prognostic Engine State Models Analysis

We discuss on how the performance of D-LSTM given the feature map representation in this part. As D-LSTM is one kind of black-box model, we need to evaluate the model performances based on the given loss function. This study applied a binary cross-entropy function since this study's problems are considered as binary classification tasks. Overall, we depicted the loss function of all nine features of both engines on Fig. 5.

Changepoint representation for Rawmill engine generally turned to either overfit or underfit phenomena after 30 epoch. More specifically, at time interval $t = 10$ s, it performed horrible since its loss value blasted over 20% at early stage (before 10 epoch, see on Fig. 5 (a) Rawmill - changepoints). These results are in conformity with the previous findings on the annotation analysis. Altogether, the model converges slower during the learning process when the time interval sets larger since the epoch number at a time interval $t = 10$ s is the smallest than the other time intervals. In this case, the model is difficult to recognize the averaging sensor information. Furthermore, both raw and last-fired representations mostly displayed similar convergent patterns on the learning curves. To seek more evidence the model evaluation, we carry on to analyze the models towards the unseen data (or testing data) in the next discussion.

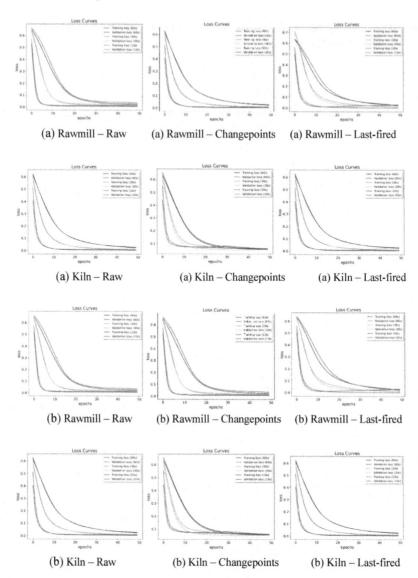

Fig. 5. The evaluation of loss function for (a) diagnostic and (b) prognostic engine state models.

4.4 Testing Model Performances Analysis

We finally analyze the models with the testing data on the prediction step. We also provide the comparison analysis with a traditional machine learning method. This study brought Gaussian Naïve Bayes (GNB) as the baseline method. In additions, this study utilized accuracy metric in order to evaluate the models, as follows:

$$\text{acc}(\%) = \frac{\# - \text{lengine state of } \mathcal{T} \text{ predicted correctly}}{|\mathcal{T}|} \times 100; \qquad (6)$$

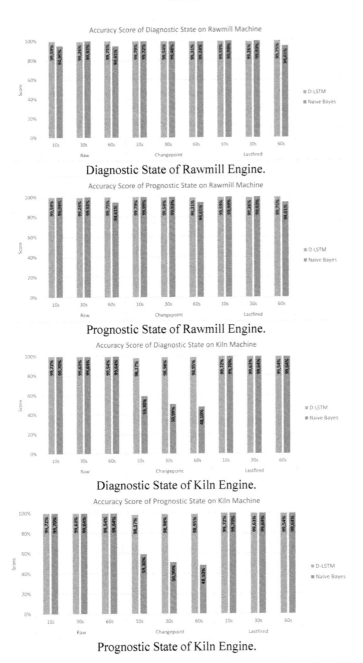

Diagnostic State of Rawmill Engine.

Prognostic State of Rawmill Engine.

Diagnostic State of Kiln Engine.

Prognostic State of Kiln Engine.

Fig. 6. The Accuracy Comparisons of D-LSTM and GNB Methods on Rawmill and Kiln Engines.

where, $|T|$ is the cardinality of testing set.

As this study focused on binary classification case and its data is huge, both D-LSTM and GNB can achieved high accuracy for all engine state predictions as depicted on Fig. 6. However, D-LSTM performed more stable than GNB in average since we know that GNB will neglect the temporal dependencies or even the dependency of each feature. That is why, we captured the accuracy of GNB slightly decrease as the time interval is larger for Kiln engine. Also, GNB is seen difficult to learn the changepoint feature as we only consider the changing state of engine. The accuracy GNB for changepoint is dramatically dropped at less than 60% level and also it went down again when we increased the average time. For Rawmill engine, it similarly happend at the time interval $t = 60$ s with decreasing at most 5 percent. However, GNB still reached almost perfectly accurate only when the time interval at $t = 10$ s.

As we expected from our idea, we obtained the similar accuracy on diagnostic and prognostic engine states of D-LSTM since we found the dependencies between two states. Unlike GNB for Rawmill engine, D-LSTM showed the highest accuracy at time interval $t = 60$ s since D-LSTM is considered as sequence model. In general, we suggest to set either raw or last-fired features with time interval $t = 60$ s for Rawmill engine. Even though, we can see that changepoint feature reached the highest accuracy but we found overfitting on the learning curves (see on Fig. 5). It means that we highly meet with the uncontrollable situation. For prognostic engine, we choose to have either raw or last-fired features with time interval $t = 10$ s.

5   Conclusion

We proposed a smart predictive maintenance scheme to classify the diagnostic and prognostic engine states of Rawmill and Kiln engines of cement factory in Indonesia. In the proposed scheme, we presented labeling method based on feature map representation and clustering approach. Next, we classified the diagnostic and prognostic engine states using deep learning technique. The proposed method can successfully annotated both diagnostic and prognostic states over Rawmill and Kiln engines with the similarities to the ground-truth are very close. Also, we can predict the future diagnostic and prognostic states of both engines accurately. We believe that the proposed method can be applied to other industries, such as, ship or train engines and etc. Additionally, we progress to forecast the error and failure times of Rawmill and Kiln engines in the near future.

References

1. Ran, Y., Zhou, X., Lin, P., Wen, Y., Deng, R.: A survey of predictive maintenance: systems, purposes and approaches. arXiv preprint arXiv:1912.07383 (2019)
2. Farsi, M.A., Zio, E.: Industry 4.0: some challenges and opportunities for reliability engineering. Int. J. Reliab. Risk Saf.: Theory Appl. **2**(1), 23–34 (2019)
3. Susanto, B., Kurniati, N.: Multi sensor-based failure diagnosis using the mahalanobis taguchi system. In: IOP Conference Series: Materials Science and Engineering, vol. 847, no. 1, p. 012036. IOP Publishing (2020)

4. Ku´Smirek, W., Szmur lo, A., Wiewi´Orka, M., Nowak, R., Gambin, T.: Comparison of KNN and k-means optimization methods of reference set selection for improved CNV callers performance. BMC Bioinform. 20(1), 1–10 (2019)
5. Zonta, T., da Costa, C.A., da Rosa Righi, R., de Lima, M.J., da Trindade, E.S., Li, G.P.: Predictive maintenance in the Industry 4.0: a systematic literature review. Comput. Ind. Eng. 106889 (2020)
6. Wu, Y., Yuan, M., Dong, S., Lin, L., Liu, Y.: Remaining useful life estimation of engineered systems using vanilla LSTM neural networks. Neurocomputing **275**, 167–179 (2018)
7. Zheng, S.: Long short-term memory network for remaining useful life estimations. In: 2017 IEEE International Conference on Prognostics and Health Management (ICPHM). IEEE (2017)
8. Aydin, O., Guldamlasioglu, S.: Using LSTM networks to predict engine condition on large scale data processing framework. In: 2017 4th International Conference on Electrical and Electronic Engineering (ICEEE), pp. 281–285. IEEE (2017)
9. San Kim, T., Sohn, S.Y.: Multitask learning for health condition identification and remaining useful life prediction: deep convolutional neural network approach, pp. 1–11 (2020)
10. Çinar, Z.M., Abdussalam Nuhu, A., Zeeshan, Q., Korhan, O., Asmael, M., Safaei, B.: Machine learning in predictive maintenance towards sustainable smart manufacturing in Industry 4.0. Sustainability **12**(19), 8211 (2020)
11. Cho, S., et al.: A hybrid machine learning approach for predictive maintenance in smart factories of the future. In: Moon, I., Lee, G.M., Park, J., Kiritsis, D., von Cieminski, G. (eds.) APMS 2018. IAICT, vol. 536, pp. 311–317. Springer, Cham (2018). https://doi.org/10.1007/978-3-319-99707-0_39
12. Li, Z.: Deep learning driven approaches for predictive maintenance: a framework of intelligent fault diagnosis and prognosis in the Industry 4.0 era (2018)
13. Dario Bruneo, F.D.V.: On the use of lstm networks for predictive maintenance in smart industries. In: 2019 IEEE International Conference on Smart Computing (SMARTCOMP). IEEE (2019)
14. Wu, H., Huang, A., Sutherland, J.W.: Avoiding environmental consequences of equipment failure via a LSTM-based model for predictive maintenance. Procedia Manuf. **43**, 666–673 (2020)
15. van Kasteren, T.L.M., Englebienne, G., Kröse, B.J.A.: Human activity recognition from wireless sensor network data: Benchmark and software. In: Chen, L., Nugent, C., Biswas, J., Hoey, J. (eds.) Activity Recognition in Pervasive Intelligent Environments. Atlantis Ambient and Pervasive Intelligence, vol. 4, pp. 165–186. Springer, Heidelberg (2011). https://doi.org/10.2991/978-94-91216-05-3_8

Data Analytics and Technologies

Optimal Portfolio Construction of Islamic Financing Instrument in Malaysia

Muhammad Firdaus Hussin[1], Siti Aida Sheikh Hussin[2(✉)], and Zalina Zahid[2]

[1] Zuzu Hospitality Solutions, 59200 Bangsar South, Kuala Lumpur, Malaysia
[2] Universiti Teknologi MARA, 50400 Shah Alam, Selangor, Malaysia
sitiaida@tmsk.uitm.edu.my

Abstract. Conventional financing and Islamic financing have the same mechanism of operation, although their nature is different. This difference creates different risk, which requires a proper investigation and observation. Thus, the conventional portfolio manager cannot use the same set of proportion or strategy when they want to invest in Islamic financial product. This study determines the optimal portfolio combination and its proportion for Islamic bank financing which include several contracts (Murabahah, Mudharabah, Musharakah, Bai Bithaman Ajil, Ijarah, Ijarah Thumma Al Bai, Istisna'). We apply single index model (SIM) since SIM enables precise calculation of the composition of each asset (financing) by identifying the value of Excess Return to Beta (ERB) as well as the cut-off point based on the acquisition of equivalent rate of profit sharing for each financing. The results show that the optimal composition of portfolio consist of Ijarah or leasing (59.93%), Musharakah or joint venture (29.18%) and Murabahah or sales (10.89%). The portfolio expected return is 1.20% with portfolio risk of 1.67% .

Keywords: Financial analytics · Portfolio optimisation · Islamic finance

1 Introduction

Shariah compliance or halal investments are subject to Islamic foundations that include sharing of profit, prohibition of exploitative gains (riba'), prohibition of gambling, investing in responsible/lawful/ethical activities or organizations and upholding moral values at all times. These foundations reduce the flexibility of investment. Thus, narrowing the choice of investments portfolio.

Every investor, portfolio manager or any financial institutions wants as high return as they could while having the lowest risk possible in their portfolio by including the right assets in a portfolio. Determining how much to invest on a particular asset is very challenging.

Portfolio construction is an ultimate key for an investor to formulate a successful investment. The significance of having well-constructed portfolio by not "putting all eggs into a basket" is strongly related to risk diversification concept. A well-diverse portfolio eliminates the unsystematic risk without affecting the expected gain of an investment. Investors strive to maximize their expected return with the minimum possible risk. The

© Springer Nature Singapore Pte Ltd. 2021
A. Mohamed et al. (Eds.): SCDS 2021, CCIS 1489, pp. 121–132, 2021.
https://doi.org/10.1007/978-981-16-7334-4_9

application of optimal portfolio construction towards Islamic Financing is intriguing, as the characteristics differs with the conventional.

Islamic financing instruments can be categorized into two types of financing, equity based financing (Mudharabah and Musharakah) and Debt based Financing (Murabahah, Bai Bithaman Ajil, Ijarah, Ijarah Thumma Al-Bai and Istisna') [11]. As the scope of the research is specific to Islamic financing, we include the definition of each financing instrument. The definition is as follows:

1.1 Equity Financing Instruments: Profit Sharing Contracts

Mudharabah (Trustee Profit and Loss Sharing)
Mudharabah is a form of partnership in which the investor (Rabbul- mal) provides 100% of the project capital, while the entrepreneur (Mudarib) oversees the venture by utilizing his/her ability. Earning from the investment shared based on the pre-agreed profit-sharing ratio, but losses carried by the provider of the funds except for real cases of carelessness by the Mudarib.

Musharakah (Partnership or Joint Venture)
Musharakah is similar to Mudharabah contract, but all partners contribute some of their capital to the investment. Profits shared in a pre-agreed ratio between partners, but losses shared in the exact proportion of the money invested by the party.

1.2 Debt Financing Instruments

Murabahah (Cost Plus Sales)
This is one of the most widely used methods of financing by Islamic financing instruments. Consumers, property, and industry use this instrument, Murabahah to finance the purchase of raw materials, machinery or equipment. It is however, most commonly used in short-term commercial finance, including the financing of credit letters. Murabahah refers to the sale of goods at a price that includes an earning margin agreed by the parties concerned. What makes the transaction Murabahah Islamically legitimate in Fiqh, is that the bank acquires the asset for resale to get a profit so that a product sold for money, and the trade is not an exchange of money for money.

Al-Bai Bithaman Ajil
Bithaman Ajil is a delayed sale whereby the entrepreneur capitalizes on his or her earnings in the sale of the property to the customer who is obliged to pay a fixed amount until the end of the tenure. In the Bithaman Ajil contract, the entrepreneur sells the house to the customer at a mark- up the price, the subject matter of which is the cost

price plus the mark-up the entrepreneur wants to make, for example, over a specified funding period of 20 years.

Al-Ijarah
Ijarah is a contract under which the tenant leases to the customer equipment, building or other facilities at an agreed rental fee(s) or fee(s), as decided by both parties.

Al-Ijarah Thumma al Bai
Financing products that allow customers to lease assets from Islamic investment and banking institutions with the option of purchasing leased assets at the end of the lease.

Al-Istisna' (Commissioned Manufacturer)
In Istisna' one party purchases the good and the other party undertakes to produce them as specified. This production undertaking includes all manufacturing, construction, assembly, and packaging processes. It is also a pre-shipment financing instrument and a contract in which the deal can be referred to something that does not exist at the time the agreement is concluded.

2 Literature Review

William Sharpe took the initiative to remodel the Markowitz model and produced Sharpe's Single Index Model. SIM simplifies the form of Markowitz model and able to define the correlation between the return of individual security with market index return by dividing the return of security into two components. Return that does not affected by market return is represented by alpha and vice versa of the previous relationship will be represented as beta [10]. SIM is suitable to apply when deciding the optimal portfolio because it analyses how and why securities should be included. [7] justified that the reason he applied Sharpe Single Index Model is due to the method requires fewer inputs as compared to, Markowitz Mean-Variance model.

SIM construction is less time consuming even when large number of securities are included in the portfolio. SIM helps in the construction of optimal portfolio by determining the security to be assimilated into the portfolio based on Excess Return to Beta Ratio (ERB). ERB ratio defines as a measure of excess return relative to one unit of risk that cannot be dispersed known as beta. These proportions tell the association between determinant of asset venture which is uncertainty and returns. The ratio between the excess return to beta ratio used to conclude which security can be integrated into the portfolio in order to build an optimal portfolio while the cut-off point will determine the security that will be included in the optimal portfolio based on the ERB low and high point [6]. Moreover, cut off rate plays vital role in constructing an optimal portfolio, which eases the investor to make decisions [8].

One of the hurdles in establishing this optimal financing portfolio is the need to assess the return behaviour of each Islamic financing instrument, that is, the expected return of each instrument when combined with other instruments (combination of more than one instrument). This is to ensure the expected return of the established financing portfolio is profitable, at the same time, avoiding high risk financing contracts with investors [1].

Most of the Islamic products offered are comparable to conventionally available products even though interest and speculation are banned in the products and services. Since there are too many products available generally, this research is focusing on Islamic financing instruments that are commonly used and has a significant influence in Malaysia market.

According to [5], the risks in Islamic banks are influenced by three factors. The first factor is the principle of the instruments or products they offer. For instance, equity-based financing increases the risk to Islamic banks. Secondly, Shariah's compliance constraint. Since the operations of Islamic banks are compelled by the Shariah principle, specific risk management strategies such as hedging and credit derivatives are not allowed in Islamic banks. Lastly is failure to standardize and regulate the Islamic banking.

Since conventional banks face three significant risks, namely credit risk, market risk, and operational risk, Islamic banks are also equally faced with these risks. The perception that Islamic banks are risk-free is incorrect and understated.

3   Methodology

In this study we are interested in the share proportion of Islamic Financing instruments in Malaysia. The data are obtained from Bank Negara Malaysia for the duration of five years from year 2014 until 2018. The selected tools or instruments are Mudharabah, Musharakah, Murabahah, Istisna', Ijara, Ijarah Thumma Al-Bai, and Bai Bithaman Ajil. The flow chart of the methodology is shown in Fig. 1.

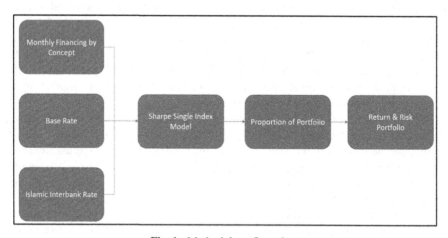

Fig. 1. Methodology flow chart

Monthly financing by concept (Mudharabah, Musharakah, Murabahah, Istisna', Ijara, Ijarah Thumma Al-Bai, and Bai Bithaman Ajil) in million (RM), base rate or risk free rate, and Islamic Interbank rate are the independent variables or input used to determine the proportion or weight of the optimal portfolio constructions. Return and risk of the optimal portfolio are calculated based on the results of the optimal portfolio of financing instruments from Sharpe Single Index Model.

3.1 Capital Asset Pricing Model

The fundamental underlying the single index model is:

$$R_i = \alpha + \beta_i R_m + e_i \tag{1}$$

Where,

R_i: return of Islamic financing instrument i
α: return of financing instrument i that is not influenced by market performance
β_i: sensitivity of return of financing instrument i to the changes in market return
R_m: market index return
e_i: error term

According to SIM, the return of any stock decomposed into the expected excess return of the individual stock due to firm-specific factors, commonly denoted by its alpha coefficient (α), which is the return that exceeds the risk-free rate. While the return due to macroeconomic events that affect the market, and the unexpected microeconomic events that affect only the firm denoted as beta coefficient (β_i) [2].

The term $\beta_i R_m$ represents the stock's return due to the movement of the market modified by the stock's beta (βi), while e_i represents the unsystematic risk of the security due to firm-specific factors. In our research i is Islamic financing instruments [4].

3.2 Construction of Optimal Portfolio

The creation of an optimal portfolio of Islamic financing instrument assuming the formation of optimal portfolio is only based on Excess Return Beta (ERB) ratio and no short sales are allowed. All formula used to construct an optimal portfolio are based on [1, 3, 6, 9]. The steps in creating the optimal portfolio are as follows:

1. Calculate the monthly return of each Islamic financing instrument using Eq. 2.
2. Calculate the the market return.

$$R_i = \frac{P_t - P_{t-1}}{P_{t-1}} \tag{2}$$

Where,

R_i: Monthly return financing instrument
P_t: Current month financing amount
P_{t-1}: Previous month financing amount

3. Calculate Beta for Each Financing Instrument Using Eq. 3.

$$\beta_i = \frac{\sigma_{im}}{\sigma_m^2} \tag{3}$$

Where,

β_i: Beta for each financing instrument
σ_{im}: Covariance between financing instrument and market
σ_m^2: Variance of market

4. Measure unsystematic risk, σ_{ei}^2

$$\sigma_{ei}^2 = \sigma_i^2 - \beta_i^2 * \sigma_m^2 \tag{4}$$

Where,
σ_i^2: Variance of Islamic financing instrument

5. Compute the ERB using Eq. 5 and rank each financing instrument from the highest value of ERB to the lowest.

$$\frac{\overline{R}_i - R_f}{\beta_i} \tag{5}$$

6. Calculate the cut off rate ratio for each financing instrument (Ci). This is to determine the cut-off ratio (C*). C* is the highest value of Ci estimation of each financing that had been previously ranked based on ERB ratio from most elevated to least. C* is a benchmark used to decide whether to include or exclude the financing instrument as a candidate of the portfolio.

$$C_i = \frac{\sigma_m^2 \sum_{i=1}^{i} \frac{(\overline{R}_i - R_f)\beta_i}{\sigma_{ei}^2}}{1 + \sigma_m^2 \sum_{i=1}^{i} (\frac{\beta_i^2}{\sigma_i^2})} \tag{6}$$

Where,

σ_m^2: Variance of market benchmark.
β_i: Beta of individual financing instrument.
σ_{ei}^2: Variance error of each financing instrument.
σ_i^2: Variance of individual financing instrument.
C_i: Cut off rate of individual financing instrument.
$\overline{R}_i - R_f$: Excess Return.

7. When the financing that shapes the portfolio has been resolved, the next stage is to decide the weight or proportion from each chosen financing in the portfolio using Eq. 7 and Eq. 8.

$$w_i = \frac{Z_i}{\sum_{i=1}^{i} Z_i} \tag{7}$$

$$Z_i = \frac{\beta_i}{\sigma_{ei}^2}(\frac{\overline{R_i} - R_f}{\beta_i} - C^*) \tag{8}$$

Where,

β_i: Beta of individual financing instrument
σ_{ei}^2: Variance error of each financing instrument
C^*: Highest value of Ci estimation
$\overline{R_i} - R_f$: Excess Return

3.3 Portfolio Return and Risk

Lastly, expected return and expected risk level of optimal portfolio is evaluated by applying Eq. 9 and Eq. 10.

$$\text{Portfolio Return} = \sum w_i \overline{R_i} \tag{9}$$

Where,

w_i: Weightage of each financing instrument
$\overline{R_i}$: Expected return on financing instrument

Portfolio Risk

$$\sqrt{(\sum w_i \beta_i)^2 * \sigma_m^2 + \sum w_i \sigma_{ei}^2} \tag{10}$$

Where

w_i: Weightage of each financing instrument.
$\overline{R_i}$: Expected return on financing instrument

4   Results and Discussion

4.1 A Composition of Islamic Financing Instrument (IFI)

Table 1 shows the amount of financing in Ringgit Malaysia for each instrument from 2014 until 2018. On average, the highest amount of financing for the last 5 years is Murabahah while the lowest amount of financing is Mudharabah.

4.2 Comparing Mean Return, Variance and Risk of IFI

Theoretically, mean return is the expected value of all the likely return of investments. In this case, two instruments are recognized as producing negative mean return as shown in Table 2, which are Bai Bithaman Ajil and Mudharabah. In general, positive return financing instrument means it is feasible as an alternative for investing the instrument that gives the highest mean return is Murabahah (2.34%). Whereas, Istisna' (10.37%) has the highest coefficient of variation implying that it is the riskiest instrument among the rest instrument while Musharakah (0.83%) indicates it is the least risky instrument.

Table 1. Amount of Financing for each instrument yearly

Year	2014	2015	2016	2017	2018	Average
Bai Bithaman Ajil	82174.08	77549.30	71204.64	67012.18	63318.97	72251.83
Ijarah	7473.12	9052.73	9828.70	10330.34	9992.87	9335.55
Ijarah Thumma Al-Bai	66270.10	71568.79	70382.98	70767.81	72164.65	70230.87
Murabahah	67215.67	107935.66	143910.43	176542.43	209728.03	141066.44
Musharakah	19618.69	25757.94	34866.97	44715.99	51126.30	35217.18
Mudharabah	91.24	78.45	73.27	66.56	53.58	72.62
Istisna'	1694.95	2033.54	2224.93	1998.50	1972.96	1984.98
Total Financing	**244537.84**	**293976.40**	**332491.93**	**371433.81**	**408357.36**	**330159.47**

Table 2. Composition of mean, risk and coefficient of variation of IFI

Instrument	Mean Return	Variance	Risk (Standard deviation)	Coefficient of variation
Bai Bithaman Ajil	**−0.48%**	**0.04%**	**1.90%**	**−3.99%**
Ijarah	0.61%	0.02%	1.53%	2.51%
Ijarah Thumma Al-Bai	0.22%	0.00%	0.50%	2.33%
Murabahah	2.34%	0.06%	2.44%	1.04%
Musharakah	1.98%	0.03%	1.65%	0.83%
Mudharabah	−1.71%	0.51%	7.11%	−4.16%
Istisna'	0.33%	0.12%	3.43%	10.37%

4.3   Capital Asset Pricing Model of IFI

Table 3 shows the regression yield for each financing instrument. $\hat{\alpha}$ represents the estimate return of financing instrument, which is not influenced by the performance of the market. Positive values show that the investment performance is better than expected investment return and vice versa. Based on Table 3, there are five financing instruments which the value of $\hat{\alpha}$ is positive, that are Ijarah, Ijarah Thumma Al-Bai, Murabahah, Musharakah and Istisna' while two instrument produced negative value, which are Bai Bithaman Ajil and Mudharabah.

$\hat{\beta}$ represents the estimate sensitivity measure of expected return depending on the market movement or changes. Based on Table 3, only two financing instrument, Ijarah Thumma Al-Bai (11.97%) and Istisna' (46.84%) show positive value while the rest are

Table 3. Capital asset pricing model of IFI

Instrument	Alpha ($\hat{\alpha}$)	Beta ($\hat{\beta}$)	Residual (e)
Bai Bithaman Ajil	−0.47%	−4.02%	1.93%
Ijarah	0.62%	−6.96%	1.55%
Ijarah Thumma Al-Bai	0.19%	11.97%	0.48%
Murabahah	2.36%	−11.13%	2.48%
Musharakah	2.05%	−41.56%	1.58%
Mudharabah	−1.65%	−33.82%	7.22%
Istisna'	0.25%	46.84%	3.43%

negative. This indicates that one percent increase of market return (Islamic Interbank rate) will increase or decrease the expected return of IFI.

4.4 Excess Return to Beta Ratio and Ranking Procedure

According to [7], negative expected return omitted in the selection of optimal portfolio as recommended by Sharpe since it will face investment risk. Hence, this study divides the construction optimal portfolio accordance to two groups of financing instrument; without negative mean instrument (NMI) return included and negative mean instrument (NMI) return included.

Table 4. Ranking based on ERB (NMI included) and (NMI excluded)

Instrument	ERB	Instrument	ERB
Bai Bithaman Ajil	1.0976	Ijarah	0.4783
Ijarah	0.4783	Murabahah	0.1438
Mudharabah	0.1670	Musharakah	0.0472
Murabahah	0.1438	Istisna	−0.0771
Musharakah	0.0472	Ijarah Thumma Al-Bai	−0.3111
Istisna	−0.0771		
Ijarah Thumma Al-Bai	−0.3111		

Based on Table 4, the highest ERB value is Bai Bithaman Ajil for NMI included while the highest value of ERB is Ijarah for NMI excluded (drop Bai Bithaman Ajil and Mudharabah). Negative ERB value is likely to be excluded in the formation of optimal portfolio development since excess return refer to return received above risk free investment return.

4.5 Selecting Instrument from Cut-Off Rate

Positive ERB value of financing instrument expected to be included in the creation of the optimal portfolio but certification to be accepted based on the cut off ratio. Cut off ratio (C*) is the maximum value of C_i whether to accept or reject the instruments into the portfolio candidate.

Table 5. Cut-off ratio (NMI included)

Instrument	$\dfrac{(\overline{R_i}-R_f)\beta_i}{\sigma_{ei}^2}$	$\dfrac{\beta_i^2}{\sigma_i^2}$	C_i	Status
Bai Bithaman Ajil	4.9358	3931	0.0009	Pass
Ijarah	14.8626	25.1686	0.0028	Pass
Mudharabah	18.6540	47.7702	0.0034	Pass
Murabahah	21.6533	68.5515	0.0040	Pass
Musharakah	55.6170	702.8339	0.0092 (C*)	Pass
Istisna	40.7052	889.5976	0.0065	Fail
Ijarah Thumma Al-Bai	−155.5310	1454.1051	−0.0228	Fail

Table 6. Cut-off ratio (NMI excluded)

Instrument	$\dfrac{(\overline{R_i}-R_f)\beta_i}{\sigma_{ei}^2}$	$\dfrac{\beta_i^2}{\sigma_i^2}$	C_i	Status
Ijarah	9.9268	20.6755	0.0018	Pass
Murabahah	12.9260	41.4568	0.0024	Pass
Musharakah	46.8897	675.7392	0.0078 (C*)	Pass
Istisna	31.9780	862.5030	0.0051	Fail
Ijarah Thumma Al-Bai	−164.2582	1427.0104	−0.0242	Fail

Based on Table 5, five instruments pass and should be included in the portfolio, while in Table 6, three instruments pass as candidates to be included in the optimal portfolio.

4.6 Proportion of Each Selected Instruments

Once all the preferred and selected instruments have been decided and tested, the proportion of the instruments calculated to give the most favourable portfolio return and risk (Table 9).

Table 7 shows the respective weightage of Bai Bithaman Ajil (33.50%), Ijarah (38.50%), Mudharabah (2.92%), Murabahah (6.95%) and Musharakah (18.14%) for NMI while Table 8 illustrates the financing weightage of Ijarah (59.93%), Murabahah (10.89%) and Musharakah (29.18%) for NMI excluded.

Table 7. Weightage of financing instruments (NMI included)

Selected instruments						
Instrument	β	σ_{ei}^2	ERB	Ci	Zi	w_i
Bai Bithaman Ajil	−0.0402	0.0004	1.0976	0.0009	−121.66	(33.50%)
Ijarah	−0.0696	0.0002	0.4783	0.0028	−139.86	(38.50%)
Mudharabah	−0.3382	0.0050	0.1670	0.0034	−10.59	(2.92%)
Murabahah	−0.1113	0.0006	0.1438	0.0040	−25.23	(6.95%)
Musharakah	−0.4156	0.0002	0.0472	0.0092	−65.87	(18.14%)

Table 8. Weightage of financing instruments (NMI excluded)

Selected instruments						
Instrument	β	σ_{ei}^2	ERB	Ci	Zi	w_i
Ijarah	−0.0696	0.0002	0.4783	0.0018	−140.28	(59.93%)
Murabahah	−0.1113	0.0006	0.1438	0.0024	−25.49	(10.89%)
Musharakah	−0.4156	0.0002	0.0472	0.0078	−68.30	(29.18%)

4.7 Portfolio Return and Risk

Once the weightage of each financing instrument has been decided, the portfolio return and risk are calculated. Based on Table 9, the portfolio return is 0.55% and the portfolio risk is 2.11% from the five financing instruments formed.

Table 9. Market referral return and risk

Portfolio				
Return	Systematic risk	Unsystematic risk	Variance	Risk
0.55%	0.0003%	0.0442%	0.0446%	2.11%

Table 10 shows that by excluding the mean value instruments, the portfolio return increases by 0.65% (1.20–0.55) and reducing the risk associated with it by 0.44% (2.11–1.67). Henceforth, maximising the return and minimizing the risk of portfolio thus the optimal portfolio obtained.

Table 10. Portfolio return and risk (NMI included)

Portfolio				
Return	Systematic risk	Unsystematic risk	Variance	Risk
1.20%	0.0006%	0.0275%	0.0280%	1.67%

5 Conclusion

It can be concluded that excluding the negative mean instrument tremendously increase the return and risk performance of portfolio. The weightage of Islamic financing instrument portfolio without the negative mean instrument is Ijarah (59.93%), Murabahah (10.89%) and Musharakah (29.18%). The establishment of an optimal portfolio formed with a Sharpe Single Index Model yields a portfolio return of 1.2% with a risk rate of 1.67%. This is a proof that diversification in various financing contracts not only increases the profits of Islamic banks; it can also reduce the level of risk at hand.

By conducting this research Islamic banks will be able to determine the main priority in the distribution of financing. The optimal portfolio can be used as a guide for Islamic bank to determine the target of financing disbursement, as financing is also the main business.

Based on the results, Islamic bank should focus more on productive financing or investment which generally use Ijarah, Murabahah and Musharakah.

Acknowledgments. Authors would like to acknowledge the Faculty of Computer and Mathematical Sciences, Universiti Teknologi MARA (UiTM) Malaysia for funding this publication.

References

1. Anggraeni, R.T.: Optimizing financing sharia bank through the formation of optimal portfolio with single index model. In: International Conference on Islamic Finance, Economics and Business,KnE Social Sciences, no. 13, pp. 255–275 (2018)
2. Chauhan, A.A.A.: Study on usage of sharpe's single index model in portfolio construction with reference to Cnx Nifty. Glob. J. Res. Anal. **3**(10), 92–94 (2014)
3. Elton, E.J., Gruber, M.J., Brown, S.J., Goetzmann, W.N.: Modern Portfolio Theory and Investment Analysis, 9th Ed. (2009)
4. Eugene, F.F., Kenneth, R.F.: The capital asset pricing model: theory and empiricism. Econ. J. **18**(3), 25–46 (2004)
5. Han, C.M.: Optimal Portfolio Construction : a Case in Bursa Malaysia (2015)
6. Marlina, R.: Formation of stock portfolio using single index model (case study on banking shares in the Indonesia stock exchange). Int. J. Bus. Econ. Law **8**(1), 67–73 (2015)
7. Nandan, T., Srivastava, N.: Construction of optimal portfolio using sharpe's single index model : an empirical study on nifty 50 stocks. J. Manag. Res. Anal. **4**(2014), 74–83 (2017)
8. Poornima, S., Remesh, A.P.: A study on optimal portfolio construction using NSE. Natl. J. Adv. Res. **2**(3), 28–31 (2016)
9. Ravichandra, T.: Optimal portfolio construction with nifty stocks. Int. J. Interdiscip. Multidiscip. Stud. (IJIMS) **1**(4), 75–81 (2014)
10. Sen, K., Fattawat, C.A.D.: Sharpe's single index model and its application portfolio construction : an empirical study. Glob. J. Finan. Manag. **6**(6), 511–516 (2014)
11. Tatiana, N., Igor, K., Liliya, S.: Principles and instruments of Islamic financial institutions. Procedia Econ. Finan. **24**(1), 479–484 (2015)

Analytics-Based on Classification and Clustering Methods for Local Community Empowerment in Indonesia

Dyah Yuniati[1,2] and Kristina Pestaria Sinaga[2(✉)]

[1] Dkatalis, Jakarta, Indonesia
dyah.yuniati@binus.ac.id
[2] Department of Master in Information System Management, Bina Nusantara University,
Jakarta, Indonesia
kristina.sinaga@binus.edu

Abstract. West Papua is reportedly the second-most populous province in Indonesia. The United Nations International Children's Emergency Fund (UNICEF) highlights Papua's performance in selecting the Sustainable Development Goals (SDG) indicators compared to other provinces in the country. The data shows that food, nutrition, health, education, housing, water, sanitation, and protection are defined as multidimensional child poverty. Population statistics and poverty figures show that inter-provincial equity in Indonesia needs to be re-measured. In 2008, the Regional Governments of Papua and West Papua Provinces implemented a Community Empowerment Program called "PNPM RESPEK", which provided direct community assistance for IDR 100 million per village. To determine the people's level of understanding and perception towards this program, PNPM RESPEK, in collaboration with the Central Statistics Agency, conducted an integrated PNPM RESPEK Evaluation Survey in July 2009. Based on the survey results, this paper identifies a model (pattern) of understanding the people of Papua and West Papua towards the program and finds the best method to build this model through classification techniques. Then the data model was also tested using unsupervised learning, the clustering method. The experimental results show that the J48 decision tree produces the highest accuracy compared to the others. As for clustering, the clustering hierarchy provides the best accuracy. Decision Tree J48 has the best accuracy with an accuracy of 97.31%. In this case, 97.31% of the people of Papua and West Papua who receive direct community assistance meet the level of understanding and perception of the PNPM RESPEK Program.

Keywords: SDG · Poverty · Equality · Machine learning · J48 Decision Tree · Hierarchical clustering

1 Introduction

Indonesia's easternmost provinces of Papua and West Papua generally referred to as Papua, are the country's most violent and resource-rich areas [1]. However, health care

© Springer Nature Singapore Pte Ltd. 2021
A. Mohamed et al. (Eds.): SCDS 2021, CCIS 1489, pp. 133–145, 2021.
https://doi.org/10.1007/978-981-16-7334-4_10

standards are lower in West Papua than in other regions of Indonesia [2]. World Health Organization (WHO) reported that poverty is a significant cause of ill health and a barrier to accessing health care when needed. This relationship is financial: the poor cannot afford to purchase things needed for good health, including sufficient quality food and health care. However, the relationship is also related to other factors related to poverty, such as lack of information on appropriate health-promoting practices or lack of voice needed to make social services work for them [3].

In 2007, the Government of Indonesia launched the Mandiri National Program for Community Empowerment (PNPM), which aims to reduce poverty, strengthen local government and community institutions' capacity, and improve local government governance. In 2008, this program covered approximately 40,000 villages in Indonesia and was expected to cover nearly 80,000 villages by 2009 [4]. In line with this, the regional governments of Papua and West Papua Provinces in 2008 implemented a Community Empowerment Program called "PNPM RESPEK." RESPEK is funded by the Provincial Expenditure Budget (APBD Propinsi), and it provides 100 million IDR directly to every village in the province [1]. The Regional Governments of Papua and West Papua provide direct community assistance (Indonesian: Bantuan Langsung Masyarakat) of IDR 100 million per village for 3,923 villages in 388 sub-districts. Meanwhile, the Ministry of Home Affairs provides more than 1,000 facilitators through PNPM [4].

The main component of PNPM is its approach called Community-Driven Development (CDD) [5]. Adopting a community-driven development (CDD) approach and with technical financial assistance from the International Bank for Reconstruction and Development, the PNPM is now a national program covering all villages and cities in the country [6, 7]. To determine the level of understanding and perception of Papua and West Papua's people towards the PNPM RESPEK Program, PNPM RESPEK, in collaboration with BPS-Statistics Indonesia, conducted the PNPM RESPEK Evaluation Survey, which was integrated through the National Socio-Economic Survey (SUSENAS) in July 2009. This research aims to identify a model (pattern) for understanding the people of Papua and West Papua towards the program and find the best method for building this model through experimental classification and clustering techniques. The primary goals of this research are to help the PNPM RESPEK improving the remote area from a data perspective and understanding principles of extracting valuable knowledge from data.

2 Methodology

Data Mining is the process of extracting and identifying patterns from large sets of data to produce output in the form of useful information or knowledge that was not previously known manually on the raw data. Data mining is carried out using statistical methods, mathematical algorithms, artificial intelligence or machine learning. In general, the stages carried out in Data Mining include data selection, pre-processing data, transformation data, modeling, and interpretation data [8].

2.1 Classification

Classification is the supervised learning technique in data mining. In supervised learning, the data label has already been defined. Classification is used to classify each item in a data set into one of a predefined set of classes or groups. The data analysis task classification is where a model or classifier is constructed to predict class labels. So, the classification technique will assign items in a collection to target categories or classes. The goal of classification is to accurately predict the target class for each case in the data. Algorithms for classification include J48 [9] and Logistic Regression [10].

- The J48 algorithm is an algorithm derived from C4.5 [10]. This algorithm generates decision trees based on rules to classify. Each aspect of information is divided into several small subsets to form the basis of decisions. J48 looks at standard data, which results in the separation of information by selecting attributes [11]. Mathematically, J48 algorithm uses the concept of entropy and information gain (IG). The information gain rate (IGR) is the splitting criterion (SplitInfo) to make the J48 decision tree. The IG, SplitInfo, and IGR are formulated as follows

$$IG(S, j) = Entropy(S) - Entropy(S|j) \tag{1}$$

$$SplitInfo_j(S) = - \sum_{k=k_0}^{k_c} \left(\frac{|S_j(k)|}{|S|} + \log_2 \frac{S_j(k)}{S} \right) \tag{2}$$

$$IGR(j) = \frac{IG(S, j)}{SplitInfo_j(S)} \tag{3}$$

where S is a parent node, j represent the j-th attribute of sample x in one class label, $Entropy(S) = - \sum_{j=1}^{d} x_j \log_2 x_j$, $Entropy(S|j)$ is the conditional entropy with $Entropy(S|j) = \sum_{k=k_0}^{k_c} \frac{|S_j(k)|}{|S|} \cdot Entropy(S_j(k))$, and $S_j(k) = \{x \in S | x_j = k\}$.

- Logistic regression is an approach to creating predictive models using equations that describe the relationship between two or more variables [13]. The dependent variable for logistic regression has a dichotomy scale. The dichotomy scale is a nominal data scale with two categories: Yes and No, Success and Failure or High and Low [12]. We often named these two categories as binary-valued labels which the correct label y values is denoted either 0 or 1 $\left(y^{(i)} \in \{0, 1\}\right)$. Mathematically, the probability that data samples belong to the "Yes" class versus the probability that it belongs to the "No" class defined as follows

$$P(y = Yes|x) = h_\theta(x) = \frac{1}{1 + \exp(-\theta^T x)} \equiv \sigma\left(\theta^T x\right) \tag{4}$$

$$P(y = No|x) = 1 - P(y = Yes|x) = 1 - h \tag{5}$$

where $\sigma(r) = \frac{1}{1+\exp(-r)}$ is the sigmoid or logistic function, and $\theta^T x \in [0, 1]$ is the gradient for linear regression. The cost function for a set of training examples with binary labels $\left\{(x^{(i)}, y^{(i)}) : i = 1, 2, \ldots, n\right\}$ to measure how close a given h_θ to the correct output y is expressed as below

$$J(\theta) = - \sum_{i=1}^{n} \left(y^{(i)} \log\left(h_\theta\left(x^{(i)}\right)\right) + \left(1 - y^{(i)}\right) \log\left(1 - h_\theta\left(x^{(i)}\right)\right)\right) \tag{6}$$

If we plug in the definition of $h_\theta(x) = \sigma(\theta^T x^{(i)})$ into (6), we will get the loss function as below

$$J(\theta) = -\sum_{i=1}^{n} \left(y^{(i)} \log\left(\sigma\left(\theta^T x^{(i)}\right)\right) + \left(1 - y^{(i)}\right) \log\left(1 - \sigma\left(\theta^T x^{(i)}\right)\right) \right) \quad (7)$$

To be noted, the smaller the values of cost function the better the model. In this sense, the model with bigger cost function clearly predict the un-great solution of $y^{(i)}$.

2.2 Clustering

Clustering is a powerful tool in data analysis. It is used for discovering the cluster structure in data sets with the most remarkable similarity within the same cluster but the most noteworthy dissimilarity between different clusters. Generally, cluster analysis became a multivariate statistical analysis branch, and it is an unsupervised learning approach to machine learning [13, 14]. Algorithms for clustering include K-Means [15], Hierarchical clustering (HCA) [16, 17], and DBSCAN algorithms [18].

- K-Means
  K-means is the simplest and most common clustering method. It is because K-means can classify large amounts of data with fast and efficient computation time. K-Means divides n data points in d dimensions into a number of k clusters where the clustering process is carried out by minimizing the sum squares distance between the data and each cluster center [15]. In its implementation, the K-Means method requires three parameters that are entirely user-defined, namely the number of clusters (# of k), cluster initialization and system distance. The objective function of K-Means is formulated as

$$J_{K-Means}(U, V) = \sum_{k=1}^{c} \sum_{i=1}^{n} \mu_{ik} \sum_{j=1}^{d} \left(x_{ij} - v_{kj}\right)^2 \quad (8)$$

$$s.t., \ \mu_{ik} \in \{0, 1\}, \ i = 1, \ldots, n, \ k = 1, \ldots, c \quad (9)$$

The objective function in (8) is optimized by using the Lagrange multipliers and obtained the updating equations of μ_{ik} and v_{kj} as follows

$$v_{kj} = \sum_{i=1}^{n} \mu_{ik} x_{ij} \bigg/ \sum_{i=1}^{n} \mu_{ik} \quad (10)$$

$$\mu_{ik} = \begin{cases} 1 & \text{if } \sum_{j=1}^{d} \left(x_{ij} - v_{kj}\right)^2 = \min_{1 \leq k \leq c} \left(\sum_{j=1}^{d} \left(x_{ij} - v_{kj}\right)^2\right) \\ 0, & \text{otherwise.} \end{cases} \quad (11)$$

- Hierarchical clustering
  Hierarchical clustering (HCA) groups data through a hierarchical chart [16]. In the initial step, the hierarchical clustering identifies the data that has the closest distance,

then associated it into one cluster. Furthermore, hierarchical clustering calculates the distance between the clusters [17]. There are seven hierarchical clustering methods including single link, complete link, group average link, McQuitty's method, median, centroid, and Ward's method. These seven hierarchical clustering methods defines a new relation from datasets to hierarchies by using different *Lance-Williams dissimilarity update formula*. However, the suitable iteration among these seven hierarchical clustering methods similar to each other, they are all carried out until all are connected. Mathematically, if points x and y are agglomerated into cluster $x \cup y$, then the *Lance-Williams dissimilarity update formula* is expressed as below:

$$d(x \cup y, k) = \alpha_x d(x, k) + \alpha_y d(y, k) + \beta d(x, y) + \gamma |d(x, k) - d(y, k)| \quad (12)$$

where α_x, α_y, β, γ define the agglomerative criterion, α_y with index y is defined identically to coefficient α_x with index x. The formulation of *Lance-Williams dissimilarity* in (12) can be expressed as follow

$$d_{x \cup y, k} = \alpha_x d_{xk} + \alpha_y d_{yk} + \beta d_{xy} + \gamma |d_{xk} - d_{yk}| \quad (13)$$

- Density-Based Spatial Clustering of Application with Noise
  Density-Based Spatial Clustering of Application with Noise (DBSCAN) is a clustering algorithm developed by density-based. DBSCAN separates high-density clusters from low-density clusters. This algorithm will start by dividing the data into d dimensions, then iteratively count the number of data points close to each other [18]. The DBSCAN relay on two parameters, called MinPts and Epsilon ($\varepsilon - neighborhood$). The $\varepsilon - neighborhood$ is a distance measure that will be used to locate the points or to check the density in the neighbourhood of any point x, formulated as

$$N_\varepsilon(x) = \{y \in d \,|\, \|y - x\| \leq \varepsilon\} \quad (14)$$

Here points x is a points inside of the cluster (MinPts) if the $\varepsilon - neighborhood$ $N_\varepsilon(x)$ of point x greater than or equal to the least number of neighbors v, denoted as $|N_\varepsilon(x)| \geq v$. A point x is directly density-reachable from a point y with respect to $\varepsilon - neighborhood$ and the minimum number of points required to form a dense region if $x \in N_\varepsilon(y)$, and $|N_\varepsilon(y)| \geq \text{MinPts}$.

3 Data Set

This research takes a case study to identify a model for the understanding of the people of Papua and West Papua towards the National Program for Community Empowerment, Strategic Plan of "Kampung" Development (PNPM RESPEK) [7]. The dataset was obtained from the results of the PNPM RESPEK survey in collaboration with BPS-Statistics Indonesia to conduct the PNPM Evaluation Survey, which was integrated through the National Socio-Economic Survey (Susenas) July 2009. The source dataset is openly accessible at https://microdata.worldbank.org/index.php/catalog/1801/ study-description.

   This data initially contains 3937 Papua and West Papua people who received support from PNPM RESPEK. Since there are 2041 missing values, the data we used in this research only contains 1896 samples of Papua and West Papua people who have benefited from the PNPM RESPEK project, with 31 attributes. Table 1 shows the data type for each attribute. As our goal is to identify the accuracy of classifiers in predicting people who are likely to get the understanding and perception towards the PNPM RESPEK program, we measured the performances by only using a single evaluation metric, called accuracy rates. It is a common known that accuracy rate is devoted to simultaneously visualize and associate the structure of data based on their similarities. Furthermore, we notice that the high accuracy rate is more important than the resources. The calculation of accuracy rate is based on the percentage of error, expressed as

$$Error\ rate = 1 - \sum\nolimits_{k=1}^{c} n(c_k)\big/n \qquad (15)$$

where $n(c_k)$ is the number of training data that obtain correct classification/clustering. All the procedures for classification and clustering including pre-processing and processing final input data will be done by using Waikato environment for knowledge analysis (WEKA).

Table 1. Data type of the attributes

No	Characteristics		
	Attribute	*Data Type*	*Values*
1	Age	Numeric	Min = 11 Max = 99
2	Gender	Nominal	1 = Male 2 = Female
3	Marital Status	Nominal	1 = Single 2 = Married 3 = Divorced 4 = Death divorce
4	Heard about PNPM	Nominal	1 = Yes, spontaneous 2 = Yes, after the interviewer explained 3 = No
5	Has been a PNPM actor	Nominal	1 = Yes 2 = No
6	Previously worked at PNPM	Nominal	1 = Yes 2 = No
7	Present at the PNPM meeting	Nominal	1 = Yes 2 = No
8	Become an SPP member	Nominal	1 = Yes 2 = No

(*continued*)

Table 1. (*continued*)

No	Characteristics		
	Attribute	*Data Type*	*Values*
9	Know the number of funds budgeted	Nominal	1 = Yes 2 = No
10	Information from "Dusun" or "RT" meetings	Nominal	1 = Yes 2 = No
11	Information from the community of mothers	Nominal	1 = Yes 2 = No
12	Information from community group meetings	Nominal	1 = Yes 2 = No
13	Information from friends/neighbors by verbal	Nominal	1 = Yes 2 = No
14	Information from the village apparatus met on the road	Nominal	1 = Yes 2 = No
15	Information from "Kampung" officials who came to the house	Nominal	1 = Yes 2 = No
16	Information from community leaders	Nominal	1 = Yes 2 = No
17	Information from religious figures	Nominal	1 = Yes 2 = No
18	Information from project companion	Nominal	1 = Yes 2 = No
19	Information from sub-district employees	Nominal	1 = Yes 2 = No
20	Information from district/city employees	Nominal	1 = Yes 2 = No
21	Announcement	Nominal	1 = Yes 2 = No
22	Information boards	Nominal	1 = Yes 2 = No
23	Written report	Nominal	1 = Yes 2 = No
24	Do you get information about the use of the "Kampung" budget?	Nominal	1 = Yes 2 = No
25	Do you get information about the use of self-help funds?	Nominal	1 = Yes 2 = No
26	Do you get info on the use of other project budgets?	Nominal	1 = Yes 2 = No

(*continued*)

Table 1. (*continued*)

No	Characteristics		
	Attribute	*Data Type*	*Values*
27	Was there a meeting the last year?	Nominal	1 = Yes 2 = No 3 = Do not know
28	How many times are these meetings?	Numeric	Min = 1 Max = 12
29	What level is the meeting?	Nominal	0 = Do not know 1 = "Desa/ Kelurahan" level 2 = "Dusun" level 4 = "RT" level 8 = Others
30	Are you looking for info?	Nominal	1 = Yes 2 = No
31	Understand the program objectives (Do you know what the funds are used for?)	Boolean	1 = Yes 2 = No

4   Result and Discussion

4.1   Classification (Supervised Learning)

We use the PNPM RESPEK 2009 data as a study case by Decision Tree J48 and Logistic regression. Our target variable is to predict whether Papua and West Papua people who received direct community assistance meet or do not meet the level of understanding and perception of funds purposes. The results of classification using Logistic regression and Decision Tree J48 are shown in Table 2.

Table 2. Class Distribution

	Class 0	Class 1
Original Data	1835	61
Logistic Regression	1886	10
Decision Tree J48	1860	36

As can be seen in Table 2, the original distribution of Papua and West Papua people who met and who did not meet the level of understanding and perception towards the PNPM RESPEK Program is *1835:61*. Here "Class 0" is defined as people who met the level of understanding and perception towards the PNPM RESPEK Program. While "Class 1" is defined as people who did not meet the level of understanding and perception towards the PNPM RESPEK Program. From Table 2, we can see the Logistic Regression and Decision Tree J48 predictions did not produce perfect results. In this sense, both classifiers suggested some false negative (FN) and false positive (FP). The FN and FP refers to people that incorrectly classified as "Class 0" or "Class 1". Specifically, there are people who was originally met the level of understanding and perception predicted as the opposite and vice versa.

The accuracy and error rates of Logistic regression and Decision Tree J48 are shown in Table 3. Table 3 demonstrated that the Decision Tree J48 is superior to the Logistic Regression with 97.31% accuracy. In contrast, Logistic Regression accuracy's 96.78%, with a 3.22% error rate.

Table 3. Classification performance results

	Accuracy	Error Rate
Logistic Regression	0.9678	0.0322
Decision Tree J48	**0.9731**	**0.0269**

The modeling result of the J48 Decision Tree method is shown in Fig. 1. Figure 1 represents that the meeting number is the most crucial variable to build people's understanding of the program. Information from sub-district employees also helps Papua and West Papua people to understand the program. Another variable that affected Papua and West Papua people's understanding and perception towards the PNPM RESPEK Program are information from the announcement, have been PNPM actors, became SPP members, received information from Dusun or RT meetings, information from the community of mothers, and previously worked at PNPM program. In comparison, people's understanding is not influenced by age and gender factors.

4.2 Clustering (Unsupervised Learning)

Because Decision Tree J48 does not make apparent immediately how they can be used for unsupervised learning, we further used the trick is to call the data of Papua and West Papua people who met the level of understanding and perception towards the PNPM RESPEK Program as "Class 1" and people who did not meet the level of understanding and perception as "Class 2." We used the Papua and West Papua people who received direct community assistance data to demonstrate these unsupervised learning of K-Means, single-linkage hierarchical clustering, and DBSCAN clustering. Since DBSCAN clustering requires two input parameters, we set the value of *Epsilon* = 2.0 and *minPts* = 35. The results of these three clustering techniques are presented in Table 4.

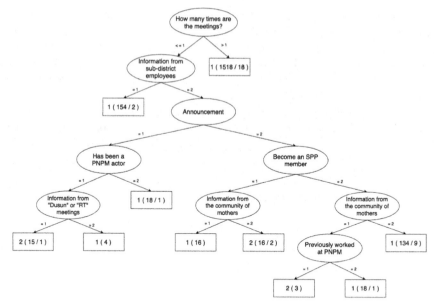

Fig. 1. Decision tree model

Table 4. Clustering performance results

	# of c	Class 1	Class 2	Accuracy	Error Rate
K-Means	2	1197	699	0.6245	0.0375
HCA	**2**	**1895**	**1**	**0.9673**	**0.0327**
DBSCAN	2	1270	626	0.9293	0.0707

Table 4 represented that the Hierarchical clustering (HCA) technique is superior to K-means and DBSCAN clustering techniques, with 96.73% accuracy. Since misclassifying a minority class instance is usually more severe than misclassifying a majority class one, it is clear that class imbalance does not affect the performance of hierarchical clustering (HCA). Figure 2, using HCA, demonstrated that 1895 of Papua and West Papua people who received direct community assistance met the level of understanding and perception towards the PNPM RESPEK. In contrast, Hierarchical clustering (HCA) represented 1 Papua and West Papua people who received direct community assistance did not meet the level of understanding and perception towards the PNPM RESPEK.

Figure 3 visualizes the distribution of Papua and West Papua people who received direct community assistance who met and did not meet the level of understanding and perception towards the PNPM RESPEK Program generated by the K-Means technique. K-means clustering obtained 62.45% accuracy with 1197 of Papua and West Papua people who received direct community assistance met the level of understanding and perception towards the PNPM RESPEK. In contrast, K-means represented 699 Papua

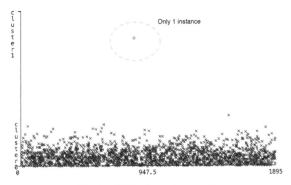

Fig. 2. HCA clustering result

and West Papua people who received direct community assistance did not meet the level of understanding and perception towards the PNPM RESPEK.

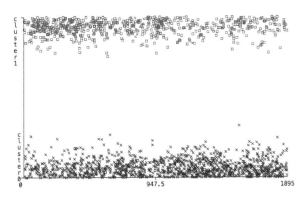

Fig. 3. K-means clustering result

Meanwhile, using the DBSCAN technique, the distribution of instances is shown in Fig. 4. DBSCAN represented 1270 out of 1896 of Papua and West Papua people who received direct community assistance met the level of understanding and perception towards the PNPM RESPEK Program. If we analyze Fig. 4 deeply, the red color distributions are well-separated. In this sense, these points that belong to the red color can be distinguished into two different clusters. These two clusters are 533 un-clustered people, and 93 Papua and West Papua people who received direct community assistance did not meet the level of understanding and perception towards the PNPM RESPEK Program. However, DBSCAN obtained a competitive accuracy of 92.93%, as shown in Table 4.

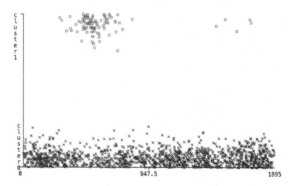

Fig. 4. DBSCAN clustering result

5 Conclusion

Conclusions are drawn based on the output of supervised (classification) and unsupervised learning (clustering). The best performance of classification techniques is generated by the J48 Decision Tree method with 97.31% accuracy. On the other hand, the best performance of clustering techniques is generated by hierarchical clustering (HCA) with 96.73% accuracy. These results are relatively high. In this sense, all 30 variables work satisfactorily in measuring whether 1896 of Papua and West Papua people who received direct community assistance met or did not meet the level of understanding and perception towards the PNPM RESPEK Program. However, these k-means and DBSCAN output can be used as a consideration to address the poverty issues in Papua and West Papua. As DBSCAN represented 533 of Papua and West Papua people who received direct community assistance are still questionable, it is recommended to evaluate this phenomenon to improve decision-making in the future. As data can help accelerate a high performance, we encourage the government to investigate these 2041 out of 3937 original data (known as missing values). These 2041 partial data are essential in providing the right insights to drive better strategic, scenario, and situational decisions.

Overall, we have implemented machine learning techniques to Papua and West Papua people who received support from PNPM RESPEK by simultaneously using two supervised and three unsupervised learning based on data collected from National Socio-Economic Survey (Susenas) July 2009. Future work is intended to conduct the update data so that an optimal result will be form appropriately. We also consider further analysis based on more supervised learning approaches to generate comprehensive results.

References

1. BPS, B.P.S. (2013): Indonesia - Survei Evaluasi Program Nasional Pemberdayaan Masyarakat Rencana Strategis Pembangunan Kampung (2009)
2. Anderson, B.: Papua's Insecurity: State Failure in the Indonesian Periphery. East-West Center, Honolulu (2015)
3. Diani, H.: Health, a specter for Irian Jaya. The Jakarta Post 2000, 21 Aug 5. http://www.lib rary.ohiou.edu/indopubs/2000/08/20/0022.html. Accessed Nov 2008

4. World Bank: Poverty and Health (2014). https://www.worldbank.org/en/topic/health/brief/poverty-health
5. Akatiga: A technical evaluation of PNPM-RESPEK infrastructure built by the barefoot engineers technical facilitator training program in Papua (2015). https://www.akatiga.org/wp-content/uploads/2018/05/Barefoot-Technical-Evaluation-Final-Report-2015.pdf
6. Susilo, A., Trisnanto, A.: The Indonesian national program for community empowerment (PNPM)–Rural: decentralization in the context of neoliberalism and world bank policies. International Institute of Social Studies, **2**(1) (2012)
7. World Bank: Indonesia: Evaluation of the Urban Community Driven Development Program: Program Nasional Pemberdayaan Masyarakat Mandiri Perkotaan (PNPM-Urban) (2013)
8. Rodrigues, I.: CRISP-DM methodology leader in data mining and big data (2020). https://towardsdatascience.com/crisp-dm-methodology-leader-in-data-mining-and-big-data-467efd3d3781. Accessed 13 Feb 2021
9. Irwansyah, E.: Clustering. https://socs.binus.ac.id/2017/03/09/clustering/. Accessed 6 Mar 2021
10. Quinlan, J.R.: C4. 5: Programs for Machine Learning. Elsevier (2014)
11. Saravanan, N., Gayathri, V.: Performance and classification evaluation of J48 algorithm and Kendall's Based J48 algorithm (KNJ48). Int. J. Comput. Trends Technol. **59**(2), 73–80 (2018). https://doi.org/10.14445/22312803/ijctt-v59p112
12. Cabrera, A.F.: Logistic regression analysis in higher education: an applied perspective. High. Educ. Handbook theory Res. **10**, 225–256 (1994)
13. Hidayat, A.: Regresi Logistik (2015). https://www.statistikian.com/2015/02/regresi-logistik.html
14. Jain, A.K., Dubes, R.C.: Algorithms for Clustering Data. Prentice-Hall, Inc. (1988)
15. Kaufman, L., Rousseeuw, P.J.: Finding Groups in Data: An Introduction to Cluster Analysis, vol. 344. Wiley, Hoboken (2009)
16. MacQueen, J.: Some methods for classification and analysis of multivariate observations. In: Proceedings of the Fifth Berkeley Symposium on Mathematical Statistics and Probability, vol. 1, no. 14, pp. 281–297 (1967)
17. Johnson, S.: Hierarchical clustering schemes. Psychometrika **32**(3), 241–254 (1967). https://doi.org/10.1007/BF02289588
18. Ester, M., Kriegel, H.P., Sander, J., Xu, X.: A density-based algorithm for discovering clusters in large spatial databases with noise. In: KDD, vol. 96, no. 34, pp. 226–231 (1996)

Two-Step Estimation for Modeling the Earthquake Occurrences in Sumatra by Neyman–Scott Cox Point Processes

Achmad Choiruddin$^{(\boxtimes)}$, Tabita Yuni Susanto, and Rahma Metrikasari

Department of Statistics, Institut Teknologi Sepuluh Nopember (ITS),
Surabaya, Indonesia
choiruddin@its.ac.id

Abstract. The Cox point process is highly considered for earthquake modeling. However, the complex earthquake data which involve a large number of occurrences and geological variables often require expensive computation. This study aims to propose an efficient algorithm based on the two-step procedure by constructing the first and second order composite likelihoods. We consider four Neyman–Scott Cox process models and apply them to fit the earthquake distribution in Sumatra. We conclude that the Cauchy cluster process performs best.

Keywords: Big data problem · Cluster point process · Disaster risk reduction · Earthquake modeling · Spatial point pattern

1 Introduction

Statistical modeling based on spatial or spatio-temporal point process has become one of the standard tools for the analysis of earthquake occurrences [e.g. 1–4]. The methodology offers a promising tool which covers some important aspects related to seismology.

For a small number of events with an assumption of independence and stationary models, there is no major theoretical and computational issue. However, the current problems usually include a large number of earthquakes [e.g. 2,5], assume cluster models [e.g. 3,6], involve marks such as depths and magnitudes [e.g. 1,2] and include geological variables [4,7]. Albeit the methodology is still able to handle such current problems, the computational issue arises mainly due a large number of parameters to estimate with complex structure. For example, for a class of Neyman–Scott Cox process [8,9], there are two groups of parameters: a group related to account for the effect of geological variables (the size depends on the number of variables) and another group for clustering (the dimension depends on the type of model). One way is to consider Markov Chain Monte Carlo to estimate all the parameters simultaneously. Although such an approach produces a good estimate, it consumes heavy computation [8].

© Springer Nature Singapore Pte Ltd. 2021
A. Mohamed et al. (Eds.): SCDS 2021, CCIS 1489, pp. 146–159, 2021.
https://doi.org/10.1007/978-981-16-7334-4_11

In this paper, we adapt the two-step estimation proposed by Waagepetersen and Guan [10], where the estimation is conducted to obtain each group of estimates at a time to gain computational efficiency. In the first step, the Poisson score is employed to estimate the first group of parameters and in the second step, minimum contrast estimation is used to estimate cluster parameters. Our approach is similar in the sense that we follow the two-step estimation strategy. However, our approach differs in employing both the first and second steps. In particular, we consider in the first step the weighted estimating equation [11] to improve the estimator efficiency of the first step of Waagepetrsen and Guan [10] and then use in the second step the second order composite likelihood estimation [12]. When available, likelihood-based approach is more appropriate and does not require tuning parameters which are sometimes hard to determine.

The remainder of the paper is organized as follows. Section 2 describes the study area and data description. Section 3 details the methodology. We discuss the results in Sect. 4 and provide conclusion in Sect. 5.

2   Study Area and Data Description

The Indonesian archipelago is situated at the junction of the three major sea plates: Eurasian, Indo-Australian, and Pacific Sea plates (Fig. 1 top). In addition, there exists so-called the pacific ring of fire, where volcanoes line up in Sumatra, Jawa, Bali, Nusa Tenggara, and Maluku (Fig. 1 top). These lead Indonesia to have a high risk of earthquake occurrences. The Sumatra is an island located at the west of Indonesia with $D = [104.517, 119.295] \times [-6.518, 7.378]$ $(100 \, \text{km}^2)$. It is known as an active tectonic area since: (1) there is an area of collision between Indo-Australian and Eurasian sea plates in Hindia ocean and (2) along the Sumatra land, there exist Sumatra fault and Bukit Barisan mountains lasts around 1650 km [13].

This study focuses on the analysis of 6900 observed major-shallow earthquakes (magnitude $\geq$4M, depths $\leq$60 km) in the Sumatra during 2004–2018 and involves three geological variables (Fig. 1 bottom). On 2004–2018, there were 35 highly major earthquakes (magnitude >6.5M), for which the top five occurred in Aceh (2004, 9.1M), Nias (2005, 8.6M), Palung Sunda (2007, 8.4M), and in Kepulauan Mentawai (2007, 7.9M) and (2016, 7.8M). The major-shallow earthquakes cause a great loss. For examples, earthquake in Aceh (2004, 9.1M) caused loss approximately US$ 4.7 million with more than 200 thousand deaths while earthquake in Padang (2009, 7.6M) caused more than US$ 2.2 million loss [14].

Figure 1 bottom depicts the distribution of the major-shallow earthquakes in the Sumatra along with the volcanoes, faults, and subduction. Most of the shallow earthquakes occur near to the subductions and some of them even trigger tsunami. Some earthquake-prone regions in Sumatra are Mentawai, Nias, Simeulue, Padang, and Bengkulu. In addition, some of the shallow earthquakes in the mainland of Sumatra tend to be close to the faults and volcanoes line. Aceh, Jambi, and Sumatra Utara are the regions with a high risk of great earthquakes in Sumatra mainland.

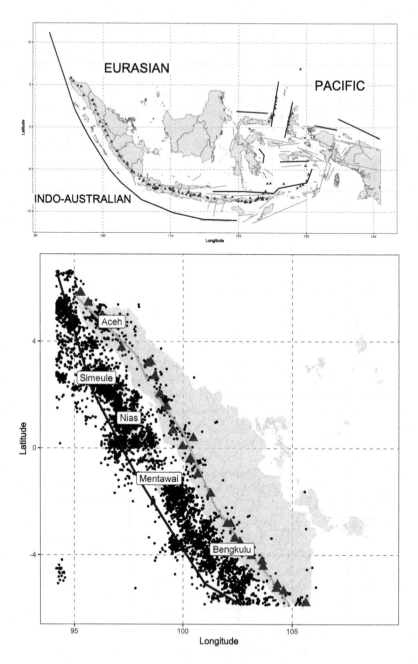

Fig. 1. Top: Map of active tectonics in Indonesia. Bottom: Locations of earthquake occurrences in Sumatra (black dot) with M ≥ 4 and depths ≤ 60 km during 2004–2018 along with locations of subduction zone (blue line), fault (orange line), and volcano (red triangle). (Color figure online)

3   Methodology

Let $\mathbf{x} = \{x_1, \ldots, x_m\}$ be a set which represents a collection of m earthquake coordinates observed on a bounded observation domain D with an area $|D|$. We assume the process underlying the earthquake occurrences is a Neyman–Scott Cox process $\mathbf{X}$ on $\mathbb{R}^2$. Such a model has been considered for earthquake modeling e.g. by [3,4]. We study this model in Sect. 3.1 and introduce an efficient algorithm for model fitting in Sect. 3.2.

The Neyman–Scott Cox process is also known as the doubly stochastic Poisson process where it could be seen as a Poisson process with non-negative random intensity Λ. Since it is fully determined by its intensity λ and second order product density $\lambda^{(2)}$ (or its derivation such as the pair correlation and K-functions), it is therefore important to assess and make inference regarding these densities.

The intensity function λ and pair correlation function g of a Neyman–Scott Cox process are

$$\lambda(u) = \mathbb{E}\Lambda(u), \quad g(u,v) = \mathbb{E}[\Lambda(u)\Lambda(v)]/\{\mathbb{E}\Lambda(u)\mathbb{E}\Lambda(v)\}, \quad u,v \in D, \quad (1)$$

where the intensity measures the probability of observing a new earthquake occurrence in a small area u, and where the pair correlation function measures the interaction between points which could detect clustering. An alternative summary function to measure the spatial interaction among earthquake occurrences is the K-function, which could be defined as

$$K(r) = 2\pi \int_0^r sg(s)\mathrm{d}s, \quad r = \|u - v\|.$$

Further details on spatial point processes could be found in [8,15].

3.1   Neyman–Scott Cox Processes (NSCP)

The Neyman–Scott Cox process model is formed first by evoking unobserved mainshocks located according to the stationary Poisson point process $\mathbf{C}$ with intensity $\kappa > 0$. Second, given the mainshock process $\mathbf{C}$, each of them generates a random number of aftershocks $\mathbf{X}_c, c \in \mathbf{C}$ according to Poisson point process with intensity $\lambda_c(u), u \in D$, distributed around the mainshock. Then $\mathbf{X} = \cup_{c\in\mathbf{C}}\mathbf{X}_c$ is a Neyman–Scott point process with mainshock process $\mathbf{C}$ and aftershock processes $\mathbf{X}_c, c \in \mathbf{C}$ driven by random intensity $\Lambda(u) = \sum_{c\in\mathbf{C}} \lambda_c(u)$ [e.g. 4,9,16].

To account for geological factors effect on the distribution of earthquake occurrences, we consider a log-linear intensity [e.g. 10,16,17] of the aftershock process $\mathbf{X}_c$

$$\lambda_c(u; \boldsymbol{\beta}) = \exp(\zeta + \boldsymbol{\beta}^\top \mathbf{z}(u))k(u - c; \omega), \quad (2)$$

where $\mathbf{z}(u) = \{z_1(u), \ldots, z_p(u)\}^\top$ is a covariates vector containing p-geological variables, $\boldsymbol{\beta} = \{\beta_1, \ldots, \beta_p\}^\top$ is the corresponding p-dimensional parameter, and

k is a probability density function determining the distribution of aftershock points around the mainshocks parameterized by ω. By having (2), the NSCP $\mathbf{X}$ is driven by a random intensity

$$\Lambda(u) = \exp(\zeta + \boldsymbol{\beta}^\top \mathbf{z}(u)) \sum_{c \in \mathbf{C}} k(u - c, \omega).$$

Therefore, by applying (1), the intensity of $\mathbf{X}$ is

$$\lambda(u; \boldsymbol{\beta}) = \mathbb{E}\Lambda(u) = \kappa \exp(\zeta + \boldsymbol{\beta}^\top \mathbf{z}(u)) = \exp(\beta_0 + \boldsymbol{\beta}^\top \mathbf{z}(u)),$$

where $\beta_0 = \zeta + \log \kappa$ represents the intercept parameter and $\boldsymbol{\beta}$ accounts for geological variables effect.

To determine the distribution of aftershocks around mainshocks k and assess the interaction between events, we consider four NSCP models: Thomas, Matérn, Cauchy, and variance-gamma cluster models. By these models, the interaction parameter $\psi = (\kappa, \omega)^\top$ is represented by the pair correlation and K-functions. Note that the intensity and pair correlation functions are now regarded as parametric functions of $\boldsymbol{\beta}$ and ψ.

Thomas Cluster Process. The aftershock occurrences are assumed to be distributed around mainshock locations according to bivariate Gaussian distribution $\mathcal{N}(\mathbf{0}, \omega^2 \mathbf{I}_2)$ [8]. The density function k is of the form

$$k(u; \omega) = (2\pi\omega^2)^{-1} \exp(-\|u\|^2/(2\omega^2)).$$

A stronger clustering effect is determined by smaller values of ω (tighter clusters) and κ (fewer mainshocks). The pair correlation and the K-functions of a Thomas cluster process are respectively

$$g(r; \psi) = 1 + \frac{1}{4\pi\kappa\omega^2} \exp\left(-\frac{r^2}{4\omega^2}\right), \qquad r = \|u - v\|,$$

and

$$K(r; \psi) = \pi r^2 + \frac{1}{\kappa}\left\{1 - \exp\left(-\frac{r^2}{4\omega^2}\right)\right\}.$$

Matérn Cluster Process. In the Matérn cluster process, the aftershocks are uniformly distributed in a circle of radius ω centered around each of the mainshocks. The density is

$$k(u; \omega) = 1/(\pi\omega^2), \text{ if } \|u\| \leq \omega.$$

The pair correlation and the K-functions of the Matérn cluster process are

$$g(r; \psi) = 1 + \frac{1}{\pi^2\omega^2\kappa} a\left(\frac{r}{2\omega}\right),$$
$$K(r; \psi) = \pi r^2 + \frac{1}{\kappa} f\left(\frac{r}{2\omega}\right),$$

where $a(z) = 2(\cos^{-1} z - z\sqrt{1 - z^2})$ for $z \leq 1$ and $a(z) = 0$ for $z > 1$, and where $f(z) = 2 + 1/\pi\{(8z^2 - 4)\arccos z - 2\arcsin z + 4z\sqrt{(1 - z^2)^3} - 6z\sqrt{1 - z^2}\}$ for $z \leq 1$ and $f(z) = 1$ for $z > 1$ [15].

Cauchy Cluster Process. An alternative is to consider a bivariate Cauchy density [18] with k

$$k(u; \omega) = \frac{1}{2\pi\omega^2}\left(1 + \frac{\|u\|^2}{\omega^2}\right)^{-\frac{3}{2}}.$$

The pair correlation and K-functions of the Cauchy cluster process are

$$g(r; \psi) = 1 + \frac{1}{8\pi\kappa\omega^2}\left(1 + \frac{\|r\|^2}{4\omega^2}\right)^{-\frac{3}{2}},$$

$$K(r; \psi) = \pi r^2 + \frac{1}{\kappa}\left(1 - \frac{1}{\sqrt{1 + \frac{r^2}{4\omega^2}}}\right).$$

Variance-Gamma Cluster Process. A variance-gamma density [18] is of the form

$$k(u; \omega) = \frac{1}{2^{q+1}\pi\omega^2\Gamma(q + 1)}\left(\frac{\|u\|}{\omega}\right)^q B_q\left(\frac{\|u\|}{\omega}\right),$$

where Γ is the Gamma function, B is a modified Bessel function of the second kind of order q and $q > -1/2$. We fix $q = -1/4$ and treat q as a fixed parameter following the default option of the spastat R package [15]. The function B is in particular $B(a) = \exp(-a)(1 + O(1/a))/\sqrt{2a/\pi}$.

The pair correlation and K-functions of the variance-gamma cluster process are

$$g(r; \psi) = 1 + \frac{1}{4\pi\kappa\omega^2 q}\frac{(\|r\|/\omega)^q B_q(\|r\|/\omega)}{2^{q-1}\Gamma(q)},$$

$$K(r; \psi) = \int_0^r 2\pi s g(s; \psi) ds.$$

3.2   Parameter Estimation

The objective function with respect to β and ψ is complex and hard to evaluate, so estimating both β and ψ simultaneously using e.g. Markov Chain Monte Carlo [8] would be computationally expensive. Therefore, we employ the two-step procedure where at each step a group of parameters is estimated. In particular, in the first step we build a weighted first-order composite likelihood [11] and maximize it to obtain $\hat{\beta}$. In the second step, we construct the second-order composite likelihood [12] given $\hat{\beta}$, and maximize it to have $\hat{\psi}$. This strategy offers more efficient computation and satisfies the theoretical properties [10]. We do not provide its convergence nor efficiency results as the results by [10–12] directly follow.

The estimation is performed using the kppm function of the spatstat R package [15]. Model assessment is conducted using the Akaike information criterion (AIC) and envelope test (Sect. 3.3) by applying the logLik.kppm and envelope.kppm functions.

Step 1: β Estimation. The first-order composite likelihood function for β [9,11] is

$$\mathrm{CL}_1(\beta) = \sum_{u \in \mathbf{X} \cap D} w(u) \log \lambda(u; \beta) - \int_D w(u) \lambda(u; \beta) \mathrm{d}u, \tag{3}$$

which coincides with the weighted likelihood function of a Poisson point process. To maximize (3), we employ the Berman-Turner approximation so that we could link (3) to the likelihood of weighted Poisson generalized linear models. See, e.g. [11,15] for the details.

Step 2: ψ Estimation. The computationally efficient method to estimate ψ is to maximize the second-order composite likelihood of ψ [12,15]

$$\mathrm{CL}_2(\psi) = \sum_{u,v \in \mathbf{X}}^{\neq} \mathbf{1}\{\|u - v\| < R\} \Big\{ \log g(u, v; \psi)$$

$$- \log \int_D \int_D \mathbf{1}\{\|u - v\| < R\} g(u, v; \psi) \mathrm{d}u \mathrm{d}v \Big\}, \tag{4}$$

where $R > 0$ is an upper bound on correlation distance model and $g(u, v; \psi)$ is the pair correlation functions of any of the Neyman–Scott Cox point processes described in Sect. 3.1.

3.3   Model Assessment

We perform model selection using the envelope test and Akaike information criterion (AIC). Envelopes are the critical bounds of the statistical test of a summary function such as K-function which validates suitability point pattern data to point process model [15]. To simulate envelopes, we first estimate the inhomogeneous K-function for the earthquake data by

$$\hat{K}(r) = \frac{1}{W|D|} \sum_{u,v \in \mathbf{X} \cap D}^{\neq} \frac{\mathbf{1}\{\|u - v\| \leq r\}}{\hat{\lambda}(u)\hat{\lambda}(v)} e(u, v; r) \tag{5}$$

where $|D|$ is the area of Sumatra, $e(u, v; r)$ is an edge correction, and $W = \frac{1}{|D|} \sum_{u \in \mathbf{X} \cap D} \hat{\lambda}(u)^{-1}$. Next we generate n point patterns as realizations of the null model and compute their K-function estimates, denoted by $\hat{K}^1(r), \ldots, \hat{K}^n(r)$. The upper and lower envelopes are then obtained by resp. determining the maximum and minimum of the n K-function estimates. We reject the null hypothesis with significance level $\gamma = 2/(n+1)$ if the estimated inhomogeneous K-function (5) lies outside the interval bound. The envelope test is useful to determine the most suitable model among list of candidate models.

As an alternative, we compare models by evaluating the maximum values of (4) and the Akaike information criterion (AIC) values. The best model is selected based on the maximum values of likelihood and minimum AIC. The AIC [17] is defined by

$$\text{AIC} = -2\text{L}_{\max} + 2k,$$

where $\text{L}_{\max}$ is the maximum of (4) and k is the number of parameters.

4 Result

4.1 Spatial Trend and Clustering Detection

We detect the spatial trend by the `quadrat.test` function of `spatstat` R package [4,15]. By $\chi^2 = 43445$ with the degree of freedom equals to 103 (or p-value $\leq 2.2e-16$), we conclude that there exists a strong evidence of spatial trend. Considering the geological variables (Figs. 1 bottom and 2), we could illustrate the relationship between earthquake occurrences and geological variables. The distance between subduction zone and Sumatra mainland is within 200 km, where major earthquakes frequently occur in this area such as Simuelue, Nias, and Mentawai. Meanwhile, the faults and volcanoes line up along the west part of Sumatra. This is in agreement with the earthquake occurrences in Aceh, Padang, Bengkulu, and Lampung.

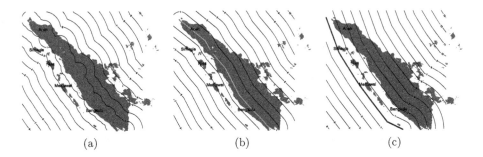

(a)   (b)   (c)

Fig. 2. Contour plot representing the distance per 100 km (black lines) of each of the geological variables in the Sumatra: (a) volcanoes, (b) faults, (c) subduction zones.

Next, the inhomogeneous K-function (Fig. 3) is employed to investigate clustering using the `Kinhom` function [15]. The baby blue dashed line shows the

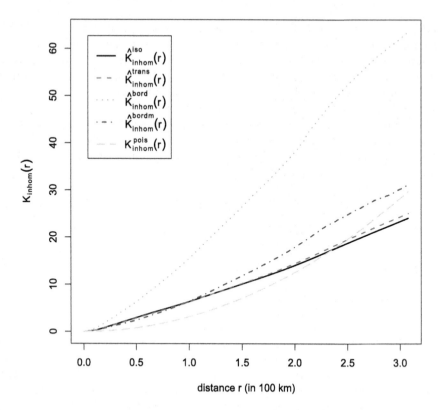

Fig. 3. Inhomogeneous K-function for the earthquake data in the Sumatra during 2004–2018.

K-function of an inhomogeneous Poisson process. In general, all the edge corrected inhomogeneous K-functions for the earthquake data indicate the presence of clustering, especially when the radius is less than 200 km. These findings suggest to use an inhomogeneous cluster point process such as Neyman–Scott Cox point process.

4.2   Model Comparison

The resulting estimators are reported in Table 1. We only present the estimates corresponding to the clustering parameters, i.e. ψ. The β estimates are identical for all models, so it is not useful for model comparison. We detail the interpretation of β estimates in Sect. 4.3.

The Cauchy model captures the highest clustering effect (smallest $\hat{\kappa}$ and $\hat{\omega}$) while the Matérn model detects the weakest clustering (largest $\hat{\kappa}$ and $\hat{\omega}$). The clustering effect by Thomas and variance-gamma models are in between the previous two models but with different behaviour. For example, the Thomas model allows for a higher intensity of the mainshock process ($\hat{\kappa}$) than that of the variance-gamma model, but with a smaller $\hat{\omega}$.

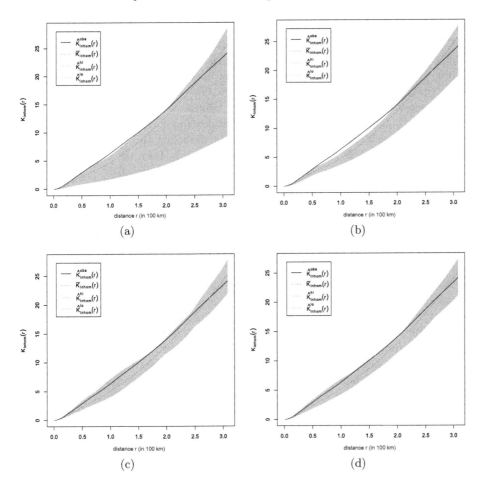

Fig. 4. Envelopes K-functions for the earthquake data in Sumatra based on: (a) Thomas, (b) Matérn, (c) variance-gamma, and (d) Cauchy cluster models.

Table 1. The clustering estimates $\hat{\psi} = (\hat{\kappa}, \hat{\omega})$, the maximum composite likelihood L_{max} and the AIC values for each of fitted model to the Sumatra earthquake data.

	Thomas	Matérn	Var-Gamma	Cauchy
$\hat{\kappa}$	0.71429	0.73337	0.53001	0.35975
$\hat{\omega}$	0.17973	0.34633	0.21760	0.14989
L_{max} $(\times 10^5)$	396.25099	396.09864	396.39394	396.48341
AIC $\times 10^5$	-792.50188	-792.19717	-792.78777	-792.96672

In general, all the four Neyman–Scott Cox models perform well. In addition to the mean of the envelopes K-function (dashed red line) that are close to the empirical ones (solid black line), the empirical K-functions fall inside the envelopes interval (grey area) for all models (Fig. 4). We notice variance-gamma and Cauchy models perform best by producing sharper envelopes (Fig. 4c–d). The comparison based on the L_{max} and AIC values show a similar message (Table 1). The variance-gamma and Cauchy have larger L_{max} and smaller AIC than that of Thomas and Matérn models. One of the main features of the variance-gamma and Cauchy models over Thomas and Matérn models is that the two earlier models allow the aftershock scattered very distant around the mainshock, which could be advantageous for modeling the earthquake data in Sumatra. The Cauchy model performs slightly better than the variance-gamma models and is used for interpretation and prediction (Sect. 4.3).

4.3   Model Interpretation and Prediction

By the Cauchy model, there are 74 estimated number of mainshocks (or 74 clusters). In addition, the aftershocks are scattered around each of the mainshock with scale of 15 km (Table 1).

To quantify the effect of each of the geological factors, we consider β estimators reported in Table 2. We first find that only fault and subduction zone are included in the model. The volcano is eliminated since it exhibits strong correlation with fault in Sumatra (See Fig. 1 bottom where the locations of volcanoes and fault are close together). Second, the β estimates for both fault and subduction are negative (Table 2), meaning that the closer the area to the fault or subduction, the more likely the major-shallow earthquake to occur. In more detail, in the area with a distance 100 km closer to a subduction zone (resp. a fault), the risk for a major earthquake occurrence increases by 3.29 (resp. 1.57) times. This also indicates that an area close to the subduction is two times (or 100%) riskier for a major earthquake occurrence than the one close to the fault. Since Simuelue, Nias, and Mentawai are the closest area to the subduction zones, more massive mitigation should be conducted in this area to prevent major risk due to earthquakes.

Figure 5 a presents the predicted maps of major earthquake risk in the Sumatra by the inhomogeneous Cauchy cluster model. We have three main points. First, the riskiest place is Aceh due to its proximity to both the Aceh subduction and two faults (segment Aceh and segment Seulimeum). During 2004–2018, there are 86 major earthquakes in Aceh. Second, due to its closeness to the subduction zones, Simuelue, Nias, and Mentawai are the second riskiest area. 3691 major earthquakes (almost 50% of all major earthquakes in Sumatra) are observed in these areas during 2004–2018. Third, the last riskiest area is in the tail of the meeting area between subduction zone and fault in Sumatra (i.e. in Bengkulu), where 151 major earthquakes occur during 2004–2018. Our general impression is the Cauchy cluster model fit well the distribution of major and shallow earthquake in Sumatra (Fig. 5a–b).

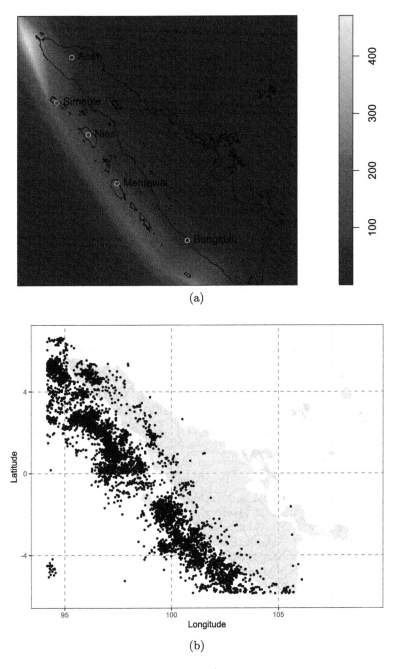

Fig. 5. (a) Prediction map of the earthquake risk in the Sumatra based on the Cauchy cluster model, and (b) Distribution of the locations of earthquakes in Sumatra.

Table 2. The important spatial covariates and their corresponding regression β estimates.

Parameter	Estimate ($\hat{\beta}$)	$\exp(\hat{\beta})$	$1/\exp(\hat{\beta})$
Intercept $\left(\hat{\beta}_0\right)$	6.44385	628.8231	
Fault $\left(\hat{\beta}_1\right)$	-0.45329	0.63553	1.57349
Subduction $\left(\hat{\beta}_2\right)$	-1.19106	0.30390	3.29053

5   Conclusion

In this study, we implement the Neyman–Scott Cox process for complex earthquake data in the Sumatra, which involve a large number of earthquake occurrences and geological covariates. We modify the two-step estimation proposed by Waagepetersen and Guan [10] to obtain a more efficient estimator using likelihood-based approach. Among all the four Neyman–Scott models, the Cauchy cluster model performs best. The subduction zone has a huge impact on increasing the earthquake risk, hence mitigation in the areas close to subduction zones in the Sumatra is of major concern. The locations of faults and volcanoes in Sumatra are close and therefore are highly correlated, so we decide to only choose one variable to avoid multicolinearity problem. Study to improve the model which takes multicolinearity into account would be interesting for future study such as regularization methods for point processes [16,19].

Acknowledgements. The research is supported by Institut Teknologi Sepuluh Nopember grant number 1292/PKS/ITS/2021. We thank the two reviewers for the comments.

References

1. Ogata, Y.: Statistical models for earthquake occurrences and residual analysis for point processes. J. Am. Stat. Assoc. **83**(401), 9–27 (1988)
2. Zhuang, J., Ogata, Y., Vere-Jones, D.: Stochastic declustering of space-time earthquake occurrences. J. Am. Stat. Assoc. **97**(458), 369–380 (2002)
3. Türkyilmaz, K., van Lieshout, M.N.M., Stein, A.: Comparing the Hawkes and trigger process models for aftershock sequences following the 2005 Kashmir earthquake. Math. Geosci. **45**(2), 149–164 (2013)
4. Choiruddin, A., Aisah, Trisnisa, F., Iriawan, N.: Quantifying the effect of geological factors on distribution of earthquake occurrences by inhomogeneous Cox processes. Pure Appl. Geophys. **178**(5), 1579–1592 (2021)
5. Aisah, I.N., Choiruddin, A.: On the earthquake modeling by using Bayesian mixture Poisson process. Int. J. Adv. Sci. Technol. **29**(7s), 3350–3358 (2020)
6. Mukhti, T.O., Choiruddin, A., Purhadi: Generalized additive Poisson models for quantifying geological factors effect on the earthquake risk mapping. J. Phys. Conf. Ser. **1863**(1), p. 012030 (2021)

7. Siino, M., Adelfio, G., Mateu, J., Chiodi, M., D'Alessandro, A.: Spatial pattern analysis using hybrid models: an application to the Hellenic seismicity. Stoch. Environ. Res. Risk Assess **31**(7), 1633–1648 (2016). https://doi.org/10.1007/s00477-016-1294-7

8. Møller, J., Waagepetersen, R.P.: Statistical Inference and Simulation for Spatial Point Processes. CRC Press (2003)

9. Møller, J., Waagepetersen, R.P.: Modern statistics for spatial point processes. Scand. J. Stat. **34**(4), 643–684 (2007)

10. Waagepetersen, R.P., Guan, Y.: Two-step estimation for inhomogeneous spatial point processes. J. Roy. Stat. Soc.: Ser. B (Stat. Methodol.) **71**(3), 685–702 (2009)

11. Guan, Y., Shen, Y.: A weighted estimating equation approach for inhomogeneous spatial point processes. Biometrika **97**(4), 867–880 (2010)

12. Guan, Y.: A composite likelihood approach in fitting spatial point process models. J. Am. Stat. Assoc. **101**(476), 1502–1512 (2006)

13. Natawidjaja, D.H.: Tectonic setting indonesia dan pemodelan sumber gempa dan tsunami. Geoteknologi-LIPI (2007)

14. Amri, M.R., et al.: Risiko bencana Indonesia. Badan Nasional Penanggulangan Bencana, Jakarta (2016)

15. Baddeley, A., Rubak, E., Turner, R.: Spatial Point Patterns: Methodology and Applications with R. CRC Press (2015)

16. Choiruddin, A., Coeurjolly, J.-F., Letué, F., et al.: Convex and non-convex regularization methods for spatial point processes intensity estimation. Electron. J. Stat. **12**(1), 1210–1255 (2018)

17. Choiruddin, A., Coeurjolly, J.-F., Waagepetersen, R.P.: Information criteria for inhomogeneous spatial point processes. Aust. New Zealand J. Stat. **63**(1), 119–143 (2021)

18. Jalilian, A., Guan, Y., Waagepetersen, R.P.: Decomposition of variance for spatial Cox processes. Scand. J. Stat. **40**(1), 119–137 (2013)

19. Choiruddin, A., Coeurjolly, J.-F., Letué, F.: Adaptive lasso and Dantzig selector for spatial point processes intensity estimation. arXiv preprint arXiv:2101.03698 (2021)

Construction of Optimal Stock Market Portfolios Using Outlier Detection Algorithm

Gee-Kok Tong[1], Keng-Hoong Ng[1]($\boxtimes$), Wun-She Yap[2]($\boxtimes$), and Kok-Chin Khor[2]($\boxtimes$)

[1] Faculty of Computing and Informatics, Multimedia University, 63100 Cyberjaya, Selangor, Malaysia
{gktong,khng}@mmu.edu.my
[2] Lee Kong Chian Faculty of Engineering Science, Universiti Tunku Abdul Rahman, 43000 Kajang, Malaysia
{yapws,kckhor}@utar.edu.my

Abstract. It is not easy for investors to trade in stock markets as building stock portfolios requires financial knowledge and consumes much time. Thus, this study aims to construct optimal stock market portfolios for investors using a LOF-based methodology. We used an outlier detection algorithm called Local Outlier Factor (LOF) to identify outperforming stocks from a stock pool. We then constructed two portfolios using these outperforming stocks, namely, tangency and equal-weighted portfolios and compared their performance against the benchmark portfolios, namely, the market portfolio and the cash market. It was followed by using Mean-Variance Portfolio Optimisation (MVPO) to measure the performance of the portfolios and determined whether they were efficient. To identify the most efficient portfolio, we used the Sharpe ratio. In general, the results showed that both tangency and equal-weighted portfolios gave better returns than the benchmark portfolios. To conclude, the LOF-based methodology helps to build and identify profitable stock portfolios for investors.

Keywords: Stock portfolios · Local outlier factor · Mean-variance portfolio · Sharpe ratio

1 Introduction

Stock markets are complex and dynamic systems. Therefore, it is always not easy to generate lucrative returns from stock investments. The investors are attempting different stock analysis techniques aiming to beat the stock markets. Hence, the research into stock markets remains a vital interest to many financial researchers.

Amateur or even professional investors understand that it is always a Herculean task to build outperforming stock portfolios. Building stock portfolios aim to gain optimal returns by constructing a combination of good performing stocks with the proper resource allocation for each of them. Identifying stocks with outstanding financial performance for portfolio selection is always on the investors' "radar". It is because such stocks often outshine their peers in capital returns. However, it is hard to uncover such stocks from, i.e., Bursa Malaysia with around 1000 listed companies (as of 2020).

© Springer Nature Singapore Pte Ltd. 2021
A. Mohamed et al. (Eds.): SCDS 2021, CCIS 1489, pp. 160–173, 2021.
https://doi.org/10.1007/978-981-16-7334-4_12

Using data mining methods for portfolio selection is prevalent nowadays. The research community utilises not only supervised learning algorithms for stock portfolios but also clustering algorithms. Clustering algorithms, including the parametric and non-parametric types, are widely used in the stock research domain. It is popular to use data mining methods because of market volatility and complexity, causing investors difficulty to make reasonable estimations.

Investors construct stock portfolios based on their predefined stock lists. For instance, a study conducted by [1] used 90 highly capitalised US stocks for 1997–2007. To optimise the stock portfolio, investors commonly use Mean-Variance Portfolio Optimisation (MVPO) [2]. Markowitz proposed MVPO in 1952, and it is still influential in portfolio analysis. MVPO aims for optimal risk-return trade-offs via portfolio diversification. Besides, MVPO is also used to measure the performance of different stock portfolios. To determine one efficient portfolio out of the portfolios constructed, investors can use the Sharpe ratio [3].

This study aims to assist investors in building profitable stock portfolios by using an outlier detection algorithm. To our best knowledge, there are very few research on building stock portfolios using outlier detection algorithms. Instead of letting the investors set a predefined list of outperforming stocks by themselves, we propose to construct stock portfolios using the LOF-based approach. It is then followed by using MVPO to measure the performance of the portfolios. Lastly, we assess the portfolios using the Sharpe ratio and determine the most efficient one.

We organise the remainder of the paper as follows. In the next section, we shall provide an overview of the data mining methods used in the stock market research. Section 3 gives a detailed explanation of how the LOF-based approach was utilised in this study. The analysis results shall then be discussed in Sect. 4, and the whole study shall be concluded in Sect. 5.

2 Related Work

2.1 Utilising Data Mining Methods

Portfolio construction is one of the primary decision-making problems for investors in stock markets. Researchers have proposed various data mining methods, including non-parametric type, to solve the problem. Using non-parametric data mining methods, researchers need not assume any underlying data distribution and not limited by a set of parameters [4]. The examples of non-parametric data mining methods are discussed as follows.

There is always the interest in utilising clustering algorithms for constructing stock portfolios. A study by [5] used K-means to mine investment information about stock category clusters. Before using K-means, the authors used the Apriori algorithm to illustrate knowledge patterns and rules for proposing stock category associations and possible stock category investment collections. Another similar study by [6] used a clustering method, Self-Organising Map (SOM), to build balanced stock portfolios.

Besides clustering algorithms, researchers also used other data mining methods. A study by [7] used the Minimax Probability Machine and Support Vector Machines to build a short-term portfolio management model. They found that both methods are suitable for selecting stocks to be traded. In the study, they derived the stock selections from the volatility around the earning announcements. The recent advancement involves combining deep neural networks and MVPO for adaptive investment portfolio management [8]. Another recent work also used a neural network model for solving the min-max optimisation problem in portfolio selection [9].

Instead of using clustering methods and other data mining methods discussed above in handling stock portfolios, we attempted an outlier detection algorithm. To date, there are very few works on stock portfolios utilising outlier detection algorithms. The algorithms are capable of identifying outperforming stocks by observing stocks with unusual data patterns. In this study, we used an outlier detection algorithm called LOF to identify outperforming stocks. The advantages and detail of LOF shall be discussed in Sect. 3.3.

2.2 Utilising MVPO and Sharpe Ratio

The early research discussed in the previous section utilised data mining methods, mostly clustering, for portfolio building. However, these works did not consider the risk inherited by the portfolios formed. It would be advantageous if the portfolios can be assessed using specific risk measures, i.e., MVPO and Sharpe ratio.

MVPO optimises a stock portfolio by measuring its risk to ensure its maximum expected return or minimise its risk given returns. Besides, the measure can also be used to measure the performance of different portfolios. MVPO has been applied to various fields, from investment in stock markets to derivative markets and other investment decision makings.

On the other hand, the Sharpe ratio has become the most widely used measure in the financial world to measure the risk-adjusted return. It can be used to identify efficient stock portfolios.

The recent research uses not only clustering algorithms but also various data mining methods incorporating risk measures. A study by [10] utilised clustering methods in portfolio selection. Besides, they minimised the portfolio risk by incorporating MVPO. K-means, SOM and Fuzzy C-means were used to construct portfolios, and K-means turned out to be the outperforming algorithm. A study by [11] improved MVPO so that the optimisation problem that consisted of the NIKKEI 500 stocks and all 1,100 stocks transacted in the Tokyo Stock Market can be solved linearly and in a short amount of time. Another study [12] applied MVPO to calculate portfolio risks. An optimal portfolio had been obtained from the combination of equity, bond, and commodity. [13] used a two-stage optimisation algorithm to solve a multi-objective portfolio optimisation model. The algorithm serves as a tool for an oil company in planning its investment strategies.

Marvin (2015) used K-means to group stocks and picked a stock from each group to form a diversified portfolio [14]. Sharpe ratio, which assessed the risk-adjusted return, was used to select a stock from each group. Another two studies, The studies by [15] and [16], also used the Sharpe ratio, but it was used as the fitness function in the genetic algorithm for selecting stocks. A study by [17] used the Sharpe ratio to develop a new portfolio efficiency that maximises the unconditional Sharpe ratio of excess returns. The forward-looking Sharpe ratios are used by [18] to construct a dynamic portfolio of stocks and corporate bonds that outperforms a static portfolio on a risk-adjusted basis.

Looking at the advantages of incorporating risk measures for portfolio optimisation, we thus used MVPO and Sharpe ratio in this study. Using both measures in this research is mainly to measure the performance and identify the most efficient stock portfolio.

3 Methodology

3.1 Preparing Stock Data

We chose the construction sector of Malaysia because of its importance in the country's economic development. Every year, the government provides a substantial amount of funds to support mega infrastructure projects. Further, in the recently announced 2020 Malaysia Budget, the construction sector remains the economic driving force [19].

We identified outstanding performance stocks from the construction sector in Bursa Malaysia. It started with the preparation of the experimental data sets. The five-year stock data sets (the financial year 2011–2015) were formed using the 12 normalised financial ratios of the construction stocks (Algorithm 1, phase 1). These financial ratios were derived and gathered from the five ratio categories, as shown in Table 1.

Firstly, *activity ratios* evaluate a company's operating performance. Total asset turnover is one example of activity ratios, and it measures the revenue generated from a company's assets. A company with a high total asset turnover implicitly reflects the efficiency of top management in utilising their company's assets to maximise income generation.

Secondly, *liquidity ratios* reveal a company's capability to repay its current debt. We used the cash ratio to gain an insight into the repayment capability of a company. A company with a high cash ratio has a much low risk of default. Therefore, a company with a high cash ratio than its peers is always preferable.

Thirdly, the *leverage ratios* are useful indicators to track the financial "healthiness" of a company. To finance a company operation, the company requires a combination of equity and debt. The leverage ratios are useful to evaluate whether or not the company is capable of repaying its debt in due time. Two financial ratios that are categorised under the leverage ratios are debt ratio and equity turnover. These two financial ratios indicate how a company uses loans or credit facilities to expand their business or acquire assets.

Fourthly, the *market value ratios* provide hints to the investors on the current stock price, whether or not it is overvalued. Four financial ratios, i.e., price-earnings ratio, price to book ratio, dividend yield and earnings yield, are examples of market value ratios.

Finally, the *profitability ratios* demonstrate the proportion of profits generated from a company's business or investment. A high profitability ratio of a company explicitly suggests that a superior management team runs the company. Return on assets, return on equity, net profit margin, and operating margin are appropriate financial ratios to demonstrate how well a company utilises its existing assets to generate profit.

Table 1. The financial ratios used in this study to find outstanding stocks.

Ratio category	Financial ratios	Formula
Activity ratio	Total asset turnover	Sales ÷ Total Assets
Liquidity ratio	Cash ratio	Cash ÷ Current Liabilities
Leverage ratios	Debt ratio	Total Debt ÷ Total Assets
	Equity turnover	Sales ÷ Equity
Market value ratios	Price earnings ratio	Price per share ÷ Earnings per Share
	Price-to-book ratio	Price per share ÷ Book Value per Share
	Dividend yield	Dividend per Share ÷ Price per Share
	Earning yield	Earnings per Share ÷ Price per Share
Profitability ratio	Return on assets	Net Income ÷ Total Assets
	Return on equity	Net Income ÷ Total Equity
	Net profit margin	Net Income ÷ Sales
	Operating margin	Profit Before Tax ÷ Sales

3.2 Preparing Proxy Data

A proxy is a broad representation of an overall market. For example, the FTSE Bursa Malaysia KLCI (FBMKLCI) is one of the Malaysian stock market's market proxies. The FBMKLCI consists of the top 30 companies by market capitalisation on Bursa Malaysia [20].

We collected the daily data from (i) the construction index (the market proxy), (ii) Malaysia inter-bank 3-month interest rates (the cash market proxy), and (iii) the individual outstanding stocks from the construction stocks of Bursa Malaysia. All the proxy data were derived from the DataStream database (Algorithm 1, phase 1).

3.3 Identifying Outliers

After data preparation, the subsequent step was to identify outliers in the construction stocks data sets using a selected outlier detection algorithm called Local Outlier Factor (LOF) [21] (Algorithm 1, phase 2). In this context, an outlier refers to a construction stock with either excellent or inferior financial performance compared to its peers. LOF was the selected outlier detection algorithm in this study due to the two justifications [22]. Firstly, it can detect outliers in multi-dimensional data sets. Secondly, it can handle clusters with different densities that are common in many real data sets. LOF calculates and assigns each instance in the data set with a value. Typically, a non-outlier instance has a LOF value approximates 1.0. A high LOF value indicates that the instance is possibly an outlier. Any construction stock with its LOF score equal to or greater than 1.5 is tagged as an outlier [22]. Before the outlier detection process, min-max normalisation was applied to the financial ratios of the data sets. This vital step ensured that each financial ratio in the data sets was treated equally by the LOF algorithm.

Algorithm 1 The proposed methodology

1: **procedure** proposedMethodology(*con_sec, lof, mvpo, sharpe*)

2: // Phase 1: data preparation and pre-processing

3: // *con_sec* are construction stocks listed on Bursa Malaysia

4: *startFY*←2011, *endFY*←2015

5: **for** *yr*←*startFY* to *endFY* do

6:     extract *yr* financial data from *con_sec* stocks using DataStream

7:   transform the financial data into 12 financial ratios and form *db_yr* data set

8: **end for**

9: normalise the _financial ratios in the 5 data sets with the min-max

    method ▶ values are [0,1]

10: //Phase 2: outlier identi_cation with lof

11: **for** each data set *db_yr* **do**

12:     perform *lof* algorithm on the *db_yr*

13:     **for** each stock, *sk* ∈ *db_yr* **do**

14:         **if** *LOFvalue(sk)* > 1.5 then

15:             tag *sk* as outlier, *skout*

16:         **end if**

17:     **end for**

18: **end for**

19: // Determine outstanding stocks from the pool of outliers

20: **for** each identified outlier, *skout* in *db_yr* **do**

21:         **if** *skout.currentYrEarning* > *skout.precedingYrEarning* and

skout.currentYrBookValue > *skout.precedingYrBookValue* then

22:             label *skout* as outstanding stock

23:         **end if**

24: **end for**

25: // Phase 3: portfolio optimisation and assessment

26: **for** outstanding stocks in each financial year 2011 to 2015 **do**

27:         measure the 1-year capital return performance with the market proxy and the cash market

            proxy, using *mvpo* and *sharpe*

28: **end for**

29: **end procedure**

The LOF algorithm is discussed in brief here. LOF computes the *k*-distance using Euclidean distance between all the possible pairs of stocks in the construction stock pool, *sp*. LOF then finds the *k*-distance neighbourhood for each stock, covering neighbour stocks with a distance less than the *k*. Subsequently, the local reachability distance *lrd* of each stock s is computed, as shown in Eq. 1, where $\|sp_k(s)\|$ denotes the number of *k*-nearest neighbours to *s*, and *rd(s'←s)* is the reachability distance *rd* between *s* and its *k*-nearest neighbour. Finally, a LOF score is computed for each *s*. The formula to compute the LOF score is as shown in Eq. 2.

$$lrd(s) = \frac{\|sp_k(s)\|}{\sum_{s' \in sp_k(s)} rd(s' \leftarrow s)} \tag{1}$$

$$LOF(s) = \frac{\sum_{s' \in sp_k(s)} lrd(s')/lrd(s)}{\|sp_k(s)\|} \tag{2}$$

The execution of the LOF algorithm on the five construction stock data sets (the financial year 2011–2015) had identified 26 outliers in total, as shown in Table 2. LOF detected seven outliers in the year 2011 data set. The algorithm also identified five outliers in each of the 2012 and 2013 data sets, respectively. In the 2014 data set, two construction stocks had been marked by LOF as outliers. The number of outliers that LOF had identified in the last data set, 2015, was seven. However, these identified outliers can either be outstanding or poor performing stocks. Therefore, they were differentiated manually based on the evaluation of two accounting summary measures, i.e., *earnings* and *book value*. These two accounting measures are effective in equity valuation; this claim was supported by a study [23], where both accounting measures were evaluated and proven to have a significant positive impact on stock prices and stock returns. In this case, an outlier is interpreted as an outstanding stock when its current earnings and book value surpasses the past financial year values. Otherwise, the outlier shall be classified as poor-performing stocks. As a result, ten were considered outstanding stocks after the screening (refer to Table 2). The remaining 16 were considered poor-performing stocks.

Table 2. The 10 outstanding stocks (indicted using bold text) were selected using two accounting measures: earnings and book value, from the outliers from the 26 outliers identified using LOF.

Data set	Outliers	No. of companies in the data set
2011	**ARK, KERJAYA**, LEBTECH, PESONA, PUNCAK, PRTASCO, WCEHB	40
2012	**PUNCAK, WCEHB**, BPURI, HOHUP, JETSON	40
2013	**ARK**, IREKA, MLGLOBAL, WCEHB, ZELAN	41
2014	MTDACPI, WCEHB	41
2015	**BENALEC, PESONA, PRTASCO, SUNCON, WCEHB**, BPURI, MTDACPI	44

Stock portfolios were then built using the outstanding outliers identified by LOF. Take the example of 2011; we constructed a portfolio using the outstanding outliers, namely, ARK and KERJAYA (refer to Table 2). We then used MVPO to measure the performance of the stock portfolio. The portfolio performance was then compared with the performance of the construction index in Bursa Malaysia and the cash market of the following year, 2012.

3.4 Measuring Performance of the Stock Portfolios Using MVPO

After constructing the portfolios, they were then assessed using MVPO and Sharpe ratio (Algorithm 1, phase 3). MVPO is vital in investment decision making [2]. It is particularly useful for risk-averse investors to build a stock portfolio by maximising their expected return given the risk or, conversely, to minimise the risk given the return.

Rational risk-averse investors prefer the less risky option when they are given portfolios that offer the same expected return. Investors are only willing to take high risks if higher expected returns compensate them. Thus, high levels of risk are usually associated with high potential returns. However, different investors possess different risk aversion characteristics.

There are specific criteria to be satisfied when carrying out portfolio optimisation. Firstly, to minimise the risk of a proxy (by the standard deviation of stock return). Secondly, to match or exceed the return of a proxy (by the mean or expected stock return). Portfolios satisfying these criteria are *efficient portfolios*. The *expected return* of a portfolio is given by the following Eq. 3,

$$E(r_p) = \sum w_i E(r_i) \tag{3}$$

where r_p is the return of the portfolio and r_i is the return on stock i, and w is the weightage of i. The returns are solely based on capital appreciation. On the other hand, the portfolio return variance is given by the following Eq. 4,

$$\sigma_p^2 = \sum w_i^2 \sigma_i^2 + 2 \sum \sum w_i w_j \sigma_i \sigma_j p_{ij} \tag{4}$$

where σ is the sample *standard deviation* of the daily returns (risk proxy) and p_{ij} is the correlation coefficient between the returns on stock i and j.

Using MVPO, investors can reduce their portfolio risk by holding a combination of stocks that are not positively correlated. The purpose of such diversification is to achieve a good portfolio return with a lower risk. However, in our study, stocks involved are in the same construction industry. Diversification is achieved by investing in different stocks in the same industry. This enables the performance comparison between the portfolios or the individual stocks and the benchmark construction index. Note that portfolio optimisation can be applied in the same industry or sector. A study by [24] proposed the Sectoral Portfolio Optimization by focusing on optimising stocks within the same sector based on financial analysis on four financial ratios: returns on asset, debt-assets ratio, current ratio, and price-earnings ratio.

Stocks may carry a certain level of risks, unlike *risk-free security*, such as government bonds that guarantee returns. All the possible combinations of stocks (or stock portfolios) can be plotted in this risk-expected return space, as shown in Fig. 1. Portfolios are the points from a feasible set of outstanding stocks. The set of a feasible portfolio is necessarily a nonempty, closed, and bounded set. In the figure, the hyperbola shows the attainable portfolio. The upper edge of the hyperbola is the *efficient frontier*. The efficient frontier is a composition of all the possible risk-carrying stocks in the absence of risk-free securities, which give investors efficient stock portfolios. Using the efficient frontier of an optimal portfolio, investors maximise the expected return for a given risk level. If risk-free assets are included, then the efficient frontier is the tangent line (the dotted line in Fig. 1) to the hyperbola at the tangency point and the rest of the upper edge of the hyperbola. The tangent line is also regarded as the *capital allocation line*.

Investors may maximise their investments as long as they invest in the portfolio composition located along the efficient frontier. However, if they wish to identify the exact portfolio composition that gives them the best investment reward, they need to utilise the Sharpe ratio.

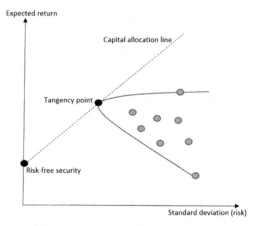

Fig. 1. Mean-variance portfolio optimisation (MVPO).

3.5 Assessing the Stock Portfolios Using the Sharpe Ratio

Sharpe ratio was introduced by [3] to measure the performance of mutual funds. It is defined as the average return earned in excess of the risk-free rate per unit of volatility or total risk. The Sharpe ratio is given by the following Eq. 5,

$$Sharpe\ ratio = \frac{r_p - r_f}{\sigma_p} \tag{5}$$

where r_p is the return of the portfolio, r_f is the risk-free rate, and σ_p is the standard deviation of the portfolio's excess return.

From the above equation, the risk-free rate is subtracted from the portfolio return and thus, allow investors to isolate better the profits associated with risk-taking activities. Hence, the Sharpe ratio can help investors understand the return of an investment compared to its risk. In general, the higher the value of the Sharpe ratio, the more attractive the risk-adjusted return.

From Fig. 1, the *tangency point* represents the most *efficient portfolio (tangency portfolio)* as it gives the best return per unit risk or maximises the Sharpe ratio. Thus, an advanced investor may invest at this point to get the highest return per unit risk with the assumption that he can accept the risk level at the tangency point.

4 Results and Discussion

Figure 2 illustrates the performance of the individual stocks, equal-weighted portfolio, tangency portfolio, market portfolio (proxy by construction index), and cash market (proxy by Malaysia interbank 3-month interest rates). Bear in mind that not all investors, especially novice investors, acquire the portfolio combination skill. Thus, this study used equal-weighted portfolios to facilitate investors. In equal-weighted portfolios, every identified outstanding stock was invested with the same amount of money. Thus, they do not require any skill in finding the portfolio combination.

An equal-weighted portfolio was constructed based on the selected outliers of the year 2011 in Table 2. Investors can easily beat the market performance in 2012 using this portfolio that comprises KERJAYA and ARK. This portfolio managed to earn a 41.36% return with a 60.44% risk. On the other hand, the market annualised return is −6.04% and yet incurred a 12.67% risk; its performance is even worse than the cash market (risk-free market) performance that earns 3.164% return with a risk of 0.0125% (virtually risk-free asset).

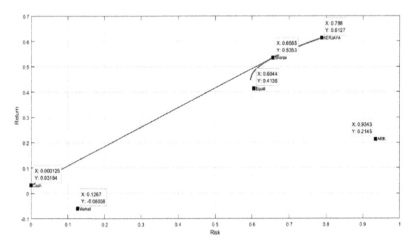

Fig. 2. The portfolio optimisation for the year 2012 based on the outliers selected in the year 2011.

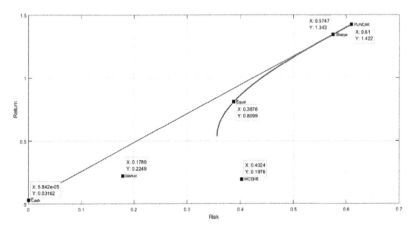

Fig. 3. The portfolio optimisation for the year 2013 based on the outliers selected in the year 2012.

The advanced investors can utilise MVPO, where the portfolio optimisation in the current year was done based on the outstanding stocks selected in the previous years. They can optimise their investment by investing in the tangency portfolio (tangency point) because this is the point that maximised the Sharpe ratio with the best risk-adjusted return. By investing in the tangency portfolio, they can earn a 53.53% return with a 65.65% risk. Such a return is also much better than the market portfolio and the cash market.

The equal-weighted portfolio beat the market portfolio and the cash market again in the year 2013. Figure 3 shows that the equal-weighted portfolio earned an 80.99% return with a 38.76% risk compared with the market portfolio that only managed to earn a 22.49% return with a 17.89% risk. The cash market gave a meagre return despite its slight risk. Investing in the tangency portfolio helped to earn a 134.3% return with a 57.47% risk.

It was unique for the year 2014. There was only one outstanding stock, ARK, as the outlier in the year 2013. Thus, the equal-weighted portfolio is the same as the tangency portfolio and the individual outlier, ARK (Fig. 4). The performance of this portfolio or ARK is much better than the market portfolio and the cash market. This portfolio generated a 69.69% return compared with the market portfolio and the cash market, which only earned returns of 0.60% (14.98% risk) and 3.46% (1.85% risk), respectively.

Since there was no outstanding outlier selected in 2014 (refer to Table 2), no stock portfolio was constructed in 2015 for performance comparison.

The year 2016 is the only year that the market performance is slightly better than the equal-weighted portfolio. Multiple factors were causing this, particularly the instability of the Malaysian stock market that was triggered by the volatility of the Malaysian forex market. Malaysia Ringgit was tumbled in the year 2016, and it seriously affected the performance of the Malaysian stock market. Figure 5 shows that the market portfolio earned a 4.15% return and the equal-weighted portfolio, on the other hand, suffered a −0.42% return. Nevertheless, with MVPO to identify the tangency portfolio, investors managed to earn a 45.24% return with a 24.09% risk. The investment performance achieved using MVPO is much better than the market portfolio and the cash market.

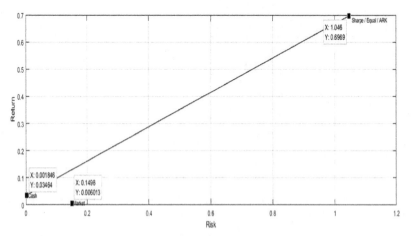

Fig. 4. The portfolio optimisation for the year 2014 based on the outliers selected in the year 2013.

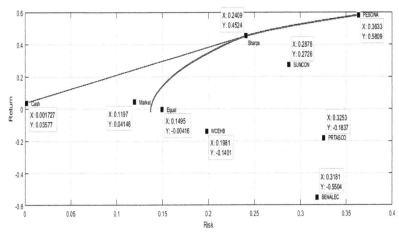

Fig. 5. The portfolio optimisation for the year 2015 based on the outliers selected in the year 2014.

The overall result of years 2012–2016 is as shown in Table 3. In conclusion, using MVPO, investors can earn the best risk-adjusted return at the tangency point that maximised the Sharpe ratio. The tangency portfolios gave the best performance among the portfolios. For novice investors, investing in the equal-weighted portfolios also resulted in returns far better than the market portfolios, particularly for 2012–2014.

Table 3. Comparison of the investment performance of different portfolios. The numbers in bold are the highest returns among the portfolios of the respective year.

		2012	2013	2014	2015	2016
Tangency portfolio	Rtn	**53.53**	**134.30**	**69.69**	No outliers selected	**45.24**
	Rs	65.65	57.47	104.60		24.0
Equal weighted portfolio	Rtn	41.36	80.99	**69.69**		−0.42
	Rs	60.44	38.76	104.60		14.95
Market	Rtn	−6.04	22.49	0.60		4.15
	Rs	12.67	17.89	14.98		11.97
Cash	Rtn	3.16	3.16	3.46		3.58
	Rs	0.01	0.00	0.18		0.17

* Rtn – Return(%); Rs – Risk(%)

5   Conclusion

This study aims to help investors construct optimal stock market portfolios using a LOF-based methodology. LOF was the outlier detection algorithm used to identify outperforming construction stocks for forming portfolios. To measure and assess the portfolios, we used MVPO and the Sharpe ratio. We constructed two portfolios using the outperforming construction stocks: (i) equal-weighted and (ii) tangency portfolios. The stocks in each equal-weighted portfolio carried equal weight, assuming that the investors invested the same amount of money in the individual stocks. On the other hand, a tangency portfolio that maximises the Sharpe ratio was constructed based on the identified tangency point on the efficient frontier. These two portfolio types were compared against (i) the cash market and (ii) the market portfolio.

We used five-year financial data (the year 2011–2015). It is obvious that the tangency portfolios gave the best performance and followed by the equal-weighted portfolios. Both portfolio types performed better than the market itself (the market portfolio) and the cash market, except for 2014, where none outperforming stock was selected.

The LOF used together with MVPO and the Sharpe ratio in this study proves its superiority in constructing optimal portfolios. However, there is room for improvement. The MVPO assumes the normal distribution for the returns of financial assets [25]. However, due to various influences concerning politics, regulations, and economics, the returns are not always normally distributed. Thus, it may violate the covariance conditions of MVPO. For better portfolio building in the future, we shall consider integrating MVPO with another data mining method that can solve the problem of the normal returns' assumption in MVPO.

References

1. Pantaleo, E., Tumminello, M., Lillo, F., Mantegna, R.N.: When do improved covariance matrix estimators enhance portfolio optimization? An empirical comparative study of nine estimators. Quant. Financ. 11(7), 1067–1080 (2011)
2. Markowitz, H.: Portfolio selection. J. Financ. 7(1), 77–91 (1952)
3. Sharpe, W.F.: The Sharpe ratio. J. Portfolio Manag. 21(1), 49–58 (1994)
4. Russell, S., Norvig, P.: Artificial Intelligence: A Modern Approach. Pearson, New Jersey (2009)
5. Liao, S.-H., Ho, H.-H., Lin, H.-W.: Mining stock category association and cluster on taiwan stock market. Expert Syst. Appl. 35(1–2), 19–29 (2008)
6. Silva, B., Marques, N.C.: Feature clustering with self-organizing maps and an application to financial time-series for portfolio selection. In: IJCCI (ICFC-ICNC), pp. 301–309 (2010)
7. Ince, H., Trafalis, T.B.: Kernel methods for short-term portfolio management. Expert Syst. Appl. 30(3), 535–542 (2006)
8. Obeidat, S., Shapiro, D., Lemay, M., MacPherson, M.K., Bolic, M.: Adaptive portfolio asset allocation optimization with deep learning. Int. J. Adv. Intell. Syst. 11(1), 25–34 (2018)
9. Nazemi, A., Mortezaee, M.: A new gradient-based neural dynamic framework for solving constrained min-max optimization problems with an application in portfolio selection models. Appl. Intell. 49(2), 396–419 (2018). https://doi.org/10.1007/s10489-018-1268-1
10. Nanda, S., Mahanty, B., Tiwari, M.: Clustering Indian stock market data for portfolio management. Expert Syst. Appl. 37(12), 8793–8798 (2010)

11. Marvin, K.: Creating Diversified Portfolios Using Cluster Analysis. Princeton University, Princeton (2015)
12. Yan, W., Sewell, M.V., Clack, C.D.: Learning to optimize profits beats predicting returns-comparing techniques for financial portfolio optimisation. In: Proceedings of the 10th Annual Conference on Genetic and Evolutionary Computation, pp. 1681–1688 (2008)
13. Cheong, D., Kim, Y.M., Byun, H.W., Oh, K.J., Kim, T.Y.: Using genetic algorithm to support clustering-based portfolio optimization by investor information. Appl. Soft Comput. **61**, 593–602 (2017)
14. Konno, H., Yamazaki, H.: Mean-absolute deviation portfolio optimization model and its applications to Tokyo stock market. Manag. Sci. **37**(5), 519–531 (1991)
15. Singhal, S.: Emergence of commodity derivatives as defensive instrument in portfolio risk hedging: a case of Indian commodity markets. Stud. Bus. Econ. **12**(1), 202–234 (2017)
16. Ximei, L., Latif, Z., Chang Feng, W., Latif, S., Khan, Z., Wang, X.: Mean-variance-kurtosis hybrid multi-objective portfolio optimization model with a defined investment ratio. J. Eng. Technol. **6**(1), 293–306 (2018)
17. Penaranda, F.: Understanding portfolio efficiency with conditioning information. J. Financ. Quant. Anal. **51**(3), 985–1011 (2016)
18. Goldberg, R.S.: A methodology for computing and comparing implied equity and corporate-debt Sharpe ratios. Rev. Quant. Financ. Acc. **44**(4), 733–754 (2015)
19. Bernama: 2020 budget: Construction sector remains economic driving force (2019). https://www.nst.com.my/news/nation/2019/10/529265/2020-budget-construction-sector-remains-economic-driving-force-says-cidb. Accessed 28 Oct 2020
20. FTSE Russell: FTSE Bursa Malaysia KLCI (2020). https://research.ftserussell.com/Analytics/FactSheets/temp/e98733f5-1671-42dd-9e71-3f4ea44acee4.pdf. Accessed 1 Dec 2020
21. Aggarwal, C.C.: Data Mining: The Textbook. Springer, New Delhi (2015). https://doi.org/10.1007/978-3-319-14142-8
22. Breunig, M.M., Kriegel, H.-P., Ng, R.T., Sander, J.: LOF: identifying density-based local outliers. In: Proceedings of the 2000 ACM SIGMOD International Conference on Management of Data, pp. 93–104 (2000)
23. Alfaraih, M., Alanezi, F.: The usefulness of earnings and book value for equity valuation to Kuwait stock exchange participants. Int. Bus. Econ. Res. J. (IBER) **10**(1), (2011)
24. Sharma, A., Mehra, A.: Financial analysis based sectoral portfolio optimization under second order stochastic dominance. Ann. Oper. Res. **256**(1), 171–197 (2016). https://doi.org/10.1007/s10479-015-2095-y
25. Sinha, P., Chandwani, A., Sinha, T.: Algorithm of construction of optimum portfolio of stocks using genetic algorithm. Int. J. Syst. Assur. Eng. Manag. **6**(4), 447–465 (2014). https://doi.org/10.1007/s13198-014-0293-7

Simulating the Upcoming Trend of Malaysia's Unemployment Rate Using Multiple Linear Regression

Nur Sarah Yasmin Mohamad Adib[(⊠)], Ahmad Asyraf Mohd Ibrahim[(⊠)], and Muhammad Hazrani Abdul Halim[(⊠)]

Universiti Teknologi MARA (UiTM) Shah Alam, 40450 Shah Alam, Selangor, Malaysia
hazrani@uitm.edu.my

Abstract. This research paper presents the approach of applying the simulation technique to predict the upcoming trend of Malaysia's unemployment rate. The recent Malaysia's unemployment rate has fluctuated at quite a high rate ever since the COVID-19 pandemic occurred. Population growth, Growth Domestic Product (GDP), inflation rate, interest rate, exchange rate, investment, government expenditure and most importantly the number of COVID-19 cases act as the independent variables in this paper. The Multiple Linear Regression (MLR) is used to determine the significance of each variable to be included in the model and also to simulate the upcoming trend of Malaysia's unemployment rate. The result of the analysis shows that the upcoming five years trend of Malaysia's unemployment rate will continue to increase in the future based on the average value of the simulations conducted.

Keywords: Unemployment rate · Multiple linear regression · Monte Carlo simulation · COVID-19 · Prediction

1 Introduction

One of the primary goals of macroeconomics is to achieve a low unemployment rate. Low unemployment rate means that the country is optimally utilizing its labour forces, and this is desirable for the prosperity of the economic growth. The rate of unemployment varies across all the countries globally. The unemployment rate in Malaysia from 2010 to 2020 ranges from 2.7% to 5.3%. The highest rate of unemployment was recorded in May 2020 when the COVID-19 pandemic hit most of the countries in the world. Other than that, there are many factors affecting the rate of unemployment such as population growth, rate of inflation, government intervention, the employee's level of education and such.

Predicting the unemployment rate is crucial as a preparation to take corrective actions as well as for decision-making purposes. The high accuracy of the prediction would be a great help for the government in guiding them to make decisions. Multiple Linear Regression (MLR) is a statistical method that can determine the significant factors affecting unemployment rate [1] as well as to predict the rate of unemployment.

© Springer Nature Singapore Pte Ltd. 2021
A. Mohamed et al. (Eds.): SCDS 2021, CCIS 1489, pp. 174–182, 2021.
https://doi.org/10.1007/978-981-16-7334-4_13

The research aims to simulate the upcoming trend of Malaysia's unemployment rate using the MLR. The analysis is being done using the Visual Basic Application (VBA) in Microsoft Excel. The upcoming trend of Malaysia's unemployment rate will give a better insight for the decision makers. The most precise prediction of the unemployment rate in Romania included the element of Monte Carlo simulation [2]. The details regarding this method will be discussed further in the latter section.

2   Literature Review

2.1   Overview and Definition of Unemployment

Unemployment happens when a person needs to work but is unable to find a job despite having the requisite experience and ability [3]. Unemployment can also be defined as an economic and social phenomenon that happens when a segment of the labour force is not active in the production and social activities [4]. Unemployment happens when the labour supply exceeds the labour demand available in the country [5]. Unemployment rate is used to determine the unemployment of the country [6]. The number of jobless people divided by the overall number of the workforce and multiplied by 100 is the calculation of the unemployment rate.

2.2   Issues of Unemployment

Unemployment has become one of the most severe economic concerns and all governments seek to implement measures that will minimise and control unemployment in their country [5]. High or uncontrollable unemployment rate will result in unstable economics of the country [7]. Almost every country in this world is coping with this macroeconomic issue, but what distinguishes them is the severity of the problem. The unemployment rate in Nigeria raised from 19.7% in 2009 to 21.1% and 23.9% for the following two consecutive years [8]. Also in Egypt, it is very worrying when the unemployment rates in that country are over 11.5% in ten years' time [9].

Youth unemployment is a greater concern than adult unemployment worldwide. According to [10], Nigeria's unemployment rate is at an all-time high, with a particularly high percentage among young people. In fact, there was a more heinous example in Italy. The percentage of young unemployment in the North was about 15%, but the rate in the South quadrupled that of the North, which was 50.5% [11]. In 2012, [3] claimed that the youth unemployment rate in Somalia is the highest in the world at 67%. Focusing on Malaysia, from 9.5% to 10.7%, it shows that the unemployment rate for young people in Malaysia has grown by 1.2% within one year and the national level of unemployment rate risen for only 0.2% from 2.9% to 3.1% [12]. Malaysia also has the third-highest rate of young unemployment in ASEAN, after the Philippines and [13]. The government should keep unemployment under control in their nation since it has ramifications for social and economic concerns.

2.3 Contributing Factors of Unemployment

William Philips stated that unemployment is negatively strong related to inflation. It is supported by the study conducted by [14] stated that inflation and unemployment are correlated whether in the short run or long run. The study also showed that the unemployment falls when the inflation rises, indicating that there is a negative relationship between inflation and unemployment [9].

Next, [9] claimed that there is a significant relationship between economic growth and unemployment. A higher GDP will result in a lower unemployment rate as stated in Okun's law [15]. It is also supported by [16] who revealed that GDP level is falling to 0.38978 due to a 1% increment in the unemployment rate. The findings are in line to those of [15], who identified a negative correlation between GDP and unemployment. [6] also mentioned that the economic performance of a country should be strengthened to lower the unemployment rate.

Furthermore, [15] showed that Foreign Direct Investment (FDI) has a negative relationship with unemployment. Plus, one unit increase in investment will reduce the unemployment rate by 0.450647 units [17]. In Somalia, a study conducted by [3] stated that external debt, population growth and gross domestic product (GDP) are positively related with unemployment rate, while gross capital information and exchange rate is significantly negatively correlated with unemployment. The unemployment and interest rate also have a positive relationship [6].

Next, [4] mentioned that Somalia's population growth gives a positive impact towards the unemployment rate. A study conducted in South Asian countries also revealed that the coefficient of 0.8082420 shows that unemployment rate and population growth have a positive relationship [17]. The high population growth is also one of the significant factors in determining the unemployment rate of a country [15].

2.4 Multiple Linear Regression

A study conducted by [1] stating that MLR is a multivariate statistical method in analysing the significant correlation between one dependent variable (Y) with two or more independent variables (X). MLR's major goal is to model the association between dependent and independent variables [18].

In predicting unemployment rate, different data types and location require different models to be used [19]. Thus, MLR can also be used for forecasting purpose. A study conducted in the East Coast of Malaysia uses MLR to estimate the youth unemployment rate by [12] claimed that MLR can be a statistical technique for predicting the result of a response variable by combining multiple explanatory factors.

2.5 Monte Carlo Simulation

Monte Carlo simulations have been used worldwide in this world where they involve the application of probability in our lives. In Romania, Monte Carlo simulation is implemented in predicting the unemployment rate and claimed that this approach has shown to be the most effective in obtaining the most exact forecasts [2]. We may repeat the simulation several times to evaluate how much variety there is in the outcomes to assess the accuracy of a Monte Carlo estimate [20]. However, the number of simulation iterations should be between 500 and 1000 to achieve the optimal value [19]. No matter what, the optimal model for forecasting the unemployment rate is still a point of contention among academics.

3 Methodology

The variables included in this research are based on past studies conducted. This research uses the secondary data obtained from several sources. The data of Malaysia's unemployment rate, government expenditure [9], GDP [8], inflation rate [14], interest rate [5], exchange rate [3], and investment growth [17] are all obtained from The Global Economy website. On the other hand, the daily number of COVID-19 cases [21] is obtained from the World Health Organization (WHO) and the data regarding Malaysia's population growth [3] is obtained from the Department of Statistics Malaysia (DOSM). The data are either in annual, quarterly or monthly form. Thus, the annual and quarterly data are all adjusted for data standardization purpose.

3.1 Multiple Linear Regression

It is vital to identify the significant factors of unemployment rate in Malaysia so that we can discover what are the root causes of this issue. We can identify the significance of the variables through stepwise regression. Through this method, the elimination or inclusion of a variable is determined based on the output of p-values obtained. The process when there are no more variables that can be included or excluded from the model.

The general form of the MLR is as follows:

$$Y = \beta_0 + \beta_1 X_1 + \beta_2 X_2 + \ldots + \beta_n X_n + \varepsilon \tag{1}$$

where,

Y: Dependent variable
β_0: Intercept of the model
β_i: Regression coefficient
X_i: Independent variables
n: Number of observations
ε: Error term of the model

There are several assumptions of the MLR model:

i.   There must be a linear relationship between independent variables and dependent variable.
ii.   The data must be free of multicollinearity
iii. The data must display homoscedasticity
iv.   The residual values are normally distributed.

3.2 Simulation of the Upcoming Trend of Malaysia's Unemployment Rate

The simulation of the upcoming trend of Malaysia's unemployment rate applies the Monte Carlo method. The analysis is run using the Visual Basic Application (VBA) in Microsoft Excel with the help of the function of a random numbers' generator.

MLR model is used to simulate the upcoming trend of Malaysia's unemployment rate for five years ahead from 2021 to 2025. The independent variables are randomized as they are the contributing factors of the unemployment rate. However, the randomization is being done only on the changes of each of the independent variables since they are normally distributed. This is also to ensure the continuity of the data from one to another.

The simulation process is improved with the user-interface element in which the users get the opportunity to key in the number of simulations as they wish. Nevertheless, [19] discovered the optimal number of simulations which is within the range 500 to 1000. Thus, considering the smooth process of the simulation, it is programmed to limit the number of simulations up to 1000 only. The simulation result is then presented in the next sheet. Based on the simulation result, their averages, maximum and minimum values are calculated and graphed.

4 Result

4.1 Multiple Linear Regression

The significance of the variables is determined using the stepwise regression. The result of the analysis is presented in Table 1 as follows:

Table 1. Regression of significance variables test

Variables	t-values	Significance
(Constant)	−0.105	0.917
Government expenditure	0.358	0.721
GDP	−3.953	0.000
Inflation rate	−1.508	0.134
Population growth	2.503	0.014
Interest rate	−10.977	0.000
Exchange rate	6.970	0.000
Investment	−3.423	0.000
COVID-19 cases	1.975	0.050

The hypotheses can be written as follows:

H0: The variable is not significant
H1: The variable is significant

If the variable has a p-value of less than 0.05, we reject the null hypothesis and conclude that the variable is significant. Based on Table 1, the p-values for the variables GDP, population growth, interest rate, exchange rate, investment and COVID-19 cases are all less than 0.05 indicating that these variables are all statistically significant. Thus, all these variables will be included in the model later.

However, the variable inflation is considered insignificant based on its p-value, but the analysis result shows that it should be included in the model as well. In resolving this issue, we run the significance test of the model using the Analysis of Variance (ANOVA) test. Two models are tested: Model 1 with all the significant variables including the inflation, while Model 2 with all the significant variables but excluding the inflation (Table 2).

Table 2. ANOVA of the regression model

Model	Residual sum of squares
1	3.563
2	3.627

The result of the analysis shows that Model 1 produces a lower Residual Sum of Squares (RSS) as compared to that of Model 2. Therefore, it is relevant to include the variable inflation in the model as well because Model 1 indicates a better fit.

4.2 Building the Regression Model

Table 3. The coefficients of significant variables

Variables	B	Std. error	t-values	Significance
(Constant)	−0.082	2.056	−0.041	0.968
GDP	−0.017	0.004	−4.134	0.000
Inflation rate	−0.053	0.035	−1.498	0.137
Population growth	0.000	0.000	2.488	0.014
Interest rate	−0.695	0.063	−11.039	0.000
Exchange rate	0.430	0.062	6.986	0.000
Investment	−0.028	0.008	−3.542	0.001
COVID-19 cases	0.000	0.000	2.102	0.038

Seven independent variables are known to be significantly affecting Malaysia's unemployment rate. Thus, these variables are included in developing the Multiple Linear Regression model. The final model equation is generated based on the value of the regression coefficients, β as follows (Table 3):

$$\hat{Y} = -0.082 - 0.017gdp - 0.053inflation + 0.0000002population\_growth$$
$$-0.695interest\_rate + 0.430exchange\_rate - 0.028investment + 0.000008covid$$

$$(2)$$

4.3   Simulation of Malaysia's Upcoming Trend of Unemployment Rate

As mentioned before, the Monte Carlo simulations technique is the most common and well-proven method for improving prediction accuracy when simulating the occurrence of an event. We use Multiple Linear Regression in simulating Malaysia's unemployment rate by using the equation in Sect. 3.2.

To simulate the unemployment rate of Malaysia for five years ahead, the independent variables that contribute to the unemployment rate are being randomized. However, to ensure that the data are normally distributed, the randomization is being done on the changes of each independent variable. This method also ensures the continuity of the data from one to another. For the research purpose, we choose 1000 as the number of simulations and the first set of data obtained are presented in this paper. We take the average, maximum value and minimum value for the simulations for every month for the upcoming five years. The results are graphed in the following Fig. 1:

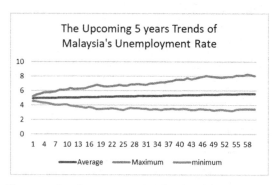

Fig. 1. The upcoming five years trends of Malaysia's unemployment rate

From the simulated results above, three alternative lines of average, maximum, and minimum unemployment rates are shown. The maximum values of unemployment rate are linearly increasing up to 8.28%. The unemployment rate for the upcoming five years also increases up to the maximum value of 5.60% for the average value of unemployment. Conversely, the unemployment rate's minimum value is decreasing gradually for five years ahead and stops at 3.25%. Therefore, we can predict minimum and maximum unemployment rates, as well as average unemployment rates, based on these numbers.

In the worst-case situation, we may anticipate that it will eventually reach the numbers found on the highest unemployment rate line. If the unemployment rate falls as a result of certain causes, it is likely to hover near the values seen on the minimal unemployment rate line. In short, in five years ahead, we can predict that the unemployment rate in Malaysia will fluctuate in the range of between 8.28% and 3.25%.

Since we are utilising random numbers in our simulation, the results will differ each time we run the simulation. As a result, in this project, we use the results obtained from the simulation when it is initially performed to be illustrated in this report. We perform 1000 simulations and find a small difference since the more simulations we do, the more accurate the outcome becomes. On the other hand, since the random numbers generated are between 0 and 1, we can notice a large difference if we perform a smaller number of simulations.

5 Conclusion

The rate of unemployment has long been a significant indication of economic progress all over the world, and this may have a greater influence on each country. Thus, this study aims to find out the significant variables that are correlated and give impact on the unemployment rate in Malaysia by using Multiple Linear Regression. This paper also focuses on simulating the upcoming five years trend of unemployment rate in Malaysia using Multiple Linear Regression and Monte Carlo simulation.

In every country, there are different factors that affect the unemployment rate. In examining which factors affect the unemployment rate in Malaysia, we found that Gross Domestic Product (GDP), inflation rate, interest rate, exchange rate, population growth, investment growth and the number of COVID-19 cases are significantly correlated with unemployment rate. This result was produced using the Stepwise technique in Multiple Linear Regression, and it demonstrates that government spending is not substantial, thus it is excluded from the model. Controlling the unemployment rate requires keeping track of several things which might have economic ramifications. It is critical for the government to plan effective ways for monitoring their country's unemployment rate in order to ensure long-term economic growth.

The final model of the Multiple Linear Regression obtained is then used in simulating the trend of Malaysia's unemployment rate for five years ahead. This Monte Carlo simulation is implemented using Visual basic Application (VBA) in Microsoft Excel. Each independent variable that gives significant impact to the unemployment rate is randomized in the simulation of unemployment trends with 1000 iterations.

The result shows that the unemployment rate in Malaysia is increasing linearly up to 5.60% for the upcoming five years. This is possible due to the COVID-19 pandemic in which more businesses were affected resulting in the escalating number of unemployed. It may take years for Malaysia to recover in terms of economy and finance. For the worst case and least case scenario, it is predicted that the unemployment rate can fluctuate between the maximum value of 8.28% and the minimum value of 3.25% respectively. All of these predicted figures may comprise of fresh graduates and the retrenched employees. Government should start on planning the necessary corrective actions in order to prevent this issue from getting worse and to avoid the unexpected rise of the unemployment rate in few years ahead.

References

1. Ansari, A.: Application of Neural Network-Support Vector Technique to Application of Neural Network Support Vector Technique to Forecast U.S. Unemployment Rate (2014)
2. Simionescu, M.: The performance of unemployment rate predictions in Romania. Strat. Improve Forecasts Accuracy Rev. Econ. Perspect. 13(4), 161–175 (2013)
3. Dalmar, M.S., Sheikh Ali, A.Y., Ali, A.A.: Factors affecting unemployment in Somalia. J. Econ. Sustain. Dev. 8(22), 200–210 (2017)
4. Yildirim, H., Basegmez, H.: Analysis and forecast of Turkey unemployment rate. Global J. Math. Anal. 5(1), 11 (2016)
5. Yüksel, S., Adalı, Z.: Determining influencing factors of unemployment in Turkey with mars method. Int. J. Commer. Financ. 3(2), 25–36 (2017)
6. Dritsaki, C.: Forecast of SARIMA models: an application to unemployment rates of Greece. Am. J. Appl. Math. Stat. 4(5), 136–148 (2016)
7. Dritsakis, N., Klazoglou, P.: Forecasting unemployment rates in USA using box-jenkins methodology. Int. J. Econ. Financ. Issues 8(1), 9–20 (2018)
8. Asif, K., Aurangzeb, D.: Factors effecting unemployment: a cross country analysis. Int. J. Acad. Res. Bus. Soc. Sci. 3(1), 2222–6990 (2013)
9. Abouelfarag, H.A., Qutb, R.: Does government expenditure reduce unemployment in Egypt? J. Econ. Admin. Sci. 37, 355–374 (2020)
10. Nkwatoh, L.: Forecasting unemployment rates in Nigeria using univariate time series models. Int. J. Bus. Commer. 1(12), 33–46 (2012)
11. Liotti, G.: Labour market flexibility, economic crisis and youth unemployment in Italy. Struct. Chang. Econ. Dyn. 54, 150–162 (2020)
12. Ramli, S.F., Firdaus, M., Uzair, H., Khairi, M., Zharif, A.: Prediction of the unemployment rate in Malaysia. Int. J. Modern Trends Soc. Sci. 1(4), 38–44 (2018)
13. Michael, E., Geetha, C.: Macroeconomic factors that affecting youth unemployment in Malaysia. Malays. J. Bus. Econ. (MJBE) 7(2), 181–205 (2020)
14. Furuoka, F., Munir, Q.: Unemployment and inflation in Malaysia: evidence from error correction model. Philippone J. Dev. 1(1), 35–45 (2014)
15. Arslan, M., Zaman, R.: Unemployment and its determinants: a study of Pakistan economy (1999–2010). SSRN Electron. J. 5(13), 20–25 (2014)
16. Kukaj, D.: Impact of unemployment on economic growth: evidence from Western Bal-kans. Eur. J. Mark. Econ. 1(1), 10 (2018)
17. Shabbir, A., Kousar, S., Zubair Alam, M.: Factors affecting level of unemployment in South Asia. J. Econ. Admin. Sci. 37(1), 1–25 (2020)
18. Abdul Shakur, E.S., Sa'at, N.H., Aziz, N., Abdullah, S.S., Abd Rasid, N.H.: Determining unemployment factors among job seeking youth in the east coast of Peninsular Malaysia. J. Asian Financ. Econ. Bus. 7(12), 565–576 (2020)
19. Morris, R.D., Young, S.S.: Determining the number of iterations for Monte Carlo simulations of weapon effectiveness (2004)
20. McDonald, R.: Derivatives Markets, 3rd edn. Pearson, London (2012)
21. Chong, T.T.L., Li, X., Yip, C.: The impact of COVID-19 on ASEAN. Econ. Polit. Stud. 9(2), 166–185 (2021)

Time Series Forecasting Using a Hybrid Prophet and Long Short-Term Memory Model

Yih Hern Kong$^{(\boxtimes)}$, Khai Yin Lim$^{(\boxtimes)}$, and Wan Yoke Chin$^{(\boxtimes)}$

Tunku Abdul Rahman University College, 53300 Setapak, Kuala Lumpur, Malaysia
kongyh-wa15@student.tarc.edu.my, {limky,chinwy}@tarc.edu.my

Abstract. Forecasting analysis is a common research topic these days. The development in this area has allowed organizations to retrieve useful information and make important decisions based on the forecast results. Different forecasting models are used to model data with different characteristics as each of the forecasting model has its own strength and weakness. As such, Hybrid Prophet-LSTM that combines Long Short-Term Memory (LSTM) and FBProphet (Prophet) is introduced. This study aims to examine the effectiveness of the hybrid model and the influence of holiday effect to the forecast result. Weighted Mean Absolute Percentage Error (WMAPE), Mean Absolute Deviation (MAD), R^2 value, and Root mean square error (RMSE) were used to evaluate the performance of the proposed hybrid model. The proposed Hybrid Prophet-LSTM is found to outperform both the standalone LSTM and Prophet, and holiday effect shows high attitude of influence to the forecast result.

Keywords: Prophet · LSTM · Hybrid forecasting · Time series forecasting · Combined forecast · Holiday effect

1 Introduction

Forecasting analysis has been a popular area of study for the past years. Numerous studies have applied the technique to forecast time series data, such as future stock movement [1], users' engagement [2], traffic matrix [3], and insurgency movement direction [4]. By establishing a connection between forecasting function and functional management, business forecasting can contribute to many functional decision-areas including both primary activities and supportive activities in a value chain. Various forecasts are needed to apply in each decision areas for different managerial decisions [5]. Forecasting is to forecast or estimate a future event or trend [6]. Forecasting model is the methodology and algorithm used to do forecasting. Different forecasting model are used to model time series data with different characteristics. According to Luan and Sudhir [7], it is important to have a reliable forecast result for a company to advertise its short lifecycle products. Besides that, some empirical research studies [7–9] have shown that it is important for marketers to have reliable market forecasts such as advertising responsiveness, sales forecast, or gross domestic product (GDP) to make any decision.

© Springer Nature Singapore Pte Ltd. 2021
A. Mohamed et al. (Eds.): SCDS 2021, CCIS 1489, pp. 183–196, 2021.
https://doi.org/10.1007/978-981-16-7334-4_14

Forecasting outcomes can be affected by various factors, especially trend, seasonality, linearity, and holiday effects of the time series data. A study [10] compared two of the most popular linear and nonlinear models in the last decade, Neural Networks (NN) and Autoregressive Integrated Moving Average (ARIMA). In the study [10], it was found that the NN performed well with their capability of generalization, but only on nonlinear data. The popular traditional model, ARIMA, on the other hand, is better to solve linear problems. The forecast accuracy of ARIMA for nonlinear data is comparative low. In general, linear and nonlinear models possess different properties that are application dependent [11].

Since a time series data rarely contains only pure linear or nonlinear information [12], particularly in marketing data [13], hybrid forecasting models [4, 12–16] that combine different methods were proposed to produce a more accurate forecast result. In spite of the enormous studies on the forecasting accuracy in time series data, little attention has been done on the impact of holidays and the correlations between irregular holidays and forecasting accuracy [17]. It was found that holiday effects in time series data are needed to be taken into consideration in the seasonal adjustment process for trend estimation [18]. This is especially important when there is a high degree of holiday features in the time series data. This is because time series data related to human mobility often exhibits a greater degree of holiday effects [15], which have a great impact to the time series data [18]. Hence, this study analyzes and examines the effect of irregular holiday schedule on the forecasting accuracy.

FBProphet (Prophet) is a newly developed time series forecasting model released by Facebook that can handle trend, seasonality, holiday effects, errors term and product reliable forecasts [19]. Whereas NN is a nonlinear forecasting model that can be used to handle residual nonlinear time series [10, 12]. Deep learning based Long Short-Term Memory (LSTM) network is a type of Recurrent Neural Network (RNN) that has the capability to address limitations of traditional time series forecasting techniques such as NN and RNN by adapting the nonlinearities of time series. Besides that, LSTM is useful to address the issue of vanishing gradient by adjusting the time scale and parameters [20]. It is therefore hypothesized that by combining Prophet and LSTM, the accuracy of the forecasting result would be improved if the time series data contains both linear and nonlinear data points, trends, seasonality, and holiday effects.

In this research, the holiday component in Prophet takes the holiday events from the time series data into consideration [19]. A hybrid model combined Prophet and LSTM consists of the strength from both models. This proposed hybrid model can capture periodic changes, nonperiodic changes, and irregular schedules by including trend, all forms of seasonality, and irregular holidays of the time series. Previous research on forecasting in different areas will be discussed in the following section. In Sect. 3, the proposed framework that combines Prophet and LSTM is explained. Section 4 consists of the evaluation results of the different models. Section 5 concludes this study with discussion of the results.

2 Literature Review

2.1 Time Series Forecasting Model

In 2017, Azzouni and Pujolle [3] proposed to use LSTM to solve the traffic matrix prediction problem. The framework was validated using real-world data from GEANT organization network. It was found that LSTM is suitable for traffic matrix prediction and outperformed several linear forecasting models such as Autoregressive Moving Average model (ARMA), Autoregressive (ARAR), Holt-Winters Model (HW), and Feed-Forward Neural Network (FFNN) by many orders of magnitude. One recent similar work on forecasting of COVID-19 epidemic uses LSTM network to forecast upcoming cases and estimate possible ending point of the outbreak [20]. The COVID-19 transmission dataset is provided publicly by Canadian health authority. The reason LSTM was chosen in the study is because the time series has a nonlinear dynamic change where the adaptation of virus over time always occurs at different magnitude.

A recent study by Yenidogan et al. [21] applied ARIMA and Prophet to Bitcoin forecasting. The dataset includes Bitcoin values of multiple currencies, was used with ARIMA and Prophet. In the study [21], time stamp conversion was performed to reduce the transaction complexity of the data, while feature selection was performed to remove weakly correlated features from dataset. The forecast result showed that Prophet outperformed ARIMA by 0.94 to 0.68 in R^2 values The authors [21] concluded that Prophet can automatically select changepoint for linear trend if the time series is having a nonlinear saturating growth or a linear trend. This finding is consistent with the study [19]. Samal et al. [22] implemented ARIMA and Prophet in air pollution forecasting. Open-source air pollution historical data was used in the research. The main strength of Prophet is its adaptability to missing data, ability of capturing the shifts in trend and outliers, and its accuracy estimating mixed data.

2.2 Hybrid Methodology

One of the most popular hybrid forecasting models [12] combined ARIMA and NN. The idea of this hybrid methodology is to combine the unique strength of ARIMA and NN in terms of their capability in linear and nonlinear modelling. The study [12] employed simulated data that contains both linear and nonlinear structure to test the performance of the hybrid model. Data transformation was used to make the time series to be stationary [12]. The trend was removed to stabilize the variance of data. The experimental results [12] indicated that the hybrid model is able to improve forecasting accuracy, which was not achievable by using either one of the models independently.

Another study [4] was carried out by combining ARIMA and Support Vector Regression (SVR). The data used is a nonstationary time series data of the insurgency that requires transformation. SVR was selected [11] because of its capability that can capture nonlinear pattern to improve forecast performance. The study found that the proposed hybrid model performed better as compared to autoregression (AR), Moving Average (MA), ARIMA, and SVR independently.

Another work [13] studied various types of linear and nonlinear forecasting methods and proposed a hybrid methodology that combines AR with NN based on the Extreme

Learning Model (ELM). The hybrid model was tested on multiple time series data including deterministic synthetic series, chaotic time series, and real data series. The deterministic synthetic series contains seasonal dependence, trend, and amplitude change, while chaotic time series contains typical chaotic time series. NN can be successfully applied to solve nonlinear domain. ELM is a type of single hidden-layer feedforward neural network (SLFN) that randomly chooses input weight and determine output weight by analyzing the input. A NN was then designed to combine AR and ELM by hybridizing the two time-series prediction algorithms.

3   Hybrid Methodology Framework

3.1   Linear Prediction

Prophet is a newly developed decomposable time series model in 2017. It is a linear model by default, and is able to model the nonlinear relationship of the time series by manipulating the parameter: change point prior scale. Its 3 main model components are trend function, seasonality component, and holiday component [19]. These 3 functions are combined in the equation: $y(t) = g(t) + s(t) + h(t) + e_t$. The trend function $g(t)$ models nonperiodic changes of the time series, aims to capture and model the patterns in time series data. $s(t)$ represents periodic changes or seasonality of the time series. The holiday function $h(t)$ represents the effects of holidays which occur differently in every year. A custom list of events or holidays can be supplemented to the model using this holiday function to includes irregular scheduled holiday effects to the time series. The error term e_t represents all identical changes which are not adapted by the other model components. Prophet also supports user to tune the seasonality mode to model time series with different seasonality such as additive component or multiplicative component. Modelling seasonality as an additive component is similar to a Generalized Additive Model (GAM) regression approach. For multiplicative seasonality, the seasonal effect will be assumed to be the factor that multiplies the trend function $g(t)$. It requires the time series data to go through log transformation for multiplicative seasonality modelling. Prophet is also able to generate decomposition result such as trend, seasonality, and holiday effects.

3.2   Nonlinear Prediction

Neural network-based models, especially LSTM has been widely used for modelling and forecasting as this model can learn complex nonlinear pattern. LSTM is an RNN-based model that solves RNN vanishing gradient and short-term memory issues. LSTM model can perform better as compared to RNN in prediction and labelling task [3].

The first step in LSTM network is the sigmoid layer called Forget gate layer. It looks at the previous state h_{t-1} and the content input x_t, and outputs 0 or 1 for each number in the cell state C_{t-1}. The sigmoid function in forget gate is $f_t = \sigma\left(W_f \cdot [h_{t-1}, x_t] + b_f\right)$. The second step is to decide the new information to store in the cell state. This involves two sigmoid layers called the input gate layer and tanh layer. The input gate layer decides which values to update, while tanh layer creates vector of new values $\tilde{C}_t$ would be added to the state. Then, these two layers would combine to update the cell state. The sigmoid function of the input gate is $i_t = \sigma\left(W_i \cdot [h_{t-1}, x_t] + b_i\right)$. The tanh function in tanh layer is $\tilde{C}_t = \tanh\left(W_C \cdot [h_{t-1}, x_t] + b_C\right)$. Combining the function from forget gate, input gate, and tanh layer, the new C_t can be computed by $C_t = f_t * C_{t-1} + i_t * \tilde{C}_t$. At the third step which is also the final step, the output gate will decide what is the output cell state of this block. The output will be based on the filtered cell state. The output gate sigmoid function $o_t = \sigma\left(W_o[h_{t-1}, x_t] + b_o\right)$ will decide which parts of the cell state to output. Then the tanh function $h_t = o_t * \tanh(C_t)$ will gives weightage to the values which passed, and deciding their level of importance ranging from -1 to 1 and multiplied with o_t.

3.3 Hybrid Time Series Forecasting Model

The hybrid forecasting model generally uses both linear and nonlinear algorithms and combine their forecasting results [13]. This author [12] gave few reasons why hybrid model can be used to solve real problems. Firstly, it is difficult to determine whether a time series is more efficient to model with linear or nonlinear prediction model. Second, the real-world environment data always contains both linear and nonlinear relationships and rarely contains purely only linear or nonlinear relationship. Hence, neither linear nor nonlinear prediction model alone can handle both linear and nonlinear patterns well.

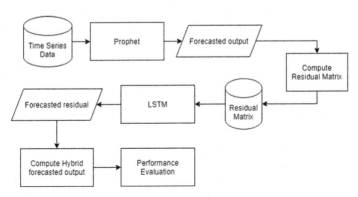

Fig. 1. Hybrid methodology framework

The forecasting model proposed is Hybrid Prophet-LSTM. This hybrid forecasting model uses both linear (Prophet) and nonlinear (LSTM) algorithms to produce a better forecasting result. Prophet can use to handle time series that have strong seasonal effects, which is not included in the other linear model. At the same time, Prophet can take holiday effects in the time series dataset into consideration. This is the major differences compared to other time series forecasting model. After Prophet models the linear relationship in time series, the residual nonlinear matrix will be handled by LSTM.

Residual matrix is computed using the formula:

$$e_{1t} = y_t - \hat{L}_t \tag{1}$$

where e_{1t} denotes the residual matrix, y_t denotes the original time series, and $\widehat{L}_t$ is the forecasted value of time series estimated by using Prophet model. The residual matrix e_{1t} is further fitted into LSTM as the target variable and the following equation is generated:

$$e_{2t} = e_{1t} - \widehat{N}_t \tag{2}$$

where e_{2t} denotes the error term, and $\widehat{N}_t$ denotes the forecasted value from LSTM.

The prediction result from LSTM needs to be converted to a useful data by applying this formula:

$$\hat{y}_t = \widehat{N}_t + \widehat{L}_t \tag{3}$$

where $\hat{y}_t$ denotes the forecasted value combined using both linear and nonlinear models, $\widehat{N}_t$ denotes the forecasted residual matrix using LSTM model. The prediction result from both Prophet and LSTM is compared to the result from hybrid model and their performance are evaluated.

In the hybrid methodology, the residual matrix from the linear model is used to the nonlinear model. Furthermore, the results of both the linear and nonlinear models are merged to produce a predicted value that is closer to the actual value. This methodology and result computation would considerably enhance the predicted outcome, making it at least better than either of the models alone.

3.4 Data

This study is to forecast customer engagement of a selected Facebook Page. Three brands in Food, Beverages, and Cosmetics were selected arbitrary. The target variable, customer engagement, and other Facebook Page metrics data can be crawled from Facebook. Facebook provides a list of page and post metrics that allows the researcher and developer to access numerous types of data [23]. In order to understand social media users' behaviors, each of the page and post metrics are studied, filtered, then selected to be crawled. Data was crawled using Facebook Graph API after selecting metrics from Facebook Page Metrics.

Users can receive metric data from Facebook by sending a request to Facebook Graph API using an access token generated with a Facebook account. Table 1 shows the main categories of Facebook Page Metrics of Insights API Version 6.0. Facebook provides a total of 217 metrics in this version and there are 15 groups of metrics.

All of the metrics were studied, and customer page engagement was chosen as the target variable. This variable shows the count of engagements from customers on one specific page. This metric is computed from user clicks, reactions, comments, shares, and many more. All other variables are crawled but remain unchanged, only customer page engagement is used for univariate analysis. In this study will only examine, analyze, and forecast a singular variable, which is the customer page engagement. Three datasets were collected from 2 different industries. Two years of daily data, which started from 1st June 2018 to 31st March 2021, were collected from the selected pages as dataset.

Table 1. Facebook Page and Post metric groups

No	Metric types
1	Page content
2	Page CTA clicks
3	Page engagement
4	Page post engagement
5	Page impressions
6	Page post impressions
7	Page post
8	Page reactions
9	Page post reactions
10	Page user demographics
11	Page views
12	Page video views
13	Page video posts
14	Page and post stories
15	Video ad breaks

Prophet's holiday component is supplemented with Malaysia Public Holiday. Comparison analysis is performed to evaluate the impact of holiday effects to the time series forecast result.

3.5 Performance Metric

The performance metric is to validate and evaluate the performance of the forecasting model. The first performance metric used is R^2 score. R^2 measures the proportion of variance that is explained by the model. A higher R^2 value generally indicates a more useful result. R^2 value can be calculated using the formula:

$$R^2 = 1 - \frac{n\left(\sum xy\right) - \left(\sum x\right)\left(\sum y\right)}{\sqrt{\left[n\sum x^2 - \left(\sum x\right)^2\right]\left[n\sum y^2 - \left(\sum y\right)^2\right]}} \tag{4}$$

where x denotes the actual value, and y is the predicted value.

Root mean square error (RMSE) or root-mean-square deviation (RMSD) is selected as one of the evaluation metrics. The formula for RMSE or RMSD is:

$$RMSD = \sqrt{\frac{\sum_{i=1}^{N}(x_i - \hat{x}_i)}{N}} \tag{5}$$

where x_i denotes the actual value, $\hat{x}_i$ is the predicted value, i is the index of each observation, and N is the total number of observations.

Mean Absolute Deviation (MAD) will also be used to evaluate the forecasting model using the sum of absolute error. A higher forecast error indicating the estimated data is further than the actual data. MAD can be computed using the following formula:

$$MAD = \frac{\sum |y_{ori} - y_{pred}|}{N} \tag{6}$$

where y_{ori} denotes the original or actual value, y_{pred} is the predicted value, and N denotes the total number of observations. RMSE and MAD value for different dataset should not be compared. The variables with huge data variation or difference cannot compared to data with small variation. As small variation led to a small difference in values and resulted smaller error percentage comparing to dataset with large variation between values.

Weighted Mean Absolute Percentage Error (WMAPE) is used to measure the average absolute percentage errors in forecasting result. The formula for WMAPE is:

$$WMAPE = \frac{\sum_{i=1}^{N} \left(|x_i - \hat{x}_i| \right)}{N} \bigg/ \frac{\sum_{i=1}^{N} |x_i|}{N} \tag{7}$$

where x_i denotes the actual value, $\hat{x}_i$ is the predicted value, i is the index of each observation, and N is the total number of observations.

4   Empirical Results

The data is normalized to a range from 0 to 1 before fitting into the model. The scale transformed data were split into training set and testing set. The forecast result was inverse scale transformed using the same scale factor before evaluation.

4.1   Prophet Decomposition Result

Before modelling, descriptive analysis was used to show and summarize the individual components of the data. The time series data was analyzed using decomposition function in Prophet. Such decomposition result was generated to observe and study various characteristic such as trend, holiday effects, and different forms of seasonality of the time series data (Fig. 2).

This decomposition result shown in Fig. 1 is generated using Dataset 3, the customer engagement of this selected page does not have a potential growth as the trend graph shown in Fig. 1(a) was not growing progressively but dropped to the minimum after a short period of growth. The trend stays in a range of growth from 0 to 175,000 engagements.

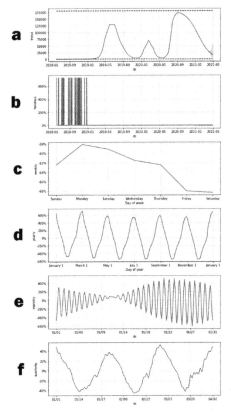

Fig. 2. Example of individual components from Prophet decomposition, a) Data trend, b) Holiday effects, c) Weekly seasonality, d) Yearly seasonality, e) Monthly seasonality, f) Quarterly seasonality

In Fig. 1(b), the holiday decomposition graph shows all the holiday effect values are not less than 0. Indicating that the holiday is bringing a neutral or positive effect to this specific dataset. Meaning when there is a holiday, the customer engagement either remain the same or increased. This is an interesting finding showing the Facebook users are be engaging to page posts during holiday. Prophet was supplied with Malaysia Public Holiday and the holiday effects were analyzed by Prophet decomposition feature. The effect of each of the holidays was identified and used to adjust the forecast value. By using the analyze result, the irregular holiday effects are also handled by Prophet during modelling.

In Fig. 1(c), weekly seasonality decomposition shows that the page's customer engagement gradually increases from Sunday and reaches the highest customer engagement on every Monday, then gradually decline on the following days. The reason why it happens may leads to further analysis to what the specific page posts to attract their customer to engage. In Fig. 1(d), the yearly seasonality of customer engagement was also decomposed and analyzed. Yearly seasonality decomposition shows the page has a stable seasonality every 2 months. The customer engagement reached a peak at every

1 month and then gradually fell on the consecutive 1 month. In Fig. 1(e), the monthly seasonality is also shown a wave pattern that is gradually stabilizing at zero until one third of the month. Then the changes are gradually increased till the end of the month, gaining back the lost value, but highly instable at the same time. The monthly seasonality is greatly affected by weekly seasonality, we can refer to Fig. 1(c) the weekly seasonality decomposition explanation for more information. In Fig. 1(f), the quarterly seasonality shows the form of seasonality for 1 quarter. The quarterly seasonality decreases gradually from the beginning of a quarter till the 14th day of the quarter, then gaining back the lost value on the next 13 days. This pattern repeats in a constant for every 10 to 14 days till the end of the quarter. The quarterly seasonality is greatly affected by monthly seasonality and weekly seasonality, we can refer to Fig. 1(c) and Fig. 1(e) the explanation for weekly seasonality and monthly seasonality for more information.

Two methods are implemented for comparison analysis using Prophet. The first method is supplemented with Malaysia Public Holiday dataset into the Prophet holiday component. The second method did not include any holiday dataset and any parameter related to its holiday feature is remain unchanged.

The selected datasets from 3 different pages are having different characteristics. Dataset 1 has an observable seasonality but mixed with an unusual pattern in each form of seasonality. Dataset 2 does not have any observable seasonality in the time series. Dataset 3 has a clear and smooth pattern in each form of seasonality. All 3 of these datasets have similar trend growth, but the percentage of the engagement value instead of raw value.

4.2 Hybrid Algorithm

Prophet was first used to model the linear pattern in time series. Then, the residual matrix was fit into LSTM and a set of forecast result will be generated based on the residual matrix (Fig. 3).

The forecast result generated by LSTM was converted from residual forecast to actual forecast result using Eq. 2. Then, the final result was recorded and the performances are measured using different performance metrics.

Table 2 compares the performance of the 3 different models. Model 1 and Model 2 are Prophet and LSTM respectively, and Model 3 uses Hybrid Prophet-LSTM. Prophet shows a weighted mean absolute percentage error of 25.19%, 52.74%, and 23.92% on 3 different datasets which are around double of LSTM error rate. LSTM outperforms Prophet by having a higher R2 value and lower error. Prophet did not show strong modelling ability in this scenario because the value adjustment from holiday component does not affect the overall result by a lot. LSTM network model has shown its outstanding performance on flexibility using the different sets of time series data. LSTM performs better than Prophet in overall as LSTM is suitable to model nonlinear dataset. It is observed that in Table 2 the comparison of WMAPE value, LSTM is able to model the nonlinear pattern in time series better than Prophet.

It is observed that the general seasonal variation can be captured by the model, but the unusual pattern captured by the model can increase error of the forecast result. Effecting the forecast result of Dataset 1 to be slightly lower compared to the forecast result of

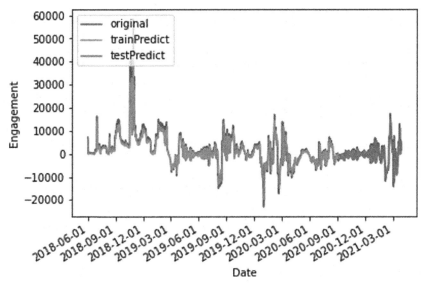

Fig. 3. LSTM modelling residual matrix

Table 2. Performance metrics for Prophet, LSTM, and Hybrid models

		Prophet	Prophet (No Holiday)	LSTM	Hybrid Prophet-LSTM	Hybrid Prophet-LSTM (No Holiday)
Dataset 1	**WMAPE**	**24.9409%**	**26.3200%**	**12.6768%**	**9.1449%**	**7.2431%**
	R2	92.1610%	91.5422%	92.8678%	99.1259%	99.3209%
	RMSE	1559.694	1619.852	1471.233	580.410	502.120
	MAD	981.001	1035.244	492.511	316.135	258.244
Dataset 2	**WMAPE**	**52.7436%**	**82.1271%**	**21.5438%**	**26.3653%**	**65.4980%**
	R2	85.9057%	54.7768%	73.391%	97.3304%	72.1865%
	RMSE	94.105	163.756	150.965	40.825	122.323
	MAD	52.242	77.635	25.776	22.218	95.391
Dataset 3	**WMAPE**	**23.8803%**	**36.3465%**	**13.5404%**	**3.87%**	**5.0921%**
	R2	93.3743%	77.3359%	97.0577%	99.8604%	99.8770%
	RMSE	6859.626	12686.764	4580.609	1142.639	1304.555
	MAD	4435.751	6751.348	2513.724	551.995	633.378

Dataset 3. For Dataset 2 forecast results, Prophet had high error rate of 52.74% when there is no observable seasonality pattern in the time series.

Table 2 also shows the comparison of forecast result of Prophet's holiday component. It is shown that the holiday can effectively increase forecast accuracy for all the 3 datasets. In Dataset 1, however, LSTM performs better in modelling the nonlinear residual matrix generated by Prophet without using the holiday component. Creating a slightly better hybrid forecast result compared to the hybrid method using holiday component. The holiday component in Prophet can significantly reduce the forecast error and increasing forecast accuracy.

In the proposed hybrid forecasting methodology, Hybrid Prophet-LSTM has higher R^2 value, lower WMAPE, lower RMSE, and lower MAD in overall compared to LSTM and Prophet on each dataset. This shows that hybrid algorithm can improve forecasting accuracy as Hybrid Prophet-LSTM has the lowest error, and higher variance are shown in overall.

5   Conclusions

In this study, Hybrid Prophet-LSTM is proposed as a new approach by combining linear and nonlinear models. The proposed model integrates features to include a custom list of events or holidays. Besides that, this work also takes seasonality and trend components into consideration during the modelling process. To validate the effectiveness of the proposed model, several experiments were conducted. The experiment result shows that the proposed Hybrid Prophet-LSTM model outperforms Prophet and LSTM, which were independently used to forecast the Facebook user engagement in terms of accuracy and robustness. The proposed model has lowest forecasting errors and work well in different scale of datasets. It is observed that the holiday component in the proposed model shows high attitude of influence on the forecasted result. Hence, it can be concluded that there are 2 important factors that should always be considered in the hybrid methodology. The first factor would be the suitability of the chosen linear model. It was found that the final outcome of the hybrid is greatly affected by the performance of the linear model. Good output in the linear model would yield a good result in the hybrid model, and vice-versa. The second factor is the feature selection. In this study, only one variable was selected to perform the univariate analysis. Analyzing multivariate data with univariate analysis would result in inconsistent error rates. The errors and generally unknown error rates caused by inconsistency can easily result misinterpretations of findings. Multivariate and univariate analysis both provide complementary result. It is best to use both univariate and multivariate methods [24] as the error rate for combination approach results is reasonably consistent [25].

References

1. Sidi, L.: Improving S&P stock prediction with time series stock similarity (2020). http://arxiv.org/abs/2002.05784
2. Srinivasan, B.V., Natarajan, A., Sinha, R., Gupta, V., Revankar, S., Ravindran, B.: Will your facebook post be engaging? In: International Conference on Information and Knowledge Management Proceedings, pp. 25–28 (2013). https://doi.org/10.1145/2512875.2512881

3. Azzouni, A., Pujolle, G.: A long short-term memory recurrent neural network framework for network traffic matrix prediction (2017). http://arxiv.org/abs/1705.05690
4. Waeto, S., Chuarkham, K., Intarasit, A.: Forecasting time series movement direction with hybrid methodology. J. Probab. Stat. **2017** (2017). https://doi.org/10.1155/2017/3174305
5. Niğde, C., Polat, Ü.: The role of forecasting and its potential for functional management: a review from the value-chain perspective. Dokuz Eylül Üniversitesi Sos. Bilim. Enstitüsü Derg. **9**(1), 373–398 (2007)
6. Magiya, J.: Introduction to Forecasting in Data Science. Towards Data Science, 19 March 2019. https://towardsdatascience.com/introduction-to-forecasting-in-data-science-676db9b55621. Accessed 22 July 2020
7. Luan, Y.J., Sudhir, K.: Forecasting marketing-mix responsiveness for new products. J. Mark. Res. **47**(3), 444–457 (2010)
8. Clements, M.P., Hendry, D.F.: The Oxford Handbook of Economic Forecasting. Oxford University Press, Oxford (2012)
9. West, D.C.: Advertising budgeting and sales forecasting: the timing relationship. Int. J. Advert. **14**(1), 65–77 (1995). https://doi.org/10.1080/02650487.1995.11104598
10. Kuvulmaz, J., Usanmaz, S., Engin, S.: Time-series forecasting by means of linear and non-linear models. In: Gelbukh, A., de Albornoz, Á., Terashima-Marín, H. (eds.) MICAI 2005. LNCS (LNAI), vol. 3789, pp. 504–513. Springer, Heidelberg (2005). https://doi.org/10.1007/11579427_51
11. Jiao, J.: A hybrid forecasting method for wind speed. In: MATEC Web Conference, vol. 232 (2018). https://doi.org/10.1051/matecconf/201823203013
12. Zhang, P.G.: Time series forecasting using a hybrid ARIMA and neural network model. Neurocomputing **50**, 159–175 (2003). https://doi.org/10.1016/S0925-2312(01)00702-0
13. Pan, F., Zhang, H., Xia, M.: A hybrid time-series forecasting model using extreme learning machines. In: 2009 2nd International Conference on Intelligent Computing Technology and Automation, ICICTA 2009, vol. 1, pp. 933–936 (2009). https://doi.org/10.1109/ICICTA.2009.232
14. Yeganeh, B., Motlagh, M.S.P., Rashidi, Y., Kamalan, H.: Prediction of CO concentrations based on a hybrid Partial Least Square and Support Vector Machine model. Atmos. Environ. **55**, 357–365 (2012). https://doi.org/10.1016/j.atmosenv.2012.02.092
15. Mcelroy, T.S., Monsell, B.C., Hutchinson, R.J.: Modeling of holiday effects and seasonality in daily time series (2018)
16. Smyl, S.: A hybrid method of exponential smoothing and recurrent neural networks for time series forecasting. Int. J. Forecast. **36**(1), 75–85 (2020). https://doi.org/10.1016/j.ijforecast.2019.03.017
17. Xu, M., Wang, Q., Lin, Q.: Hybrid holiday traffic predictions in cellular networks. In: IEEE/IFIP Network Operations and Management Symposium: Cognitive Management in a Cyber World, NOMS 2018, pp. 1–6 (2018). https://doi.org/10.1109/NOMS.2018.8406291
18. Shuja, N., Lazim, M.A., Wah, Y.B.: Moving holiday effects adjustment for Malaysian economic time series. Dep. Stat. **1**, 36–50 (2007)
19. Taylor, S.J., Letham, B.: Business time series forecasting at scale. PeerJ Prepr. 5e3190v2 **35**(8), 48–90 (2017). https://doi.org/10.7287/peerj.preprints.3190v2
20. Chimmula, V.K.R., Zhang, L.: Time series forecasting of COVID-19 transmission in Canada using LSTM networks. Chaos Solitons Fractals **135** (2020). https://doi.org/10.1016/j.chaos.2020.109864
21. Yenidogan, I., Cayir, A., Kozan, O., Dag, T., Arslan, C.: Bitcoin forecasting using ARIMA and PROPHET. In: 3rd International Conference on Computer Science and Engineering, UBMK 2018, February 2019, pp. 621–624 (2018). https://doi.org/10.1109/UBMK.2018.8566476

22. Samal, K.K.R., Babu, K.S., Das, S.K., Acharaya, A.: Time series based air pollution forecasting using SARIMA and prophet model. In: ACM International Conference Proceedings Series, pp. 80–85 (2019). https://doi.org/10.1145/3355402.3355417
23. "Insight - Pages." https://developers.facebook.com/docs/platforminsights/page. Accessed 29 Mar 2020
24. Saccenti, E., Hoefsloot, H.C.J., Smilde, A.K., Westerhuis, J.A., Hendriks, M.M.W.B.: Reflections on univariate and multivariate analysis of metabolomics data. Metabolomics **10**(3), 361–374 (2013). https://doi.org/10.1007/s11306-013-0598-6
25. Hummel, T.J., Sligo, J.R.: Empirical comparison of univariate and multivariate analysis of variance procedures. Psychol. Bull. **76**(1), 49–57 (1971). https://doi.org/10.1037/h0031323

Financial Analytics on Malaysia's Equity Fund Performance and Its Timing Liquidity

Siti Farizah Saad[1], Siti Meriam Zahari[2(✉)] ⓘ, and Muhammad Azri Mohd[2] ⓘ

[1] Standard Chartered Bank, Kuala Lumpur, Malaysia
[2] Universiti Teknologi MARA, Shah Alam, Malaysia
{mariam,azri}@tmsk.uitm.edu.my

Abstract. This study focuses on equity fund performance in the Malaysian market with three different time frames (daily, monthly, and yearly) and examines the relationship between its expected return and liquidity timing. Traditional financial ratios such as Sharpe, Jensen Alpha, Treynor index, and Capital Asset Pricing Model (CAPM) help in analyzing the performance of the equity fund whereas Trading Volume and Turnover methods were utilized to measure the fund liquidity. Deduced from the analysis, the equity fund performs differently within the group itself, depending on the time frames stated. This study also found that the liquidity timing affects the expected return where the fund manager can use the beta values from the fund analysis to increase their market exposure, prior to market timing liquidity.

Keywords: Financial ratios · Liquidity timing · Equity fund

1 Introduction

Mutual Fund comes from a pool of monies that are collected from numerous investors with similar investment objectives. The return of investing in the mutual funds can be realized when there is an increase or decrease in the price value as the value of each unit are directly correlated to it. Generally, there are two types of return on mutual fund investments, which are income distribution and capital appreciation. Many factors influence the fund manager's investing decision, such as past performance of the mutual funds, market return of the mutual funds, market volatility, and market liquidity of the mutual funds. Investing in mutual funds comes with the risk associated with it, such as market risk, liquidity risk, management risk, inflation risk, and interest rate risk, therefore the investors must consider the risks before making any investment decision. Liquidity can be defined as the ease of converting assets to cash or cash to the asset without affecting the asset's price. In the financial sector, the ability of the market to purchase or sell an asset without causing a drastic change in the asset's price is called market liquidity. Equivalently, market liquidity also refers to the extent to which a market allows the assets to be bought and sold at stable prices. Most Malaysians from all types of income classes face difficulties not only to save more but also in growing their wealth. Since the gap between savings and investing becomes larger and optimism followed by the 2021

© Springer Nature Singapore Pte Ltd. 2021
A. Mohamed et al. (Eds.): SCDS 2021, CCIS 1489, pp. 197–212, 2021.
https://doi.org/10.1007/978-981-16-7334-4_15

Budget which includes initiatives to increase the cash available to the people, it leads to the increasing interest in mutual fund investments. It is also crucial for all investors including the government, where it helps in improving the economy, the well-being of the country, and the citizens.

1.1 Mutual Fund in Malaysia

Malaysia's mutual fund industry is at the stage of large-scale growth in terms of size and choices. Investors preferred to invest in mutual funds as they can diversify their investments and risks, but at the same time, this investment also offers profitable possibilities [1]. On top of that, the mutual fund has been an investors' choice for long-term investment due to the diversified risk and investment at a low cost [2–4]. It also becomes very popular because they are well managed by financial experts and owning units of the mutual fund is a cost-effective way of diversification [5].

Malaysia's mutual fund industry had made a solid recovery and grows rapidly after the 1997 East Asian financial crisis and added that during the 1990s, the Malaysian Government took an initiative to encourage Malaysians to invest by offering the Amanah Saham Nasional. The impact is that the Net Asset Value (NAV) for mutual funds climbed by approximately 160%; the total net asset value grew from RM87.4 billion in 2004 to RM226.8 billion in 2010. It shows that the government plays a crucial role to encourage people to invest in mutual funds [6]. When the mutual fund industry is getting popular, the investment is now focused on the Fund Manager. The ability of fund managers to create value depends on market liquidity conditions which in turn introduces a liquidity risk exposure (beta) for skilled managers [5]. Associated studies with the same methods to measure liquidity are in [7–10]. There are thirty-six (36) Unit Trust Management Companies (UTMC) in Malaysia that offer various types of funds to potential investors. Every fund manager at UTMC has a distinct style of allocating funds. However, there is a scarcity of past studies that focus on the performance of equity funds in the Malaysian market and the liquidity timing ability as compared to other markets. Therefore, this study focuses on evaluating the Malaysian equity fund performance and examines one of the factors affecting the funds' performance which is timing liquidity.

2 Methodology

Equity funds are the focus in this study where only 56 funds were chosen due to its availability of complete data. Fund's Net Asset Value (NAV) is used, Kuala Lumpur Composite Index (KLCI) closing price as a market proxy, and closing price of Kuala Lumpur Interbank Offered Rate (KLIBOR) based on three different time frames: daily, monthly, and yearly as the proxy for risk-free rate. Three different time frames are applied to get a better understanding and multiple outlooks on Malaysian equity funds. The average monthly return will be used to provide the output of which equity funds perform better than the other. Meanwhile, the yearly return marks which equity funds' performance beat the other for each year as the result may differ for different years. The daily returns a month before and after Malaysia's General Election 14 (GE14) is employed to test the pre-effect and post-effect of Malaysia's political events towards equity funds performance in Malaysia. To determine the performance of equity funds in Malaysia, this study used Sharpe Ratio, Jensen's Alpha Index, and Treynor Index.

2.1 Return and Traditional Financial Ratios

The study used logarithmic or continuously compounded monthly returns for all selected equity funds using the fund's net asset value.

The same calculation goes to the monthly market return using the FBM Bursa Malaysia (KLCI) index at time t. The return of the equity funds is then compared against the market return to see the performance of the equity funds across three different time frames. Sharpe ratio is also referred to as reward to variability ratio which assumes that small investors will demand a premium for the total risk if they fully invest in the mutual fund and they do not hold any portfolio to exclude unsystematic risk. The higher the positive SR, the better. However, if the value is negative, it means that the return of investment is lower, then it is a risk-free rate. The Treynor ratio is based on systematic risk. The unsystematic risk can be eliminated when the investor holds a diversified portfolio. The higher the value of the Treynor ratio, the better the performance of the fund. Contrary to Sharpe's and Treynor ratio, Jensen alpha focused on the problem of evaluating a fund manager's ability to providing higher returns to the investors [11]. The return of the portfolio is calculated on a basis of risk-adjusted by taking the market performance as a benchmark. Moreover, it can be used to measure the performance by way of the excess return provided by the portfolio over the risk-adjusted return or the portfolio's theoretical expected return which is predicted by the capital asset pricing model (CAPM). Jensen also assumes that at least CAPM returns have been expected as a return for an investor. A positive α indicates that the mutual fund manager has a superior forecasting ability. It also shows that Jensen alpha returns are better than predicted by CAPM, meaning that the mutual fund manager is beating the market performance.

2.2 Capital Asset Pricing Model

The difference between CAPM with all the other three methods in measuring the performance of the equity funds is that CAPM is used to find the expected return based on historical data. However, both Jensen's Alpha Index and Treynor Index presume the single-factor CAPM with both Alpha and Beta values in the equation. The Alpha value is the surplus of the return while the Beta value represents the systematic risk.

$$CAPM : E(R_i) = R_f + \beta_i(R_m - R_f) \tag{1}$$

where $E(R)_i$ is the expected portfolio return, R_f is the risk-free interest rate, R_m is the return on market/benchmark portfolio, and β_i is the Portfolio's market risk (volatility of portfolio return against that of market portfolio return).

2.3 Liquidity Measures, Trading Volume and Turnover

To examine the relationship of the expected return of mutual funds and market timing liquidity, this study uses two approaches to calculate the timing liquidity whereby the values will be used in the multiple linear regression. The first approach is Trading Volume (TV) and the second approach is Turnover (TO). The similarity between these two methods is that both methods are based on trading activity. Therefore, the supply of market participants and transactions is measured to evaluate liquidity.

To measure equity market trading activity, trading volume like Financial Times Stock Exchange (FTSE) KLCI is used as a liquidity proxy. The FTSE KLCI data retrieved from Bloomberg is used to calculate monthly estimates of dollar trading volume for each stock. Then, the data across stocks is average out for a monthly aggregate liquidity measure and denoted as VOL. Specifically,

$$VOL_t = \frac{1}{N} \sum\nolimits_{i=1}^{N} P_{i,t} V_{i,t} \tag{2}$$

where $P_{i,t}$ is the closing share price for a company i in month t, $V_{i,t}$ is the number of shares of a company i traded in month t, N is the number of companies included in the sample in month t, VOL_t is the estimate of average dollar trading volume per stock during month t. The liquidity timing model in this study is then the natural logarithm of this measure. To test for mutual funds' liquidity timing ability, we use innovation in the monthly trading volume measure rather than levels data. Innovation is calculated over a 12-month moving average of the average monthly trading volume.

The second proxy for trading activity is turnover. Turnover can be defined as the number of shares traded per number of shares outstanding, averaged across all stocks traded during a month. We use FTSE KLCI data from Bloomberg and denote turnover as *TO*. Specifically,

$$TO = \frac{1}{N} \sum\nolimits_{i=1}^{N} V_{i,t}/S_{i,t} \tag{3}$$

where $S_{i,t}$ is the number of shares of a company i outstanding in month t, N is the number of companies included in the sample in month t. To test for mutual funds' liquidity timing ability, we use innovations in the monthly turnover measure rather than levels data. Innovations are calculated over a 12-month moving average of average monthly stock turnover.

2.4 Multiple Linear Regression

The regression is used to explain the relationship between the equity funds and the liquidity timing. CAPM was first used, only then the first-order Taylor series expansion will be added to the equation. To express the linear function of market beta with the market liquidity in excess of its time-series average, the Taylor series expansion was applied. The First-order Taylor series is as below.

$$\beta_{f,m,p} = \beta_{0,m,p} + \Upsilon_{mp}\left(L_{mt} - \overline{L_m}\right) \tag{4}$$

where L_{mt} is the market liquidity measure in month t, $\overline{L_m}$ is the average of the market liquidity, $\beta_{0,m,p}$ is the risk of the equity funds in absence of liquidity timing. Lastly, by substituting the Taylor expansion equation into the CAPM equation, the final equation below was applied.

$$CAPM : \alpha_p + \beta_{0,m,p}R_{mt} + \gamma_{mp}\left(L_{mt} - \overline{L_m}\right)R_{mt} + \varepsilon_{pt} \tag{5}$$

where γ_{mp} is the mutual fund managers' liquidity timing ability. The higher the value, the higher market exposure the equity funds have. In this study, all of the equations and methods used were motivated by past literatures, for instance in [12–15].

3   Analysis and Findings

3.1 Performance of Equity Funds Versus Market Return

Table 1 exhibits the average monthly return of top-five and bottom five equity funds, out of 56 mutual funds in the Malaysian market. As reflected in Table 1, Eastspring Investment Small Cap Fund, Kenanga Growth Fund, and KAF Vision Fund have the maximum monthly average return whereas Eastspring Investment Small-Cap, Kenanga Growth Fund, and KAF Vision Fund beat the market average return. Four equity funds from PMB Asset Management, and one equity fund from Pacific Mutual Fund have the lowest average monthly return as compared to the average market return. PMB Shariah Mid-Cap Fund, PMB Dana Mutiara, and PMB Dana Al-Aiman have the lowest monthly return, which is lower by 0.36%, 0.32% and 0.25% from the average market return respectively. There are nine equity funds from PMB Asset Management that are used in this study and four of them recorded the lowest average monthly return for 99 months. In addition, from all 56 funds included in the study, there are a total of 14 equity funds that performed lower than the average monthly market return, indicating that their performance is not at par with other equity funds available in the market. Considering the average market return of 0.48% as a benchmark, 71.43% of the equity mutual funds available in the Malaysian market have higher and promising returns as compared to the average monthly market return. 3.57% of the equity mutual funds perform similarly to the average monthly market return and 25% of the equity mutual funds have a lower return than the average monthly market return. Thus, if an investor is only interested to look at the average monthly return in making their investment decision, there are 40 mutual funds available to be considered as an option to invest in. Out of 14 Unit Trust Management Companies (Fund Houses) included in this study, only three of the Fund Houses like Eastspring Investment Berhad, Kenanga Investor Berhad, and KAF Investment Berhad offer a higher average monthly return as compared to the average monthly market return. Table 2 shows the association between each equity fund and the market return by looking at the movement of the one return of equity funds with the market return. If the value of the coefficient correlation is positive (negative), the movement of the mutual fund return is in the same (opposite) direction as the market return. PMB Shariah Index Fund shows the strongest correlation of 0.9593 with the market return, followed by Pacific Dana Aman Fund and AMB Dana Yakin Fund as compared to all the selected equity mutual funds. Interpac Dana Safi has the lowest rate of correlation with the market return, followed by Interpac Dynamic Equity Fund and RHB Small Cap Opportunity Fund.

From all the 56 mutual funds selected in this study, no mutual funds display negative correlations with the market return, as all the mutual funds have positive value of coefficient correlation. From Table 3, we can see three different performance methods such as Jensen's Alpha Index, Sharpe Ratio, and Treynor Index. It can be concluded that Eastspring Investment Small-Cap Fund has the highest return based on all three methods; Jensen's Alpha index or so called Jensen, Treynor index, and average return, however, it scores the second-highest for Sharpe ratio. For the Sharpe ratio, Kenanga Growth Fund has the highest rate as compared to others. Moreover, for the underperformed equity funds, all four (4) methods shown the same result when all four PMB equity funds have the lowest rate of return. Table 4 displays the expected return using the Capital Asset Pricing Model (CAPM) with risk-free rate of 0.511% and market risk premium at − 0.027%.

Table 1. Return earned by selected mutual funds.

Equity fund name	Monthly average return (%)	Compare to the market (%)
Eastspring Investment Small-Cap Fund	1.39	0.91
Kenanga Growth Fund	1.32	0.84
KAF Vision Fund	1.03	0.55
Eastspring Investment Equity Income Fund	1.00	0.52
Eastspring Investments Dana Al-Ilham	0.94	0.46
Kenanga Syariah Growth Fund	0.89	0.41
PMB Shariah Mid-Cap Fund	0.12	−0.36
PNB Dana Mutiara	0.16	−0.32
PMB Dana Al-Aiman	0.23	−0.25
PMB Dana Bestari	0.23	−0.25
Pacific Pearl Fund	0.28	−0.20

AMB Dividend trust fund has the highest expected return at 0.496% as compared to other equity funds, followed by Eastspring Investment Equity Income Fund and both PMB Shariah Premier Fund and InterPac Dynamic Equity Fund. On the other hand, KAF Vision Fund and RHB Malaysia Diva Fund both indicate the lowest expected rate of return at 0.48%. There is no negative value of the expected rate of return for all 56 equity funds. All equity funds scheme denotes higher return or at least the same value of return in comparison with the average market return.

Table 2. Coefficient correlation and correlation of determination (R square).

Fund name	Correlation coefficient	R square
PMB Shariah Index Fund	0.9593	0.9203
Pacific Dana Aman	0.9438	0.8908
AMB Dana Yakin	0.9423	0.8880
Pacific Dividend Fund	0.9352	0.8746
Eastspring Investments Dana Al-Ilham	0.9336	0.8716
InterPac Dana Safi	0.5543	0.3072
InterPac Dynamic Equity Fund	0.5642	0.3184
RHB Small Cap Opportunity Unit Trust	0.6924	0.4795
PMB Shariah Aggressive Fund	0.7054	0.4976
RHB Emerging Opportunity Unit Trust	0.7082	0.5016

Table 3. Comparison of the mutual fund performance.

Fund name	Jensen Alpha (%)	Sharpe (%)	Treynor (%)	Average return (%)
Eastspring Investment Small-Cap Fund	0.92	16.48	1.01	1.39
Kenanga Growth Fund	0.83	24.61	0.71	1.32
KAF Vision Fund	0.55	10.96	0.59	1.03
Eastspring Investment Equity Income Fund	0.51	16.79	0.26	1.00
Eastspring Investments Dana AL-Ilham	0.45	13.21	0.35	0.94
PMB Shariah Mid-Cap Fund	−0.37	−12.46	−0.53	0.12
PMB Dana Mutiara	−0.32	9.04	−0.37	0.16
PMB Dana Al-Aiman	−0.26	−9.26	−0.47	0.23
PMB Dana Bestari	−0.26	−8.71	−0.38	0.23

3.2   Performance of Equity Funds Versus Market Return: A Month Before and After the 14th General Election (GE14)

Based on Table 5, the rate of return for FBM Bursa Malaysia KLCI was at −1.42%. The negative rate of return for KLCI implying that the market condition was slightly downhill. The highest rate of the equity funds' performance for the day was AMB Index-Linked Trust Fund, which beats the market return by 1.93%. Next, RHB KLCI Tracker Fund performed as the second-best which overperformed the market return by 1.61% and next followed by PMB Shariah Index Fund. InterPac Dynamic Equity Fund performed worst and was tailed by InterPac Dana Safi and KAF Vision Fund. As their return was negative in value, this indicated that the funds brought huge losses to its investors. Table 6 shows that a month after GE14, the rate of return for FBM Bursa Malaysia KLCI was at −3.69% indicating that the market condition slightly worsened as compared to the month before. Remarkably, the most performing equity fund of the day was BIMB i Growth, which outperformed the market return by 12.86%. It can be considered as a shocking verdict since in the previous month BIMB i Growth was the worst performed

Table 4. Expected return of the mutual fund.

Fund name	Beta	Expected Rate of Return (%)
AMB Dividend Trust Fund	0.54489	0.50
Eastspring Investments Equity Income Fund	0.68711	0.49
PMB Shariah Premier Fund	0.72526	0.49
InterPac Dynamic Equity Fund	0.72678	0.49
PMB Dana Al-Aiman	0.73110	0.49
Eastspring Investments Small-Cap Fund	1.34356	0.48
Pacific Pearl Fund	1.20354	0.48
RHB Small Cap Opportunity Unit Trust	1.18247	0.48
KAF Vision Fund	1.16679	0.48
RHB Malaysia DIVA Fund	1.15764	0.48

Table 5. Equity funds return versus market return: a month before GE14.

Equity fund name	Equity fund return	Comparison to market
AMB Index-Linked Trust Fund	0.51%	1.93%
RHB KLCI Tracker Fund	0.19%	1.61%
PMB Shariah Index Fund	0.13%	1.55%
KAF Vision Fund	−12.00%	−10.58%
InterPac Dana Safi	−16.40%	−14.98%
InterPac Dynamic Equity Fund	−17.89%	−16.47%

equity fund as compared to other equity funds. Its rate of return soared by 9.56% in just a month. The second most performing equity fund was PMB Shariah Dividend which was higher than the market return by 7.5%, followed by PMB Shariah Mid-Cap Fund, beating the market return by 6.54%. AmanahRaya Islamic Equity Fund was recorded as the worst performed equity fund, followed by PMB Dana Bestari and RHB Capital Fund, which underperformed the market return by −3.4%. This indicated that investors are suffering a loss from their investment in the equity funds.

Table 7 shows the summary of all the performance for the equity funds. The rate of market returns one month before the GE14 indicated losses at −1.42%. However, on Election Day, the rate of return for the FBM Bursa Malaysia KLCI Index bounced to 0.52% before it decreased to −3.69% due to the post-Election Day aftershock. From Table 7, it can be concluded that the performance of the funds was higher on and after Election Day as compared to its rate of return before Election Day. Notice that the rate of return for BIMB i Growth reached its peak performance at 9.17% a month after the election, followed by PMB Shariah Dividend Fund and PMB Shariah Mid-Cap Fund. The equity fund with the poorest performance within this time frame (a month before and after Election Day) was InterPac Dynamic Equity Fund, followed by InterPac Dana Safi as both funds had the lowest rate of return a month before Election Day. Thus, it can be concluded that a month before the Election Day, the equity funds were not performing at their best and only started to gain traction following Election Day.

Table 6. Equity funds return versus market return: a month after GE14.

Equity fund name	Equity fund return	Comparison to market
BIMB i Growth	9.17%	12.86%
PMB Shariah Dividend Fund	3.81%	7.50%
PMB Shariah Mid-Cap Fund	2.85%	6.54%
RHB Capital Fund	−7.10%	−3.40%
PMB Dana Bestari	−7.13%	−3.44%
AmanahRaya Islamic Equity Fund	−8.51%	−4.81%

Table 7. Equity funds return versus market return on GE14.

Equity fund name	Equity fund return	FBM KLCI return	Comparison to market
AMB Index-Linked Trust Fund	0.51%	−1.42%	1.93%
RHB KLCI Tracker Fund	0.19%	−1.42%	1.61%
PMB Shariah Index Fund	0.13%	−1.42%	1.55%
KAF Vision Fund	−12.00%	−1.42%	−10.58%
InterPac Dana Safi	−16.40%	−1.42%	−14.98%
InterPac Dynamic Equity Fund	−17.89%	−1.42%	−16.47%
KAF Tactical Fund	6.06%	0.52%	5.55%
KAF Vision Fund	5.20%	0.52%	4.68%
PMB Shariah Dividend Fund	4.75%	0.52%	4.23%
RHB Malaysia DIVA Fund	0.32%	0.52%	−0.20%
RHB Smart Treasure Fund	0.11%	0.52%	−0.40%
BIMB i Growth	−0.39%	0.52%	−0.91%
BIMB i Growth	9.17%	−3.69%	12.86%
PMB Shariah Dividend Fund	3.81%	−3.69%	7.50%
PMB Shariah Mid-Cap Fund	2.85%	−3.69%	6.54%
RHB Capital Fund	−7.10%	−3.69%	−3.40%
PMB Dana Bestari	−7.13%	−3.69%	−3.44%
AmanahRaya Islamic Equity Fund	−8.51%	−3.69%	−4.81%

3.3   Relationship Between Expected Return of CAPM and Liquidity Timing

Trading Volume Liquidity Measure

Previously, Table 8 presents the Trading Volume for liquidity measure for 10 out of 56 equity funds were selected. The results are statistically significant for six out of the 10 equity funds. Three equity funds are significant at 0%, two equity funds are significant at 1% and only one is significant at 5% level. Lastly, four equity have no significant liquidity timing values at any level.

Table 8. Relationship between CAPM and trading volume liquidity measure.

Fund name	Alpha	Beta	Liquidity timing	Adjusted R-square
AMB Dividend Trust Fund	3.763E−3**	0.6946***	−6.24E−09	0.7493
Eastspring Investments Equity Income Fund	5.099E−3***	0.8056***	1.27E−10	0.8651
PMB Shariah Premier Fund	−2.77E−05	0.9974***	1.31E−08	0.8527
InterPac Dynamic Equity Fund	0.0005128*	0.8833***	5.083E−8 **	0.5177
PMB Dana Al-Aiman	−2.60E−03	0.8385***	9.70E−09	0.83
InterPac Dana Safi	9.97E−04	0.8705***	4.603E−8 *	0.4781
Kenanga Syariah Growth Fund	4.023E−3**	0.824***	2.341E−7 **	0.8338
PMB Shariah Dividend Fund	−2.08E−04	0.8761***	1.40E−08	0.6805
KAF Core Income Fund	2.03E−03	0.859***	9.41E−09	0.735
AMB Index-Linked Trust Fund	8.10E−04	0.8241 ***	−9.82E−09	0.8861

*Sig. codes: 0 '***' 0.001 '**' 0.01 '*' 0.05 '.' 0.1 ' ' 1*

Table 9 shows the adjusted beta which indicates the relative change in fund equity market exposure with a benchmark standard deviation of 0.0248. However, the values of the adjusted beta were modest for all 10 equity funds which signify that there is not a highly significant economic result. The highest adjust beta value was AMB Index-Linked Trust Fund, whereby the timing ability only increases market exposure by 1.283% depending on one standard deviation increase in the aggregate market trading volume. It can be concluded that for 10 equity funds, the relative exposure amount ranges from 1.07% to 1.3%. Nevertheless, this trading volume test further indicates that the equity funds demonstrated the liquidity timing skills with the positive adjust beta values where the equity funds adjusted to the market exposure prior to changes in the aggregate liquidity. In line with [7], the study stated that the equity funds have a timing ability prior to the market exposure. Most of the six equity funds have significant liquidity timing values, with high R-Square which indicates that there was a strong relationship between expected return and liquidity timing.

Table 9. Trading volume liquidity measure.

Fund name	Std dev	R-square	Δ Beta	Beta	Δ Beta %
AMB Dividend Trust Fund	0.0271	0.7536	0.8243	0.6989	1.1794
AMB Index-Linked Trust Fund	0.0297	0.8876	1.0642	0.8295	1.2830
Eastspring Investments Equity Income Fund	0.0291	0.8715	1.0238	0.8064	1.2697
InterPac Dana Safi	0.0421	0.5318	0.9042	0.8476	1.0668
InterPac Dynamic Equity Fund	0.0413	0.5947	0.9916	0.8584	1.1552
KAF Core Income Fund	0.0334	0.7397	0.9980	0.8538	1.1689
Kenanga Syariah Growth Fund	0.0301	0.8512	1.0344	0.8120	1.2739
PMB Dana Al-Aiman	0.0307	0.8422	1.0453	0.8341	1.2532
PMB Shariah Dividend Fund	0.0353	0.7018	1.0013	0.8698	1.1511
PMB Shariah Premier Fund	0.0305	0.8621	1.0608	0.9910	1.0705

Turnover Liquidity Measure

Table 10 presents the Turnover for liquidity measure for 10 equity funds for the explanation and illustration in this section. As recorded in Table 10, the results are statistically significant for only three (3) out of the ten (10) funds. There were two (2) equity funds that were significant at 1% namely InterPac Dynamic Equity Fund and Kenanga Syariah Growth Fund. Meanwhile, there was only one equity fund that was significant at a 5% level – InterPac Dana Safi Fund. The other seven (7) equity funds namely AMB Dividend Trust Fund, Eastspring Investment Equity Income Fund, PMB Shariah Premier Fund, PMB Dana Al Aiman Fund, PMB Shariah Dividend Fund, KAF Core Income Fund, and AMB Index-Linked Trust Fund have no significant liquidity timing values at any level. Out of three (3) equity funds that have a significant value of timing ability, only one (1) equity fund has a high R-Square of 83.72% which is Kenanga Syariah Growth Fund, while the other two namely InterPac Dynamic Equity Fund and InterPac Dana Safi Fund only have moderate R-Square values of 52.76% and 48.88% respectively. Table 11 shows the adjusted beta which indicates the relative change in fund equity market exposure with a benchmark standard deviation of 0.0248.

Table 10. Relationship of CAPM and turnover liquidity measure.

Fund name	Alpha	Beta	Liquidity timing	Adjusted R-square	R-square
AMB Dividend Trust Fund	3.76E−3 **	0.6946***	−6.24E−09	0.7493	0.7544
Eastspring Investments Equity Income Fund	5.099E−3 ***	0.8056***	1.27E−10	0.8651	0.8678
PMB Shariah Premier Fund	−2.77E−05	0.9974***	1.31E−08	0.8527	0.8557
InterPac Dynamic Equity Fund	5.128E−4 *	0.8833***	5.083–08 **	0.5177	0.5276
PMB Dana Al-Aiman	−2.60E−03	0.8385***	9.70E−09	0.83	0.8335
InterPac Dana Safi	9.97E−04	0.8705***	4.603E−08 *	0.4781	0.4888
Kenanga Syariah Growth Fund	4.02E−3 **	0.824***	2.34E−08**	0.8338	0.8372
PMB Shariah Dividend Fund	−2.08E−04	0.8761***	1.40E−08	0.6805	0.687
KAF Core Income Fund	2.03E−03	0.859***	9.41E−09	0.735	0.7405
AMB Index-Linked Trust Fund	8.10E−04	0.8241***	−9.82E−09	0.8861	0.8884

*Sig. codes: 0 '***' 0.001 '**' 0.01 '*' 0.05 '.' 0.1 ' '*

Table 11. Turnover liquidity measure.

Fund name	Std Dev	R-square	Δ Beta	Beta	Δ Beta %
AMB Dividend Trust Fund	0.0271	0.7544	0.8252	0.6946	1.1880
AMB Index-Linked Trust Fund	0.0297	0.8884	1.0652	0.8241	1.2925
Eastspring Investments Equity Income Fund	0.0291	0.8678	1.0195	0.8056	1.2655
InterPac Dana Safi	0.0421	0.4888	0.8311	0.8705	0.9547
InterPac Dynamic Equity Fund	0.0413	0.5276	0.8833	0.8584	1.0290
KAF Core Income Fund	0.0334	0.7405	0.9991	0.8590	1.1630
Kenanga Syariah Growth Fund	0.0301	0.8372	1.0174	0.8240	1.2347
PMB Dana Al-Aiman	0.0307	0.8335	1.0345	0.8385	1.2337
PMB Shariah Dividend Fund	0.0353	0.687	0.9802	0.8761	1.1188
PMB Shariah Premier Fund	0.0305	0.8557	1.0530	0.9974	1.0557

However, the values of the adjusted beta were only average for all 10 funds which signify there is not a highly significant economic result. In line with trading volume, AMB Index-Linked Trust Fund has the highest adjust beta values for turnover at 1.2925%, slightly higher than the trading volume rate which was 1.283%. Meaning that the timing ability increases market exposure by 1.2925% prior to one standard deviation increase in the aggregate market trading volume. It can be concluded that for all 10 equity funds, the relative exposure ranges from 0.9% to 1.3%.

4 Conclusion

This study discovers that there is a total of fourteen equity funds that performed lower than the average monthly market return, indicating that their 99 monthly performance period is not at par with other equity funds available in the market. In consideration of the monthly average market return of 0.48% as a benchmark, 71.43% of the equity mutual funds available in the Malaysian market have a higher and promising return as compared to the average monthly market return, 3.57% of the equity mutual funds perform similarly as the average monthly market return and 25% of the equity mutual funds have a lower return than the average monthly market return. However, the result can be quite biased since, in this time frame, only the average return was used for the period of study; 99 months. Thus, this study includes the second time frame that is a yearly basis, time frame different is to manage or check on any discrepancies from the output. On a yearly time frame, there is a mixed result where each year there are different equity funds that either outperformed or underperformed the market return. This is possible due to the fluctuating of Net Asset Value (NAV) and the fact that it was compared with different market returns as the benchmark. This study employed the third time frame to understand the performance of the equity funds on daily basis and focuses on one month before Malaysia's 14th General Election (GE14) and one month after the GE14. Based on this time frame, it can be concluded that the market return escalated towards Election Day but slightly dropped after Election Day. Although the study is divided into three different periods, the returns of each fund show little indication of profitability, and this can cause these funds to be the non-investors choice due to their uncertain price movements. Therefore, it is recommended that techniques like forecasting or chaos theory be included to forecast future movements of fund prices as the financial time series usually have a rapid variance change [16–18]. A combination of forecasting techniques with the method used in this study can also indicate fund price movements in the future and can make the analyzed fund's performance more meaningful to investors and researchers [19].

The second objective of this study was to help the Fund Managers to investigate the relationship between expected mutual fund returns and liquidity timing. Previous studies stressed that the market liquidity moves consistently with the market return, which means the Fund Managers can decide to either reduce or increase the market exposure prior to the value of the market liquidity condition certainly. In line with [4], this study found that there are significant findings that timing liquidity affects the expected return of the equity funds. However, for both the Turnover method and Trading Volume, the adjusted beta values are modest for all ten equity funds which are between 1.07% to 1.3% and between 0.9% to 1.3% respectively. This indicates that Fund Managers consider increasing the market exposure prior to one standard deviation increase in the aggregate market. Based on the result, it shows that liquidity timing is very important to the fund manager and it is also a benchmark of the ability of a fund manager. It demonstrates the fund manager's ability to balance the investment portfolio between risk and return in the event of any change in the market risk. For future research, it is recommended to explore other methods such as the Data Envelopment Analysis (DEA) model which combines performance-related factors and can be used to measure the overall performance appraisal of a fund [20] or hybrid DEA model such as the Network Data Envelopment Analysis (NDEA) that is also expected to support the liquidity timing of a fund manager [21].

Acknowledgements. The authors gratefully acknowledge financial support from the Faculty of Computer and Mathematical Science, Universiti Teknologi MARA, Shah Alam for this study.

References

1. Ming, L.-M., Hwa, L.-S.: Evaluating mutual fund performance in an emerging Asian economy: the Malaysian experience. J. Asian Econ. **21**(4), 378–390 (2010)
2. Abdullah, E.A., Zahari, S.M., Shariff, S.S.R., Rahim, M.A.I.A.: Modelling volatility of Kuala Lumpur composite index (KLCI) using SV and GARCH models. Indones. J. Electr. Eng. Comput. Sci. **13**(3), 1087–1094 (2019)
3. Don, U.A.G., Roshdi, I., Fukuyama, H., Zhu, J.: A new network DEA model for mutual fund performance appraisal: An application to U.S. equity mutual funds. OMEGA Int. J. Manag. Sci. **77,** 168–79 (2018)
4. Rajpurohit, S.: A comparative study of performance of top 5 mutual funds in India. SAMVAD **8** (2015)
5. Galagedera, D.U., Roshdi, I., Fukuyama, H., Zhu, J.: A new network DEA model for mutual fund performance appraisal: an application to US equity mutual funds. Omega **77**, 168–179 (2018)

6. Jamaludin, N., Smith, M., Gerrans, P.: Mutual fund investment choice criteria: a study in Malaysia. Int. J. Educ. Res. 1(4), 1–10 (2013)
7. Amihud, Y.: Illiquidity and stock returns: cross-section and time-series effects. J. Financ. Mark. 5(1), 31–56 (2002)
8. Pástor, Ľ, Stambaugh, R.F.: Liquidity risk and expected stock returns. J. Polit. Econ. 111(3), 642–685 (2003)
9. Sadka, R.: Momentum and post-earnings-announcement drift anomalies: the role of liquidity risk. J. Financ. Econ. 80(2), 309–349 (2006)
10. Cao, C., Chen, Y., Liang, B., Lo, A.W.: Can hedge funds time market liquidity. J. Financ. Econ. 109(2), 493–516 (2013)
11. Jensen, M.C.: The performance of mutual funds in the period 1945–1964. J. Financ. 23(2), 389–416 (1968)
12. Abdul Rahim, R., Mohd Nor, A.H.S.: A comparison between Fama and French model and liquidity-based three factor models in predicting portfolio returns. Asian Acad. Manag. J. Account. Financ. 2(2), 43–60 (2006)
13. Abdul Rahim, R., Mohd Nor, A.H.S.: The role of illiquidity risk factor in asset pricing models: Malaysian evidence. J. Pengurusan 26(2007), 67–97 (2007)
14. Cao, C., Simin, T.T., Wang, Y.: Do mutual fund managers time market liquidity? J. Financ. Mark. 16(2), 279–307 (2013)
15. Liao, L., Zhang, X., Zhang, Y.: Mutual fund managers' timing abilities. Pacific-Basin Financ. J. 44, 80–96 (2017)
16. Qamruzzaman, Md.: Comparative study on performance evaluation of mutual fund schemes in Bangladesh: an analysis of monthly returns. J. Bus. Stud. Q. 5(4), 190 (2014)
17. Shamsuddin, S., Hanafi, N., Samian, M., Amin, M.: Chaotic stochastic lee-carter model in predicting kijang emas price movements: a machine learning approach. In: Kor, L.-K., Ahmad, A.-R., Idrus, Z., Mansor, K.A. (eds.) Proceedings of the Third International Conference on Computing, Mathematics and Statistics (iCMS2017), pp. 521–528. Springer, Singapore (2019). https://doi.org/10.1007/978-981-13-7279-7_65
18. Ramli, S.F., Firdaus, M., Uzair, H., Khairi, M., Zharif, A.: Prediction of the unemployment rate in Malaysia. Int. J. Mod. Trends Soc. Sci 1(4), 38–44 (2018)
19. Johari, S.N.M., Farid, F.H.M., Nasrudin, N.A.E.B., Bistamam, N.S.L., Shuhaili, N.S.S.M.: Predicting stock market index using hybrid intelligence model. Int. J. Eng. Technol. (UAE) 7, 36–39 (2018)
20. Kamarudin, N., Ismail, W.R., Mohd, M.A.: Network data envelopment analysis as instrument for evaluating water utilities' performance. J. Qual. Meas. Anal. JQMA 14(2), 1–10 (2018)
21. Premachandra, I.M., Zhu, J., Watson, J., Galagedera, D.U.A.: Mutual fund industry performance: a network data envelopment analysis approach. In: Zhu, J. (ed.) Data Envelopment Analysis. ISORMS, vol. 238, pp. 165–228. Springer, Boston (2016). https://doi.org/10.1007/978-1-4899-7684-0_7

An Autoregressive Distributed Lag (ARDL) Analysis of the Relationships Between Employees Provident Fund's Wealth and Its Determinants

Haidah Syafi Parly[1], Siti Meriam Zahari[2]($\boxtimes$) (iD),
Muhammad Asmu'i Abdul Rahim[2] (iD), and S. Sarifah Radiah Shariff[2] (iD)

[1] Bancassurance Associates, RHB Bank, Kuala Lumpur, Malaysia
[2] Universiti Teknologi MARA, Shah Alam, Selangor, Malaysia
{mariam,asmui,radiah}@tmsk.uitm.edu.my

Abstract. This study empirically examines the potential determinants of the wealth of Malaysian Employees Provident Fund (EPF). The auto-regressive distributed lag (ARDL) bounds test approach was employed to determine the existence of a long-run relationship between EPF's wealth and its determinants, namely inflation rate (INF), gross domestic product (GDP), life expectancy (EXP), and the Gini coefficient (GINI) as a proxy for income inequality. Data for this study consisted of a set of annual time series data from 1980 to 2018. The findings indicate the existence of a long-run equilibrium relationship between the total EPF wealth and GDP, income inequality, INF, and EXP. Further, there is a relatively quick adjustment in the total EPF wealth when all of the determinants change. GDP, EXP, and GINI were found to be significantly important drivers of EPF wealth. A 10% change in GDP will result in a long-run change of 25.8% in the total EPF wealth. Meanwhile, a 1% increase in income inequality and life expectancy will reduce the total EPF wealth by 11.8% and 20.4%, respectively. One of the implications of this study's findings is that EPF wealth would grow continuously with an increase in economic prosperity and improvements in income inequality and demographic structure. Thus, this study suggests that EPF wealth can be improved by making positive changes to economic and demographic factors in ensuring the sustainability of the fund.

Keywords: Employee provident Fund · ARDL · Long-run relationship · Short-run dynamics · Income inequality

1 Introduction

Ageing is a lifelong process that affects all living creatures. Working hard during younger ages in preparation for happy retirement is the hope and aspiration of every individual. As a person reaches a particular age, they have the right to retire from their current job permanently and receive their retirement fund benefits. A retirement fund is vital for supporting retired pensioners, especially among the elderly. The elderly is acknowledged

© Springer Nature Singapore Pte Ltd. 2021
A. Mohamed et al. (Eds.): SCDS 2021, CCIS 1489, pp. 213–227, 2021.
https://doi.org/10.1007/978-981-16-7334-4_16

as one of the groups that are vulnerable to financial insecurity (Sulaiman and Mohammed 2016). The pension system in Malaysia was first developed in 1951 under Pensions Ordinance. The system was subsequently refined several times to keep up with economic developments. Reform in the pension system commenced in the 1980s, moving from the pay-as-you-go (PAYG) pension system to the defined contribution (DC) pension scheme. Since then, about 30 developing countries have partially or fully replaced their existing pension schemes with private pension schemes (Ja'afar and Daly 2016). The older pension system was unlikely to be able to provide adequate and sustainable pension benefits during retirement, thus justifying a move to a private pension system (Ja'afar and Daly 2016). Research on EPF in Malaysia have been extensively discussed in Hussin et al. (2011) and an insightful overview of the Malaysian EPF can also be found in inter alia; Hussin (2012), Hassan (2018a, b) and Hussein (2019). In Malaysia, retirement benefits are provided via both private and public sector pension schemes (Hussein). Currently, the pension system in Malaysia constitutes several distinct institutions that are classified into five distinct areas: a tax-funded defined benefit (DB) pension scheme for public or civil servants, a DC scheme for military personnel, a DC scheme for private sector workers, social insurance for private sector workers, and a DC scheme that is open to everyone. The Employees Provident Fund (EPF) is a statutory savings scheme established to cater to public and private workers. It targets to secure employees' right to retirement benefits and to increase the value of their benefits for their wellbeing and post-employment financial security (Hassan and Othman 2018a, b). The primary goal of the scheme is to provide financial protection to pensioners; however, its withdrawal purposes have been widened to allow for non-retirement withdrawals, catering to those who need funds for medical treatment, financing children's higher education, and hajj (Ja'afar and Daly 2019). Both employers and employees are obliged to contribute to the fund according to specific percentages based on income level. Members are allowed to make full withdrawal or partial withdrawal at the age of 50, even though the compulsory contributions paid after the age of 55 cannot be withdrawn until the age of 60 (Hussein 2019). EPF made the first change in 1994 to classify each contributor's account into three accounts, namely Account 1, 2, and 3 which carry 60, 30, and 10 per cent of the member's balance, thus securing the highest amount for retirement via Account 1. Members' accounts were subsequently restructured into two accounts as part of the improvements to enhance the overall efficiency and effectiveness of EPF governance (Hassan and Othman 2018a, b).

A major concern is that many Malaysians do not have sufficient retirement savings in addition to their EPF funds. It is high time for industry and community groups to investigate private retirement schemes and other ways to boost long-term savings more seriously (Gomez 2018). Asher and Bali (2015) classified pension coverage into three areas: the risk covered by the system, the number of employees enrolled in the pension programme, and the sufficiency of the pension system to alleviate poverty. Thus, the sustainability of the pension system is vital as it reflects the capacity to pay pension in both the short term and the long term. The ability to fund the current explicit debt, which is paying the current pensioners, is a short-term commitment. It requires consideration of the current liabilities and assets. Meanwhile, the long-term commitment refers to the ability to fund the implicit debt consisting of future pension payments, taking into account the future revenues and expenditure (Barr and Diamond 2010). In the Malaysian case, long-term sustainability depends on the ability of the fund institution to make pension payments to current and future retirees. However, the institution would eventually run into problems without contributions from the employees, such as problems in accumulating financial reserves for investing and facing difficulty to obtain the required returns for ceaseless payment (Lee 1997). Malaysia cannot avoid dealing with the multifaceted aspects surrounding the pension scheme. Although the scheme appears to be inclusive, unsustainability and the existing gap in covering the adequacy are issues that have yet to be addressed (Nurhisham 2019).

Thus, it is crucial to examine the potential determinants of EPF wealth. Pension scheme sustainability could be affected by demographic changes and economic uncertainties (Jaafar and Daly 2016). Blake (2000) claimed that pension savings are more important than the pension system, as savings levels would determine a decent pension system. Demographic changes, such as increased longevity, will cause a pension fund scheme to face a stability crisis. Besides, the recent proportion of the retirement age population (60 years and above) was at 10% and is expected to increase rapidly to roughly 25% over the next couple of decades (Gomez 2018). Moreover, another potential determinant of EPF wealth is nominal gross domestic product (GDP). Economists have argued that reducing the EPF wealth rate will help to boost the domestic spending in the country, and it was included as one of the measures in the 2020 Economic Stimulus Package (Chung 2020). Inflation rate is another factor that may potentially influence the stability of EPF. A study by Estrada et al. (2017) found that an increase in the inflation rate would reduce the value of EPF's wealth. Furthermore, income inequality is expected to affect EPF wealth as the larger the disparities in income, the smaller the savings are. This is because low wages and small EPF wealth cause depositors to have low savings in the pension fund to sustain their retirement.

The pension system is one of the most important elements of the social security system. It has a close relationship with sustainable economic development and social harmony. This is because a pension fund is needed by retirees to support themselves and their families after retirement. However, a change in the demographic structure and an uncertain and inconsistent economic value might affect the financial soundness of Malaysia's EPF. These factors might risk the stability of EPF's wealth in the future. Thus, it is necessary to analyse the relationship between EPF's wealth and the potential determinants that may influence the performance and stability of the fund. In other words, financial stability should be maintained by continuously assessing and reviewing the structure of EPF's scheme along with the surrounding internal and external factors. Therefore, this study employs the ARDL approach to examine the existence of a long-run relationship between the total wealth of EPF and its determinants. In Sect. 2, this study discusses the methodology of Pesaran (2001) in the ARDL setting. Next, Sect. 3 reported the analysis and findings of the study, followed by a brief discussion in the last section.

2 Methodology

Most of the macroeconomic variables are non-stationary due to time-variant in the mean and variance. Estimating a regression model that involves non-stationary variables could lead to spurious regression. This study avoided spurious results by ensuring that none of the variables is integrated in the order of two, I(2). The presence of the I(2) variable would render it impossible to interpret the values of the F-statistics provided by Pesaran et al. (2001). Thus, augmented Dickey-Fuller (ADF) and Phillips-Perron (PP) unit root tests were performed to examine the stationarity of each variable. The null hypothesis of the unit root was tested against a stationary alternative for both ADF and PP tests. Next, the study examined the presence of co-integration among the variables using the unrestricted error correction model (ECM) as the representation of the ARDL approach. The steps are explained below:

Step 1: Based on previous studies and data availability, we specified the following double-log model:

$$\ln EPF_t = c + \beta_2 \ln INF_t + \beta_3 \ln GDP_t + \beta_4 \ln EXP_t + \beta_5 \ln GINI_t + \varepsilon_t \quad (1)$$

where the disturbance term is assumed to be normally distributed, $\varepsilon_t \sim (0, \sigma^2)$. The coefficients, $\beta_i, i = 2, 3, 4$ and 5 are the elasticities of total EPF wealth with respect to inflation rate, GDP, life expectancy, and income equality, respectively. In estimating model (1) by using ordinary least squares (OLS), we applied the ARDL $(p_1, p_2, p_3, p_4, p_5)$ bounds approach using the following specified model:

$$\Delta \ln EPF_t = c + \beta_1 \ln EPF_{t-1} + \beta_2 \ln INF_{t-1} + \beta_3 \ln GDP_{t-1} + \beta_4 \ln EXP_{t-1}$$
$$+ \beta_5 \ln GINI_{t-1} + \sum_{i=1}^{p_1} \alpha_{1i} \Delta \ln EPF_{t-i} + \sum_{i=0}^{p_2} \alpha_{2i} \Delta \ln INF_{t-i} + \sum_{i=0}^{p_3} \alpha_{3i} \Delta \ln GDP_{t-i}$$
$$+ \sum_{i=0}^{p_4} \alpha_{4i} \Delta \ln EXP_{t-i} + \sum_{i=0}^{p_5} \alpha_{5i} \Delta \ln GINI_{t-i} + \varepsilon_t \quad (2)$$

where Δ is the first-order differential operator, u_t is the white noise, and p_1, p_2, p_3, p_4, and p_5 are the maximum lag orders.

Step 2: The optimal lag length was determined based on the lowest error measure criteria, such as Akaike information criterion (AIC), Bayesian information criterion (BIC), and Schwarz criteria. The model selected was tested using the autocorrelation test to ensure that the errors of the model are serially independent.

Step 3: A 'bounds test' was performed to examine the existence of a long-run relationship between total EPF wealth and its determinants. The F-test was used to determine whether co-integrating relationships exist among the variables. Referring to Eq. (2), the hypothesis is formulated as follows:

$H_0: \beta_1 = \beta_2 = \beta_3 = \beta_4 = \beta_5 = \beta_6 = 0$ (A long-run relationship does not exist)
$H_1: \beta_1 \neq \beta_2 \neq \beta_3 \neq \beta_4 \neq \beta_5 \neq \beta_6 \neq 0$ (A long-run relationship exists)

where β_i refers to long-run elasticities (multipliers). The calculated F-statistics were compared against the critical values tabulated in the statistical table CI (III) (Pesaran et al. 2001). If the calculated F-statistics are greater than the upper bounds, then H_0 is rejected. It means that the variables included in the model have long-run relationships among themselves. If the calculated F-statistics are smaller than the lower bounds, then H_0 cannot be rejected, indicating any long-run relationship. If the calculated F-statistics fall between the upper and lower bounds' values, then the decisions are inconclusive.

Step 4: Since co-integrating relationships were found among the variables, the next step is estimating a long-run model and a separate, restricted ECM. The ECM's coefficient, *ect* shows how quickly or slowly the variables return to equilibrium and it should have a statistically significant coefficient with a negative sign. The short-run dynamic model can be written as an error correction model with an additional *ect* term, as follows:

$$\Delta \ln EPF_t = c + \sum_{i=1}^{p} \alpha_{1i} \Delta \ln EPF_{t-i} + \sum_{i=0}^{p} \alpha_{2i} \Delta \ln INF_{t-i}$$
$$+ \sum_{i=0}^{p} \alpha_{3i} \Delta \ln GDP_{t-i} + \sum_{i=0}^{p} \alpha_{4i} \Delta \ln EXP_{t-i} + \sum_{i=0}^{p} \alpha_{5i} \Delta \ln GINI_{t-i} + \theta ECT_{t-1} + \varepsilon_t \quad (3)$$

Step 5: The final step involves verifying and validating the model by conducting several diagnostic tests. These tests are Jarque-Bera test for normality, Breusch-Godfrey LM test for autocorrelation, Breusch-Pagan test for heteroskedasticity, and cumulative sum of recursive residuals (CUSUM) and cumulative sum of recursive residuals squared (CUSUMSQ) tests for stability and consistency. Jarque-Bera test is to test the goodness of fit of the model based on kurtosis and skewness. Breusch and Godfrey LM test was carried out to ensure that the model errors are serially independent, such that $cov(\varepsilon_t \varepsilon_s) = 0; \forall_t \neq 0, t \neq s$. Breusch-Pagan test was also performed to test the presence of inconstancy of the error variance of the model. CUSUM and CUSUMSQ tests were applied to the residuals of the estimated ECM to affirm long-term parameter stability and consistency along with the movements of the short-term equations.

3 Analysis and Findings

This section discusses the empirical analysis of the ARDL approach and the research findings. This includes time series plot, unit root test, ARDL bound test and co-integration test.

3.1 Time Series Plots

Figures 1a) to 1e) show the time series plot for each variable, consisting of a total of 39 observations.

The time series plot of EPF wealth demonstrates an exponential trend where the curved line illustrates a rise in EPF wealth at an increasing rate over the years. It can be inferred that there is an improvement in terms of the incremental contribution to members' fund. In addition, there is no random or irregular pattern observed. The trends in Figs. 1a) and 1b) are markedly different from others. There is a significant outlier in the time series of inflation rate in 1981, which significantly affects the trend. The inflation rate in 1981 is markedly higher than the average inflation rate of 2.67% per year between 1980 and 2018. This outlier coincides with a dramatic increase in global oil prices due to the Iran-Iraq war during the 1980s. The global oil shocks led to an increase

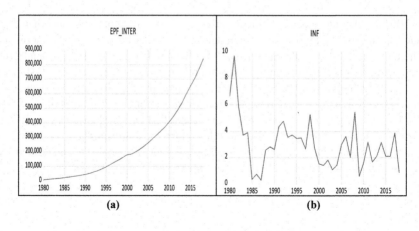

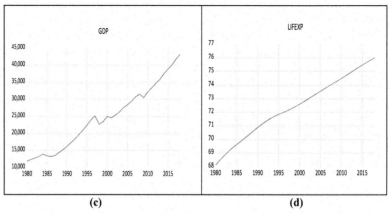

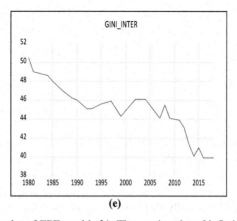

Fig. 1. a). Time series plot of EPF wealth. **b).** Time series plot of inflation rate. **c).** Time series plot of GDP. **d).** Time series plot of life expectancy. **e).** Time series plot of GINI.

in domestic retail fuel prices. Prices increased at a higher rate because of interruptions in the global energy and food supply. The shortage of global food supplies caused food prices to increase significantly. Besides, there are changes in the trend and level of the time series data in several periods, with random temporary shocks occurring between 1980 and 1990. Sudden significant drops in the inflation rate caused changes in the trend pattern in 1985 and subsequent years. In 1990, inflation increased to 2.67% and remained at the average of 2.60% until the early 2000s. In essence, there is no consistent trend over the period. The series appears to slowly wander up or down. There is no seasonality, as the time series pattern shows no regularity and changes occur every year. Figures 1c) shows that GDP has a positive upwards trend, indicating an improvement in Malaysia's economic growth. Throughout the period, the series seems to have three cycles in which it falls and rises within unspecified periods. A drop in the GDP in 1998 depicts a short-term memory effect where a decline in the output of the real economy plunged the country into its first recession, and it took about two years for the economy to return to its normal pattern in the year 2000. Figures 1d) shows an upward trend, demonstrating a steady increase in the life expectancy of Malaysians over the years, perhaps due to medical advancements and an improved lifestyle. The time series pattern shows no irregular pattern. Figures 1e) shows a downward trend in income inequality from 1980 to 2018. The pattern is quite random. Nonetheless, the downward trend indicates a positive change in income inequalities among Malaysians during the period. In general, each plot indicates a non-stationary series due to the presence of a trend and time invariance in the mean and variance of the series.

3.2 Unit Root Test

For illustration, individual significance test and join significant test can be carried out to determine whether a series is stationary. The results are reported in Figures 2a) to 2e). Given that the first five autocorrelation coefficients for EPF are 0.894, 0.795, 0.702, 0.612, and 0.527, a value would be significant (stationary) if it lies outside the interval of $(-0.3139, +0.3139)$ at the 5% level. Since each value is significant, the series is not stationary at each lag. For the joint significance test, Box-Ljung test can be used whereby the Q-statistic is compared against the critical value from the Chi-squared distribution table. The Q-statistic is 113.09, which is far greater than the tabulated $\mathcal{X}^2(5) = 11.11$ at the 5% level. Thus, the null hypothesis that the autocorrelations up to lags 5 are uncorrelated is rejected. Hence, stationary is also rejected. The correlogram plots support this statistical test result as the spikes seem to die out at a much slower rate as the number of lags increases. The results are consistent for all variables, except for INF, where the individual test shows stationarity for most of the lags since most of the spikes are lying within the confidence bands, except lags 1 and 5. This result might indicate that the INF series is stationary or nearly stationary at the level form. This result is not surprising as the series is computed in 'rates'. Hence, we should expect less variability in the series. To avoid lengthy pages, only EPF and INF correlograms are displayed in this paper. Nonetheless, more appropriate unit root tests, namely ADF and PP, were carried out to confirm this result. In addition, these tests were carried out to ensure that none of the variables is integrated of the second-order I(2), as required under the ARDL approach. Next, the structural break unit root test was examined to support the analysis.

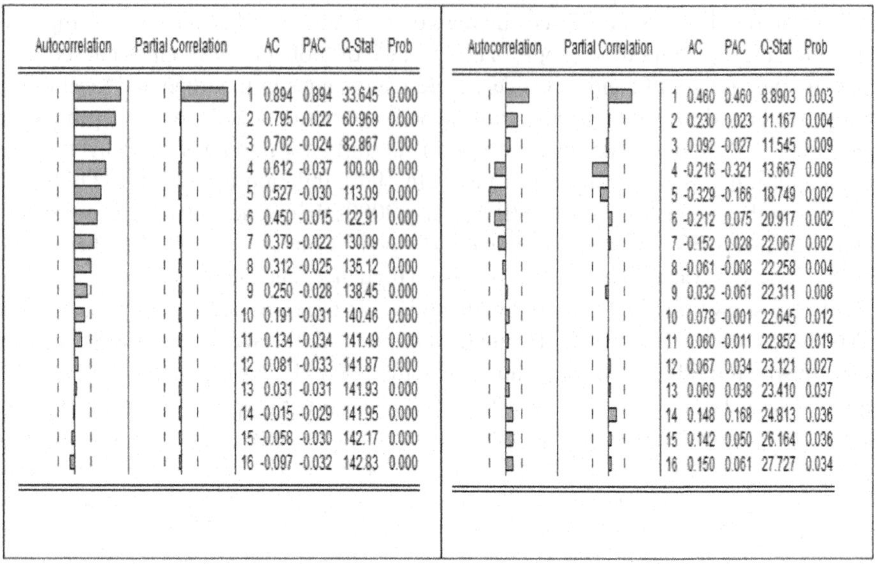

| (a) | (b) |

Fig. 2. a). Correlogram plot of EPF. b). Correlogram plot of INF.

Table 1. Probability values from ADF and PP unit root tests.

Variables	ADF				PP			
	Constant		Constant & Trend		Constant		Constant & Trend	
	Level	1st diff	Level	1st diff	Level	1st diff	Level	1st diff
LEPF	0.2093	**0.0104**	**0.0104**	0.8461	**0.0000**	0.0104	0.1559	**0.0245**
LINF	**0.0029**	0.0003	**0.0147**	0.0018	**0.0037**	0.0000	**0.0178**	0.0000
LGDP	0.8846	**0.0001**	0.6761	**0.0011**	0.8836	**0.0001**	0.5570	**0.0011**
LEXP	0.8126	**0.0291**	**0.0008**	0.4552	**0.0078**	0.0291	**0.0012**	0.3310
LGINI	0.8282	**0.0002**	0.7576	**0.0012**	0.8027	**0.0002**	0.7576	0.0013

Table 1 shows the probability value of the unit root test for each variable. At the 5% significance level, the results show that LINF is already stationary at level form, supporting the earlier findings. The other variables are only stationary after first differencing, signifying that they are integrated of order 1. The results show that there are two integration orders at I(1) and I(0), and importantly, no variable is integrated at I(2), suggesting that an ARDL model is suitable for this study.

3.3   Testing for the Presence of Co-integration

The next procedure is to find out whether there is a long-run relationship between the total EPF wealth and its determinants. Since the variables in this study are not integrated of the same order, the traditional co-integration approaches by Engle and Granger (1987) and Johansen and Juselius (1992) were not employed in this analysis. This study adopted the ARDL bounds approach by Pesaran and Shin (1999) and Pesaran (2001). To do this, first, an optimal lag length was selected based on the lowest error measure criteria: log-likelihood, likelihood ratio, final prediction error, AIC, Schwarz criteria, and Hannan-Quinn criteria as **reported** in Table 2.

Table 2. Optimal lag selection criteria.

Lag	LogL	LR	FPE	AIC	SC	HQ
0	82.25001	NA	0.000579	-4.621213	-4.349120	-4.529662
1	94.17547	18.79164*	0.000299*	-5.283362*	-4.965921*	-5.176553*
2	95.08752	1.381881	0.000302	-5.278031	-4.915242	-5.155964
3	95.47747	0.567200	0.000314	-5.241059	-4.832920	-5.103732
4	95.95998	0.672591	0.000326	-5.209696	-4.756208	-5.057111
5	96.06925	0.145692	0.000347	-5.155712	-4.656876	-4.987869

The output shows that most criteria opted for lag 1 as the optimal lag length. The next step is determining the order of ARDL $(P_1, P_2, P_3, P_4, P_5)$. Based on the analysis, ARDL $(1,0,0,1,1)$ is chosen as an optimal model due to its lowest AIC value of -5.6603 among the 16 models. The results of the estimated ARDL $(1,0,0,1,1)$ model are reported in Table 3.

This study used a correlogram plot to check for the presence of autocorrelation. The plot shows that all spikes are bounded within the confidence bands, indicating no autocorrelation in the model's residuals. This is also supported by the p-value that exceeds the 5% significance level for each lag. The F-statistic from the Breusch-Godfrey Serial Correlation LM test gives a p-value of 0.2446, which is greater than the 5% significance level. Therefore, there is no serial correlation up to the higher order of 2. Table 4 shows the F-statistic results for testing the existence of a long-run co-integration. Since the F-statistic value of 36.20 exceeds the upper bound at the 5% significance level, we can strongly reject the null hypothesis of no long-run relationship and hence, reinforce our conclusion that there is evidence of long-run relationships between LEPF and its determinants, namely LINF, LGDP, LEXP, and LGINI.

As mentioned above, the long-run relationships were estimated based on the lowest AIC value. The results in Table 4 show that there are long-run equilibrium relationships between the variables in the model. LEXP and LGINI are found to be statistically significant at the 5% significance level and have negative signs. These variables will negatively impact the total EPF wealth in the long run. In the long run, a 1% increase in life expectancy will lead to a 2.04% decrease in the total EPF wealth, other factors being held constant. The total EPF wealth will reduce by 1.18% with every 1% increase in the

Table 3. ARDL (1,0,0,1,1) output.

Variable	Coefficient	Std. Error	t-Statistic	Prob.*
LEPF(-1)	0.872308	0.013524	64.49839	0.0000
LINF	0.000717	0.003481	0.205926	0.8382
LGDP	0.329989	0.047375	6.965508	0.0000
LLIFEXP	-7.309614	4.948179	-1.477233	0.1497
LLIFEXP(-1)	7.048872	4.968183	1.418803	0.1659
LGINI	-0.623283	0.162703	-3.830797	0.0006
LGINI(-1)	0.472255	0.166973	2.828327	0.0081

R-squared	0.999907	Mean dependent var	11.79855
Adjusted R-squared	0.999889	S.D. dependent var	1.248742
S.E. of regression	0.013148	Akaike info criterion	-5.660346
Sum squared resid	0.005359	Schwarz criterion	-5.358685
Log likelihood	114.5466	Hannan-Quinn criter.	-5.553018
Durbin-Watson stat	1.935314		

*Note: p-values and any subsequent tests do not account for model selection.

Breusch-Godfrey Serial Correlation LM Test:
Null hypothesis: No serial correlation at up to 2 lags

F-statistic	1.478974	Prob. F(2,29)	0.2446
Obs*R-squared	3.517185	Prob. Chi-Square(2)	0.1723

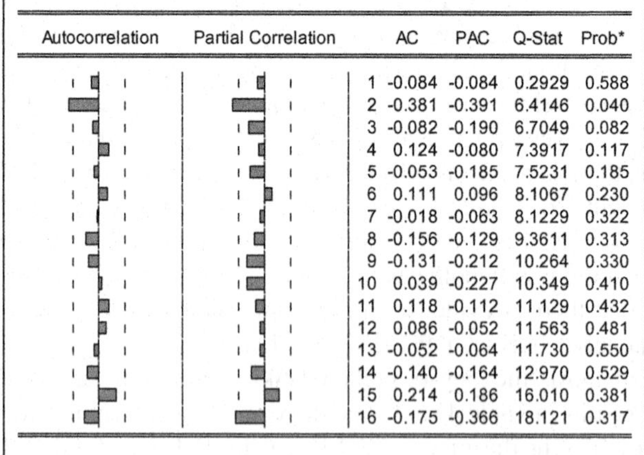

Autocorrelation	Partial Correlation		AC	PAC	Q-Stat	Prob*
		1	-0.084	-0.084	0.2929	0.588
		2	-0.381	-0.391	6.4146	0.040
		3	-0.082	-0.190	6.7049	0.082
		4	0.124	-0.080	7.3917	0.117
		5	-0.053	-0.185	7.5231	0.185
		6	0.111	0.096	8.1067	0.230
		7	-0.018	-0.063	8.1229	0.322
		8	-0.156	-0.129	9.3611	0.313
		9	-0.131	-0.212	10.264	0.330
		10	0.039	-0.227	10.349	0.410
		11	0.118	-0.112	11.129	0.432
		12	0.086	-0.052	11.563	0.481
		13	-0.052	-0.064	11.730	0.550
		14	-0.140	-0.164	12.970	0.529
		15	0.214	0.186	16.010	0.381
		16	-0.175	-0.366	18.121	0.317

Fig. 3. Correlogram of SACF of ARDL (1,0,0,1,1).

Table 4. Long-run coefficients.

— Variable	Coefficient	Std. Error	t-Statistic	Prob.
LINF	0.005614	0.027573	0.203609	0.8400
LGDP	2.584251	0.175026	14.76492	0.0000
LLIFEXP	-2.041954	0.876272	-2.330274	0.0265
LGINI	-1.182749	0.577669	-2.047452	0.0492

F-Bounds Test		Null Hypothesis: No levels relationship		
Test Statistic	Value	Signif.	I(0)	I(1)
		Asymptotic: n=1000		
F-statistic	36.19540	10%	1.9	3.01
k	4	5%	2.26	3.48
		2.5%	2.62	3.9
		1%	3.07	4.44

Gini coefficient, other factors being held constant. Thus, increases in these variables will cause the total EPF wealth to shrink considerably in the long run. LGDP is positively related to LEPF and statistically significant at the 5% significance level in the long run. In this context, a 1% increase in economic growth or GDP will increase the total EPF wealth by 3.27%, holding other factors constant. LINF is not statistically significant as the p-value is greater than the significance level. Based on the findings of this study, the total EPF wealth is significantly affected by economic growth, life expectancy, and income inequality in the long run. Thus, EPF members should consider increasing or at least maintaining their EPF wealth as their pensions will be derived from their EPF savings. A low level of EPF wealth will lead to a reduction in accumulated EPF savings. The existence of long-run relationships among these variables is also supported by other studies, such as Lee and Muhammed (2016), Estrada et al. (2017), Hassan and Othman (2018a, b).

As the variables are co-integrated over the long run as per Table 4, the next step is to estimate the short-run dynamics among the variables. This is also known as an error correction model, in which the term ect functions as the speed of adjustment parameter to bring the variables back to an equilibrium point. The results are presented in Table 5. The coefficient of the error-correction term, ect(-1), is negative, indicating convergence and a very significant value. This is what we would expect if there is co-integration between LEPF and its determinants. The magnitude of this coefficient implies that nearly 98% of any disequilibrium between LEPF and all of its determinants is corrected within one period (one year).

Table 5 shows the dynamic short-run results, where the current change in LEPF reacts to the past equilibrium error, current change in LDGP, current change in LGINI, and LEPF's past change. There is a 0.54% decline in the current change of EPF wealth with every 1% increase in the past change of income inequality in Malaysia. In contrast, a 1% increase in past change of GDP in the short run has a 0.37% positive effect on the

Table 5. Error correction model of ARDL(1,0,0,1,1).

Variable	Coefficient	Std. Error	t-Statistic	Prob.
D(LEPF(-1))	0.841442	0.075471	11.14920	0.0000
D(LINF)	-0.001810	0.002570	-0.704121	0.4870
D(LGDP)	0.373772	0.067422	5.543775	0.0000
D(LLIFEXP)	-8.881139	14.52463	-0.611454	0.5457
D(LLIFEXP(-1))	8.557158	15.54993	0.550302	0.5863
D(LGINI)	-0.543960	0.164209	-3.312608	0.0025
D(LGINI(-1))	0.215877	0.153147	1.409607	0.1693
ECT(-1)	-0.980368	0.196970	-4.977242	0.0000
R-squared	0.909112	Mean dependent var		0.115354
Adjusted R-squared	0.887173	S.D. dependent var		0.036381
S.E. of regression	0.012220	Akaike info criterion		-5.782633
Sum squared resid	0.004331	Schwarz criterion		-5.434326
Log likelihood	114.9787	Hannan-Quinn criter.		-5.659839
Durbin-Watson stat	1.703865			

current change of EPF wealth. Likewise, EPF wealth for the current year tends to partly derive some dominant characteristics from its wealth in the previous years.

3.4 Unit Root Test

The validity and robustness of the estimated equations are confirmed by employing relevant diagnostic tests such as cumulative sum (CUSUM), Breusch Godfrey serial correlation, Breusch Pagan test, and Jarque-Bera normality test. Parameter stability test, CUSUM, and cumulative sum of squares (CUSUMQ) were carried out, and the results are displayed in Fig. 4. As both plots seem to be lying within the 5% critical bound, the null hypothesis of the stability of the parameters cannot be rejected. In that sense, the estimated parameters do not have structural instability, thus they are said to be stable within the study period. As Fig. 5 shows, the p-value of the test is 0.3819, indicating that the homoscedasticity assumption is valid in the model. The normality test shows

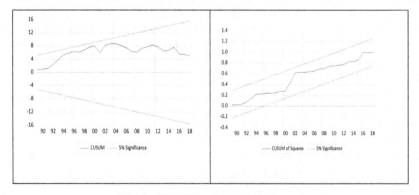

Fig. 4. CUSUM and CUSUMSQ tests.

that the estimated residuals are normal since the Jarque-Bera statistic is 0.3566, which is higher than the significance level 5%.

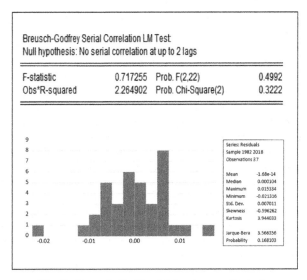

Fig. 5. Breusch-Godfrey LM and normality tests.

4 Conclusion

This study employed the ARDL bounds method to analyse EPF's wealth stability by estimating the long-run and short-run elasticity between EPF wealth and its potential determinants consisting of inflation rate, gross domestic product, life expectancy, and Gini coefficient. Each time series variable was checked using the unit root test, and it was found that they are integrated at order 0 and 1 and none is integrated at 2. Hence, the ARDL bounds test was employed in this study. The co-integration test revealed the existence of long-run and short-run relationships between EPF wealth and its determinants. The results imply that the explanatory variables, namely inflation rate, gross domestic product, life expectancy, and income inequality, collectively move together with EPF wealth to achieve a long-run steady state. GDP, GINI, and EXP are found to be important determinants of EPF wealth. Positive economic growth seems to have a positive effect on EPF wealth. On the contrary, increases in income inequality and life expectancy will adversely affect EPF wealth. For the short-run analysis, only GDP and Gini coefficient are statistically significant at the 5% level. It is worth noting that GDP and Gini coefficient are key driving factors in influencing the growth of EPF wealth. Based on these empirical analyses, this study suggests that improvement to the existing EPF scheme is important in view of the economic and demographic changes taking place in Malaysia. Sufficient pension benefits are required to maintain a certain level of comfort during retirement. For future works, it is recommended that the research can be further analysed with forecasting and using alternative enhanced ARDL models such as Multiscaled Neural ARDL (Saadaoui and Messaoud 2020) approach for comparison purposes.

Acknowledgements. The authors are grateful for the financial support from the Faculty of Computer and Mathematical Science, Universiti Teknologi MARA, Shah Alam.

References

Asher, M., Bali, A.S.: Public pension programs in Southeast Asia: an assessment. Asian Econ. Policy Rev. **10**(2), 225–245 (2015). https://doi.org/10.1111/aepr.12100

Barr, N., Diamond, P.: Pension Reform: A Short Guide. Pension Reform: A Short Guide (2010). https://doi.org/10.1093/acprof:oso/9780195387728.001.0001

Blake, D.: Does it matter what type of pension scheme you have? Econ. J. **110**(461), F46–F81 (2000)

Chung, C.: Cut in EPF can boost economy. The Star (February 2020). https://www.thestar.com.my/news/nation/2020/02/29/cut-in-epf-can-boost-economy

Engle, R., Granger, C.: Co-integration and error correction: representation estimation and testing. Econometrica **55**(2), 251–276 (1987)

Estrada, M., Khan, A., Staniewski, M., Mansor, N.: How inflation and the exchange rate affect the real value of pension plan systems : the case of Malaysia. SSRC Work.Pap. Ser. **2**(201), 1–27 (2017)

Gomez, O.C.: News:Malaysia's pension infrastructure lacks long-term sustainability, says Mercer. The Edge Markets, 26 November 2018. https://www.theedgemarkets.com/article/news-malaysias-pension-infrastructure-lacks-longterm-sustainability-says-mercer

Hassan, S., Othman, Z.: Determinants of employees provident fund in Malaysia: potential factors to jeopardize the EPF sustainability. Bus. Econ. Horiz. **14**(1), 29–42 (2018a). https://doi.org/10.15208/beh.2018.3

Hassan, S., Othman, Z.: Forecasting on the long term sustainability of the employees provident fund in Malaysia via the Box-Jenkins' ARIMA model. Bus. Econ. Horiz. **14**(1), 43–53 (2018b). https://doi.org/10.15208/beh.2018.4

Hussein, N.: The Malaysian pension system. Nomura Found. 15–20 (2019). http://www.nomurafoundation.or.jp/en/wordpress/wp-content/uploads/2019/03/NJACM3-2SP19-04.pdf

Hussin, S.S.: Employees Provident Fund (EPF) Malaysia: Generic models for asset and liability management under uncertainty. Dissertation. Brunel University, School of Information Systems, Computing and Mathematics (2012)

Hussin, S.S., Mitra, G., Roman, D., Kamaruzaman, W., Ahmad, W.: Employees' Provident Funds of Singapore, Malaysia, India and Sri Lanka: a Comparative Study. In: Mitra, G., Schwaiger, K. (eds.) Asset and Liability Management Handbook. Palgrave Macmillan, London (2011). https://doi.org/10.1057/9780230307230_8

Ja'afar, R., Daly, K. J.: Reviewing the financial soundness of the malaysia''s employees provident fund. Int. J. Humanit. Manag. Sci. (IJHMS) **4**(1), 14–18 (2016)

Ja'afar, R., Daly, K.J.: The sustainability of malaysia's defined contribution pension system: implementation of deterministic linear programming. Int. J. Innov. Technol. Explor. Eng. **8**(12S2), 97–103 (2019). https://doi.org/10.35940/ijitee.l1019.10812s219

Johansen, S., Juselius, K.: Testing structural hypotheses in a multivariate cointegration analysis of the PPP and the UIP for UK. J. Econom. **53**(1–3), 211–244 (1992)

Lee, H., Muhammed, A.K.: Is inequality in Malaysia really going down? Development Research Group (DECRG) Kuala Lumpur Seminar Series, 40 (2016)

Lee, K.C.: The Malaysian government pension scheme : whither its future direction ? Jurnal Ekonomi Malaysia **31**, 87–106 (1997)

Nurhisham, H.: The Malaysian pension system. Nomura J. Asian Cap. Mark. **3**(2), 15–20 (2019)

Pesaran, M.H., Shin, Y.: An autoregressive distributed-lag modelling approach to cointegration analysis. In: Strom, S. (ed.) Econometrics and Economic Theory in the 20th Century: The Ragnar Frisch Centennial Symposium, Cambridge University Press, pp. 371–413, Cambridge (1999). https://doi.org/10.1017/ccol521633230.011

Pesaran, M.H., Shin, Y., Smith, R.: Bounds testing approaches to the analysis of level relationships. J. Appl. Econ. **16**, 289–326 (2001)

Saadaoui, F., Messaoud, O.B.: Multiscaled neural autoregressive distributed lag: a new empirical mode decomposition model for nonlinear time series forecasting. Int. J. Neural Syst. **30**(8), 2050039 (2020)

Sulaiman, N., Mohammed, M.I.: Towards income sustainability among Elderly Malaysians : exploring the reverse mortgage option. Business (2016)

Data Mining and Image Processing

Iris Segmentation Based on an Adaptive Initial Contour and Partly-Normalization

Shahrizan Jamaludin[1]([✉]) [ID], Nasharuddin Zainal[2] [ID], W. Mimi Diyana W. Zaki[2] [ID], and Ahmad Faisal Mohamad Ayob[1] [ID]

[1] Faculty of Ocean Engineering Technology and Informatics, Universiti Malaysia Terengganu, 21030 Kuala Nerus, Terengganu, Malaysia
shahrizanj@umt.edu.my

[2] Department of Electrical, Electronic and Systems Engineering, Faculty of Engineering and Built Environment, Universiti Kebangsaan Malaysia, UKM, 43600 Bangi, Selangor, Malaysia

Abstract. Active contour is accurate for iris segmentation on the non-ideal and noisy iris images. However, understanding on how active contour reacts to the motion blur or blurry iris images is presently unclear and remains a major challenge in iris segmentation perspective. Moreover, studies on the initial contour position in the blurry iris images are infrequently reported and need further clarification. In addition, convergence or evolution speed is still a major drawback for active contour as it moves through the boundaries in the iris images. Based on the above issues, the experiment is conducted to obtain an accurate and fast iris segmentation algorithm for the blurry iris images. The initial contour is also investigated to clarify its positioning for the blurry iris segmentation. To achieve these objectives, the Wiener filter is used for pre-processing. Next, the morphological closing is applied to eliminate reflections. Then, the adaptive Chan-Vese active contour (ACVAC) algorithm is designed from the adaptive initial contour (AIC), δ and stopping function. Finally, the partly-normalization is designed where only prominent iris features near to the inner iris boundary are selected for normalization and feature extraction. The experimental results show that the proposed algorithm achieves the highest segmentation accuracy and the fastest computational time than the other active contour-based methods. The accurate initial contour position in the blurry iris images is clearly clarified. This shows that the proposed method is accurate for iris segmentation on the blurry iris images.

Keywords: Iris segmentation · Adaptive Chan-Vese Active Contour · Adaptive initial contour · Segmentation accuracy · Computational time

1 Introduction

The sales of biometric system have continuously increased to meet the global market demands. The biometric system is preferred because of non-transferrable [1], fraud resistant [2] and very convenient [3] compared to the conventional system. The contact and contactless biometric systems can provide an acceptable identification [4] and verification [5]. The contact biometric system such as fingerprint is easy to use [6] with a

© Springer Nature Singapore Pte Ltd. 2021
A. Mohamed et al. (Eds.): SCDS 2021, CCIS 1489, pp. 231–240, 2021.
https://doi.org/10.1007/978-981-16-7334-4_17

high accuracy [7]. Moreover, the contactless biometric system such as iris recognition is expected to witness a boost in demand due to the COVID-19 pandemic. Iris recognition can maintain personal hygiene [8] and can reduce virus transmission since it can be used from a distance [9]. Moreover, iris recognition has many advantages such as spoof-proof [10], difficult to forge [11], high scalability [12], accurate matching [13] and natural protection from harsh environment [14]. This system has a huge database of iris information which can be obtained and accessed from the various government agencies [15].

Recently, there are many active contour methods that have been developed for iris segmentation. Chang et al. [16] developed the geodesic active contour (GAC) with innovative algorithms to segment the accurate iris region. Pupil was segmented with GAC, to obtain the correct center and radius of pupil. Meanwhile, Abdullah et al. [17] used the gradient vector flow (GVF) active contour for iris segmentation. A new pressure force was added to the existing GVF active contour to create a fusion of expanding and shrinking active contours.

The active contour models can achieve good results in terms of segmentation accuracy and recognition accuracy, when the iris boundary is not symmetrical in the non-ideal environment. Moreover, the accurate iris region occluded by eyelash, eyelid and reflection can also be detected. However, understanding on how active contour reacts to the motion blur or blurry iris images instead of various noises is presently unclear and remains a major challenge in iris segmentation perspective. Moreover, studies on the initial contour position in the blurry iris images are infrequently reported and need further clarification. In addition, convergence or evolution speed is still a major drawback for active contour as it moves through the boundaries in the iris images.

Based on the above issues, the experiment reported in this paper is conducted to obtain an accurate and fast iris segmentation algorithm for the blurry iris images. The initial contour is also investigated to determine and clarify its positioning for the blurry iris segmentation.

The rest of the paper is organized as follows: Sect. 2 describes the proposed algorithm. Section 3 presents and discusses the obtained results. Finally, Sect. 4 draws the conclusion.

2 Methodology

2.1 Pre-processing

The proposed work is based on the Chan-Vese (CV) active contour [18] since it is less dependent on edge detection. The proposed adaptive CV active contour (ACVAC) algorithm focuses on segmenting the precise iris region in the blurry iris images.

The blurry images happen because of disturbance during image acquisition such as camera or object movement, inappropriate capturing time, scattered light and poor focus. Moreover, this phenomenon is frequently observed in the non-ideal iris images and unconstrained environment. Because of that, pre-processing is needed to improve the quality of blurry iris images.

In this work, the Wiener filter method as reported in [19, 20] is used to improve the quality of iris images. This method has a better performance than the Lucy-Richardson algorithm and blind deconvolution algorithm for image deblurring. Firstly, the point spread function is configured from the optical system light by calculating the width of blurry pixels in the iris image. Then, the constraint of point spread function is applied with the Gaussian low-pass filter. This will allow Wiener filter to produce the better iris textures since it can adapt its mechanism to the blurry environment specified by the point spread function. Moreover, the important characteristics in the image can be preserved.

2.2 Adaptive Initial Contour (AIC)

However, pre-processing alone is not able to deblur all iris textures in the blurry iris image. Because of that, the ACVAC algorithm with AIC is proposed to segment the precise iris region in the blurry iris images.

Firstly, the pupil region is segmented by calculating all pixel values in the contiguous and discontiguous regions in the iris image. The specific pupil threshold is assigned which is obtained from the pre-test on 100 iris images. The pre-test is also conducted to investigate the information of adaptable radius variable and convergence threshold. After that, the obtained information is analyzed to determine all connected components in the image. The obtained connected components are sorted to find out the precise location of pupil region in the iris image. The biggest connected component is considered as the pupil region. The centroid and radius information of pupil region are also recorded. The false detection of eyelash region as the precise pupil region can be avoided due to the assigned pupil threshold.

On the other hand, reflections in iris image are also detected by the above method. The detected reflections can be eliminated with the morphological closing. Firstly, the iris image is complemented. The reflections are represented with a dark color. The morphological closing will enlarge and connect all bright regions. Then, the dark regions are filled with pixels of their nearest neighbors, thus the reflections are eliminated.

Most of the recent active contour-based algorithms [16, 17, 21] use active contour for both pupil segmentation and iris segmentation which is a time-consuming process. Hence, the proposed method uses the centroid and radius information that are obtained previously to create an initial contour. On the other hand, the false segmentation can happen if the evolution curve reaches the prominent edges in the iris image. The upper eyelash has rich textures and prominent edges, thus the evolution curve may spread along the eyelash and stop prematurely. Because of that, the initial contour must not solely depend on the centroid of pupil region since it can intercept with the eyelash [22]. In this work, the AIC is designed where the pupil centroid is shifted away from the upper eyelash, hence the evolution curve can avoid reaching the prominent edges. The proposed AIC can be defined in (1):

$$AIC = \varepsilon r_p - \sqrt{(x - x_P)^2 + (y - y_P + k)^2} \tag{1}$$

where ε is an adaptable radius variable, which is obtained from the pre-test, r_p is the pupil radius, x_p and y_p are the coordinate of pupil centroid, and k is the y-axis value that is shifted away from the upper eyelash. The εr_p will create a region-of-interest (ROI) from the centroid of AIC, which can reduce the searching area in an iris image.

The length parameter μ is an important scaling role which can determine the length of evolution curve. This parameter must be set appropriately to prevent false iris segmentation. μ can accommodate the desired iris region by restraining the length of evolution curve. Interference evaluation factor is used in [21], where μ must be small to segment small objects such as eyelash, and μ must be large to segment large objects such as pupil and iris. Because of that, δ is introduced in this work to restrain the length of evolution curve from the upper eyelash, where μ will evolve only for the large object such as the iris region. It will not search the pupil region since it has been segmented previously. The formula of δ can be defined in (2):

$$\delta = \frac{\omega}{(N_{AIC} + \varepsilon r_p)N_{AIC}} \tag{2}$$

where ω is the number of counts of pixel 0 in the upper eyelash region, and N_{AIC} is a square ROI at the upper eyelash region based on the AIC. εr_p will ensure interceptions between the AIC and the square ROI, hence the length of evolution curve can be restrained from the upper eyelash.

The curve will evolve until it reaches the desired iris boundary. However, because of the eyelash interference, the evolution curve may not stop at the correct iris boundary, thus can produce a false segmentation. Due to that, the stopping function is designed to stop the evolution when it reaches the iris boundary. Two stopping functions are used in [21] for pupil and iris segmentations. However, in this paper, the stopping function is modified since only one active contour is used for iris segmentation. The stopping function can be defined in (3) and (4), where the curve will start evolving from εr_p.

$$\underset{c_1,c_2,C}{Inf} \left\{ F^{ACVAC}(c_1, c_2, C) \right\} \approx \delta \bullet \mu \bullet Length(C) \tag{3}$$

$$Length(C) = \frac{|Length(C(t+1)) - Length(c(t))|}{2} \tag{4}$$

The stopping function will ensure the evolution curve to stop evolving when *Length (C)* is smaller than the defined convergence threshold. The convergence threshold is obtained from the previous pre-test.

Finally, the proposed AIC, δ and stopping function are inserted into the ACVAC algorithm. The formula of ACVAC algorithm F^{ACVAC} can be defined in (5):

$$F^{ACVAC}(c_1, c_2, C) = \delta \bullet \mu \bullet Length(C) + v \bullet Area(inside(c))$$

$$+ \lambda_1 \int_{inside(C)} |\mu_0(x, y) - c_1|^2 dxdy$$

$$+ \lambda_2 \int_{outside(c)} |\mu_0(x, y) - c_2|^2 dxdy \tag{5}$$

where constants $\lambda_1 = \lambda_2 = 1$, $v \geq 0$, and iteration threshold $= 25$.

2.3 Partly-Normalization

The widely used rubber sheet model can maintain a reference to the same iris region regardless of iris constriction or dilation [22]. However, this method will capture all features in the iris region including the less important features near to the outer iris boundary, which are not prominent enough for recognition [17].

In this work, the rubber sheet model is modified where only 70% of the iris region near to the inner iris boundary is selected for normalization, while the remaining 30% is ignored. Subsequently, the number of iris feature points to be mapped from the iris region is reduced. This is because the prominent iris features are located near to the inner iris boundary, instead of the outer iris boundary. Moreover, the iris features near to the outer iris boundary can be obstructed by the upper eyelash. Finally, the iris features in the normalized iris image are extracted with 1D Log-Gabor filter [17].

3   Results and Discussion

3.1 Pre-test on Iris Images

The pre-test is conducted before the main experiment to investigate the information of pupil threshold, adaptable radius variable and convergence threshold in selected 100 iris images of CASIA v4 database. After that, the average values of these parameters are inserted into the proposed ACVAC algorithm. Next, the ACVAC algorithm is experimented on 400 blurry iris images. These iris images are blurred to simulate the motion blur environment.

3.2 Adaptive Initial Contour (AIC)

The initial contour position in the blurry iris images must be accurate since the active contour is sensitive to the initialization [23]. If its position is inaccurate, then the segmentation accuracy and convergence time can be reduced. The examples of the accurate initial contour position in the blurry iris images are shown in Fig. 1.

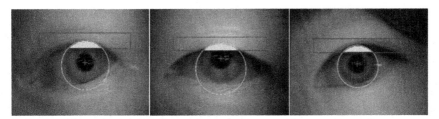

Fig. 1. The proposed AIC where the white-red circles are the proposed AIC, the red boxes are the square ROI, the green regions are the intercepted region, and the plus signs are the pupil centroid. (Color figure online)

From Fig. 1, it can be observed that the AIC is designed where its centroid has the similar x-axis value as the pupil region, while its y-axis value is shifted away from the upper eyelid region. Due to that, the AIC covers only some portions of the upper eyelid region which contains rich textures and prominent edges. In addition, the square ROI is based on the AIC centroid. The AIC will intercept with the square ROI, thus this intercepted region is excluded from the segmentation. During iris segmentation, the evolution curve will search the correct iris boundary on the lower eyelid until it reaches the intercepted region. Consequently, the curve will not spread along the upper eyelid and eyelash. This finding is in line with [22] which states that the initial contour is more accurate if it can avoid the rich edges and energy levels. The outcomes of the proposed segmentation with AIC, δ and stopping function can be observed in the next subsection.

3.3 Iris Segmentation on Blurry Iris Images

The segmentation results on selected blurry iris images are demonstrated in Fig. 2.

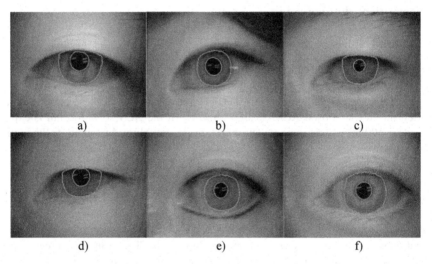

Fig. 2. The ACVAC segmentation results on selected blurry iris images.

From Fig. 2, it can be observed that the proposed algorithm manages to segment the accurate iris regions from the blurry iris images. Figure 2(a) and Fig. 2(b) show the segmentation results on the blurry and heavy eyelash occlusions. Figure 2(c) and Fig. 2(d) show the segmentation results on the blurry and heavy eyelid occlusions. Figure 2(e) and Fig. 2(f) show the segmentation results on the double eyelids occlusions. The proposed algorithm can distinguish the correct upper and lower eyelids, pupil and sclera, with the proposed AIC, δ and stopping function. The AIC position in the blurry iris images is accurate as mentioned in Subsect. 4.2. Meanwhile, δ manages to restrain the length of evolution curve, where the upper eyelash and pupil region are not included into the searching area by the interceptions of AIC and square ROI. Only the lower eyelid is included into the searching area since it has fewer rich textures. On the other hand, the stopping function manages to stop the evolution curve at the correct iris boundary, when *Length (C)* is smaller than the convergence threshold. Because of above factors, a separate eyelid detection is not needed. This finding is in line with the work reported in [17], which states that the abovementioned parameters can improve segmentation accuracy on the non-ideal and noisy iris images. However, the effect of these parameters on the blurry iris images is not being mentioned.

Moreover, the average segmentation accuracy obtained by the proposed algorithm is better than the other active contour-based methods, as shown in Table 1.

Table 1. Comparison of segmentation accuracy and computational time.

Method	Average Segmentation Accuracy (%)	Average Computational Time (s)
[16]	76.1	1.90
[17]	77.6	2.31
[21]	81.2	1.81
[22]	48.6	0.71
Proposed method	97.2	0.65

From Table 1, the proposed method achieves 97.2% segmentation accuracy. The closest one is method [21] where it achieves 81.2% segmentation accuracy. Combining the above segmentation accuracy results, several factors are discussed. The initial contour in [21] was designed with the pixel gray information. This method can achieve satisfactory results on the non-ideal and noisy iris images. However, the results are unsatisfactory if implemented on the blurry iris images since the initial contour position is less accurate to detect the blurry upper eyelid boundary. Moreover, the stopping function may not be sufficient for the blurry iris images, hence the evolution curve spreads along the upper eyelid and upper eyelash. Method [22] used the CV active contour for iris segmentation. The initial contour position is created by assuming the iris radius is three times of the pupil radius. However, it achieves the lowest segmentation accuracy since it only focuses on the lower iris region. Meanwhile, method [16] created the initial contour from the combination of GAC, Canny edge detection and Hough transform. This method can segment the non-ideal iris images. However, it can detect the hidden circles in the blurry iris image, which can produce a false iris circle estimation. Method [17] used the GVF active contour for iris segmentation. The initial contour is optimized from the initial curve sensitivity and failure detection of non-convex object. However, the initial contour position only focuses on the non-ideal iris segmentation.

Several factors may contribute to the highest segmentation accuracy of the proposed algorithm when it is implemented on the blurry iris images. Firstly, the designed AIC allows accurate initialization for the proposed ACVAC algorithm. This is in line with [24], where an accurate initialization can increase the segmentation accuracy. Secondly, the designed δ restrains the length of evolution curve, thus can create an efficient convergence mechanism. This finding is in line with [25], which states that the curve length must be controlled to obtain a smooth boundary. Thirdly, the modified stopping function allows the curve to stop evolving when it reaches the correct iris boundary.

3.4 Computational Time

Table 1 shows that the proposed method is superior in terms of computational time than the other active contour-based methods. The proposed algorithm has fewer operations for its implementation. It uses Wiener filter, morphological closing and one active contour for iris segmentation. Furthermore, the initial contour position also contributes to the fast computational time. An accurate initial contour can reduce iteration for boundary convergence. This is in line with [26], which states that the small iteration can reduce execution time for iris segmentation. On the other hand, the partly-normalization reduces points to be mapped from the segmented iris region, thus less computational time when dealing with blurry iris images. This is in line with [22], which states that fewer processes for normalization can improve the performance of algorithm.

Moreover, the circle detection algorithms such as integro-differential operator and Hough transform are time-consuming process [27]. Methods in [16, 17] used the circle detection algorithm in their implementation. Meanwhile, the proposed method does not use any circle detection algorithm, hence it is faster than those methods. On the other hand, method [21] used no circle detection algorithm. However, it used two active contours for segmentation, which requires more computational time than one active contour. Method [22] achieved almost similar time with the proposed method since it segments only small portions of iris region. It used the morphological closing and flood-fill for pre-processing, and one active contour for iris segmentation.

4 Conclusion

In this paper, the adaptive Chan-Vese active contour (ACVAC) model based on the adaptive initial contour (AIC) and partly-normalization has been proposed to segment accurate iris regions in the blurry iris images. The proposed method consists of pre-processing, AIC and partly-normalization. For pre-processing, the Wiener filter and morphological closing were applied to deblur the blurry iris images and eliminate reflections. However, pre-processing alone cannot deblur all iris textures in the blurry iris image. Because of that, the AIC, δ and stopping function were designed to improve the segmentation accuracy on the blurry iris images. The AIC was designed where it covers only some portions of rich textures and prominent edges on the upper eyelid region, hence can achieve robust segmentation and fast computational time. Then, δ was employed to restrain the length of evolution curve, where the intercepted region of AIC and square ROI is excluded from segmentation, thus can obtain a smooth iris boundary. Finally, the partly-normalization was designed where only prominent iris features near to the inner iris boundary are selected for normalization and feature extraction. This can reduce points to be mapped from the segmented iris region, hence less computational time.

The experimental results showed that the proposed algorithm achieves the highest segmentation accuracy and the fastest computational time than the other active contour-based methods. The accurate initial contour position on the blurry iris images was clearly clarified. The convergence speed was improved by employing fewer pre-processing methods, one active contour and partly-normalization for iris segmentation. However, the highly computational cost of active contour is well known. For future work, the proposed method can be designed with the graphic processing unit (GPU) to speed up its computational time.

Acknowledgement. This study uses the CASIA v4 database collected by the Chinese Academy of Sciences' Institute of Automation (CASIA).

References

1. Sarier, N.D.: Comments on biometric-based non-transferable credentials and their application in blockchain-based identity management. Comput. Secur. **105**, 102243 (2021)
2. Shrivastava, H., Tcheslavski, G.V.: On the potential of EEG for biometrics: combining power spectral density with a statistical test. Int. J. Biom. **10**(1), 52–64 (2018)
3. Hossain, M.S., Balagani, K.S., Phoha, V.V.: Effectiveness of symmetric rejection for a secure and user convenient multistage biometric system. Pattern Anal. Appl. **24**(1), 49–60 (2020). https://doi.org/10.1007/s10044-020-00899-0
4. Anne, N., et al.: Feasibility and acceptability of an iris biometric system for unique patient identification in routine HIV services in Kenya. Int. J. Med. Inform. **133**, 104006 (2020).
5. Blasco, J., Peris-Lopez, P.: On the feasibility of low-cost wearable sensors for multi-modal biometric verification. Sensors **18**(9), 2782 (2018)
6. Amreen, S., Mockus, A., Zaretzki, R., Bogart, C., Zhang, Y.: ALFAA: active learning fingerprint based anti-aliasing for correcting developer identity errors in version control systems. Empir. Softw. Eng. **25**(2), 1136–1167 (2020)
7. Alsmirat, M.A., Al-Alem, F., Al-Ayyoub, M., Jararweh, Y., Gupta, B.: Impact of digital fingerprint image quality on the fingerprint recognition accuracy. Multimed. Tools Appl. **78**(3), 3649–3688 (2018). https://doi.org/10.1007/s11042-017-5537-5
8. Jamaludin, S., Azmir, N.A., Ayob, A.F.M., Zainal, N.: COVID-19 exit strategy: transitioning towards a new normal. Ann. Med. Surg. **59**, 165–170 (2020)
9. Zhang, M., He, Z., Zhang, H., Tan, T., Sun, Z.: Toward practical remote iris recognition: a boosting based framework. Neurocomputing **330**, 238–252 (2019)
10. Kaur, B.: Iris spoofing detection using discrete orthogonal moments. Multimed.Tools Appl. **79**(9–10), 6623–6647 (2019). https://doi.org/10.1007/s11042-019-08281-x
11. Cohen, F., Sowmithran, S., Li, C.: 3D iris model and reader for iris identification. Concurr. Comput. Pract. Exp. **33**(12), e5653 (2021)
12. Shin, J., Kim, T., Lee, B., Yang, S.: IRIS-HiSA: highly scalable and available carrier-grade SDN controller cluster. Mob. Netw. Appl. **22**(5), 894–905 (2017)
13. Wang, K., Kumar, A.: Cross-spectral iris recognition using CNN and supervised discrete hashing. Pattern Recogn. **86**, 85–98 (2019)
14. Sujatha, E., Chilambuchelvan, A.: Multimodal biometric authentication algorithm using iris, palm print, face and signature with encoded dwt. Wirel. Pers. Commun. **99**(1), 23–34 (2018)
15. Chen, Y., Wu, C., Wang, Y.: T-center: a novel feature extraction approach towards large-scale iris recognition. IEEE Access **8**, 32365–32375 (2020)
16. Chang, Y.-T., Shih, T.K., Li, Y.-H., Kumara, W.G.C.W.: Effectiveness evaluation of iris segmentation by using geodesic active contour (GAC). J. Supercomput. **76**(3), 1628–1641 (2018). https://doi.org/10.1007/s11227-018-2450-2
17. Abdullah, M.A., Dlay, S.S., Woo, W.L., Chambers, J.A.: Robust iris segmentation method based on a new active contour force with a noncircular normalization. IEEE Trans. Syst. Man Cybern. Syst. **47**(12), 3128–3141 (2016)

18. Chan, T.F., Vese, L.A.: Active contours without edges. IEEE Trans. Image Process. **10**(2), 266–277 (2001)
19. Jamaludin, S., Zainal, N., Zaki, W.M.D.W.: Deblurring of noisy iris images in iris recognition. Bull. Electr. Eng. Inform. **10**(1), 156–159 (2021)
20. Baselice, F., Ferraioli, G., Ambrosanio, M., Pascazio, V., Schirinzi, G.: Enhanced wiener filter for ultrasound image restoration. Comput. Methods Progr. Biomed. **153**, 71–81 (2018)
21. Chen, Y., Liu, Y., Zhu, X.: Robust iris segmentation algorithm based on self-adaptive chan-vese level set model. J. Electron. Imaging **24**(4), 043012 (2015)
22. Jamaludin, S., Zainal, N., Zaki, W.M.D.W.: Sub-iris technique for non-ideal iris recognition. Arab. J. Sci. Eng. **43**(12), 7219–7228 (2018)
23. Duan, Y., Peng, T., Qi, X.: Active contour model based on LIF model and optimal DoG operator energy for image segmentation. Optik **202**, 163667 (2020)
24. Ding, K., Xiao, L., Weng, G.: Active contours driven by region-scalable fitting and optimized Laplacian of Gaussian energy for image segmentation. Signal Process. **134**, 224–233 (2017)
25. Jin, R., Weng, G.: A robust active contour model driven by fuzzy c-means energy for fast image segmentation. Digit. Signal Process. **90**, 100–109 (2019)
26. Fang, J., Liu, H., Zhang, L., Liu, J., Liu, H.: Active contour driven by weighted hybrid signed pressure force for image segmentation. IEEE Access **7**, 97492–97504 (2019)
27. Djekoune, A.O., Messaoudi, K., Amara, K.: Incremental circle hough transform: an improved method for circle detection. Optik **133**, 17–31 (2017)

Identifying the Important Demographic and Financial Factors Related to the Mortality Rate of COVID-19 with Data Mining Techniques

Nur Sara Zainudin[1] (ID), Keng-Hoong Ng[1]([⊠]) (ID), and Kok-Chin Khor[2] (ID)

[1] Faculty of Computing and Informatics, Multimedia University, Cyberjaya, Malaysia
khng@mmu.edu.my

[2] Lee Kong Chian Faculty of Engineering and Science, Universiti Tunku Abdul Rahman, Bandar Sungai Long, 43000 Kajang, Malaysia
kckhor@utar.edu.my

Abstract. The whole world has been greatly affected by the recent emergence of the COVID-19 pandemic since December 2019, and the death toll has reached millions. Thus, this problem needs to be addressed and mitigated immediately. In this study, the primary objective is to determine the factors affecting the mortality rate of COVID-19 in demographic and financial factors. This study utilised supervised learning methods with feature selection methods: filter and wrapper, to identify factors attributed significantly to the Case Fatality Ratio (CFR), a measure for mortality. The result showed that the wrapper method running K-Nearest Neighbour with the Sequential Forward Selection produced the feature subset that gave the best result. The feature selection results also suggest that the factor - household debt is the key to affecting the mortality rate of this infectious disease.

Keywords: COVID-19 · Mortality rate · Classification · Regression · Feature selection

1 Introduction

Countries worldwide are affected by COVID-19 caused by the new coronavirus, Severe Acute Respiratory Syndrome Coronavirus 2 (SARS-CoV-2). The illness has been listed as a pandemic by the World Health Organization (WHO) on 11th March 2020 [1]. Common symptoms of COVID-19 include but are not limited to fever, cough, sore throat, breathlessness, fatigue, etc. However, some individuals diagnosed with the disease were asymptomatic or showed no symptoms [2]. The virus spreads through droplets caused by sneezing or coughing and is caught either by inhalation or coming into contact with contaminated surfaces and later touching the T-zone, consisting of nose, mouth, and eyes [3].

The Department of Operational Support [4] offers operational support to United Nations Secretariat entities and reports that about one in every six people with COVID-19 become seriously ill and develop severe symptoms, such as shortness of breath and persistent chest pains. Severe lung damage leads to Acute Respiratory Distress Syndrome

© Springer Nature Singapore Pte Ltd. 2021
A. Mohamed et al. (Eds.): SCDS 2021, CCIS 1489, pp. 241–253, 2021.
https://doi.org/10.1007/978-981-16-7334-4_18

(ARDS), and it is the primary causal agent resulting in death [5]. Even now, the total deaths caused by the COVID-19 pandemic continue to grow, and countries worldwide have to enforce preventive and mitigation measures to stop the number from growing. As such, we must determine the factors related to this pandemic's mortality rate, which is the project's main objective. In particular, the project emphasises identifying the financial factors and demographic factors correlated to the mortality rate of COVID-19 using data mining techniques. We compiled data from reliable sources, such as World Bank, the International Monetary Fund (IMF), and Trading Economics. We employed data mining techniques, i.e., supervised learning and feature selection methods, on the compiled data to achieve the objective.

The paper is followed by the Literature Review section, which reviews the data mining techniques and related work. Then, we continue to describe the data mining techniques used in the Methodology section. The results of the experiments we obtained are discussed in the Results & Discussion section. Finally, we conclude our study in the Conclusion & Future Work section.

2 Literature Review

2.1 Related Work

Research related to the project were reviewed and summarised below. The research is closely related to factors that affect the mortality rate of COVID-19.

According to [6], the researchers believe that other than biological and epidemiological factors, socioeconomic factors also affect the outcome of the coronavirus pandemic. They collected data from 96 countries with 29 determinants that are all related to socioeconomic characteristics. Two distinct models were created using Bayesian model averaging (BMA); one model related the variables to the log of coronavirus cases per million population. The second model related the same variables to the log of coronavirus deaths per million population. The interesting model is the second model, which shows the important determinants of the log of the mortality rate due to COVID-19. BMA calculates the posterior probability that determines the most valuable variables. The results were then arranged based on their posterior inclusion probabilities (PIP) to show the importance of each variable. PIP analyses the significance of single determinants of the models. In their finding, no determinant is a significant factor to the coronavirus deaths per million population. However, population size and government health expenditure are strongly related to the number of coronavirus cases per million population.

In the study by Goutte, Péran and Porcher [7], economic structural factors, such as unemployment and poverty rates, are significant in determining the mortality rate of COVID-19. The study started with 66 features related to economic, financial and structural factors of a highly populated region in France, Île-de-France, which includes its eight administrative divisions, such as Yvelines and Paris. Principal component analysis (PCA) was used to narrow down the scope of features. After analysing, it is found that features such as unsuitable housing (e.g. overcrowded houses, household size) and economic precariousness indicators (e.g. poverty rate, no or little graduate in the workforce) are likely to have higher mortality.

In another study, Prakash et al. [8] showed that machine learning techniques could predict COVID-19 better. The authors used a data set containing seven different independent features regarding the pandemic in India. Different machine learning models were tested, i.e., Support Vector Machine (SVM), K-Nearest Neighbors using Neighbourhood Components Analysis (KNN + NCA), Decision Tree Classifier, Gaussian Naïve Bayes Classifier, Multilinear Regression, Logistic Regression, Random Forest and XGBoost Classifier. A correlation matrix was built on understanding the relationships between the features inside the data set better. Additionally, the feature importance score for each classifier was computed. From there, the classifiers were used to predict the confirmed cases of COVID-19. The Random Forest classifier and Random Forest regressor outperformed the rest of the tested machine learning models.

We believe that it is important to study possible factors from different perspectives, especially in demographic and financial factors related to the mortality rate of COVID-19. Moreover, we also took a different approach from similar work to predict the mortality rate of COVID-19. The research conducted used feature selection to identify important factors and supervised learning models for prediction.

2.2 Feature Selection

Feature selection is the process of selecting a small subset of features from a data set that contributes most to prediction. Irrelevant features create noise and lead to inaccurate results and longer training time. Feature selection provides many benefits, including improving learning accuracy, reducing training time, and achieving better model interpretability [9, 10]. According to Lotte et al. [11], the common and prominent methods in feature selection are filter, wrapper, and embedded. All three methods have their pros and cons. Selecting which method to use generally depends on the scenario and problems.

The filter feature selection method assigns a score to every feature based on statistical measures to ascertain its correlation with the outcome variable. With this, only useful features are selected while the rest are removed from the data set. On the other hand, the wrapper method finds the best set of features in a machine learning algorithm by considering its mining performance. With this method, combinations of features are created and evaluated according to their predictive accuracy. In the embedded feature selection, the core idea is that the knowledge regarding the features is already embedded in the problem's solution. Thus, the model performs feature selection during model training and shall use the pruned data set for training. Decision tree classifiers, such as ID3, include the embedded feature selection method in its algorithm.

Several search methods are used in feature selection methods to decide the feature set to evaluate, such as forward selection or backward selection. Forward selection would add good features to an empty list at each iteration, whereas backward selection removes the weak performing features from a complete set of features [12].

2.3 Supervised Machine Learning

Supervised machine learning trains models using annotated training data to make predictions [13]. Labels or classes provided in the training data help to improve the prediction of a model. Supervised machine learning is utilised for two problems which are regression problems and classification problems. Of the two problems mentioned, regression is part of supervised learning and is a process to find correlations between independent and dependent variables. The output is a continuous variable useful for certain cases, such as predicting the price of items or a person's weight. Another supervised learning algorithm is classification, whereby data gathered by a user is divided into distinct predetermined classes. There are binary classifiers that only classify two possible outcomes and multi-class classifiers that classify more than two outcomes [14].

3   Methodology

The project follows a series of experiment steps, which are similar to the data science pipeline. Figure 1 shows the overall task flow of the entire project and depicts the seven steps.

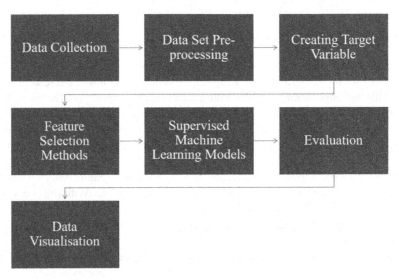

Fig. 1. The overall task flow in this study.

3.1   Data Collection

We collected the data via reliable online sources, such as World Bank, Trading Economics, and IMF. The data of 165 countries were compiled and merged in a CSV file as the final data set. For the COVID-19 data, the total number of death cases and the total number of recovered cases are up until 2nd January 2021. As we are studying the factors related to the mortality rate of COVID-19, we use the Case Fatality Ratio (CFR) as the measure for mortality. According to WHO [15], the value for CFR during an ongoing pandemic is calculated using Eq. 1. The final data set comprises the target variable -

CFR and 20 factors (features), of which 10 are demographic factors, and another 10 are financial factors (Table 1).

$$CFR(\%) = \frac{Number\ of\ deaths\ from\ disease}{Number\ of\ deaths\ from\ disease\ +\ Number\ of\ recovered\ from\ disease} \tag{1}$$

Table 1. The CFR and factors related to mortality, all numeric types, are included in the data set.

Target Variable	Variable Name
Case Fatality Ratio	CFR
Demographics Factors	**Feature Name**
Population age 65 + (% of total)	Population_65 2019
Population ages 0–14 (% of total)	Population_14 2019
Population, male (% of people ages 15 and above)	Population_Male 2019
Total Population	Total_Population 2019
Population Density (persons per square km)	Population_Density 2018
Life Expectancy	Life_Expectancy 2018
Literacy rate, adult total (% of people ages 15 and above)	Literacy_Rate 2015
Human Capital Index	HCI 2020
Prevalence of Tobacco Smoking	Smoking 2020
Prevalence of overweight (% Of Adults)	Overweight 2016
Financial Factors	**Feature Name**
Bottom 40% Income Share	Bottom_40_Income_Share 2019
Consumer Price Index	CPI 2019
Current Health Expenditure	Health_Expenditure 2017
GDP Growth (Annual %)	GDP_Growth 2019
GDP Per Capita	GDP_Per_Capita 2019
Household debt, loans, and debt securities (% of GDP)	Household_Debt 2018
International tourism, expenditures (current USD)	International_Tourism 2018
Trade (% of GDP)	Trade 2019
Transport Services (% of Service Exports, BoP)	Transport_Services 2019
Total Unemployment (% of Total Labor Force)	Unemployment_Total 2019

3.2   Data Set Pre-processing

The data pre-processing steps included filling the missing values and normalising the data using min-max normalisation. If there were a missing value, it would be replaced with the average value of the samples in the same column. Next, we applied data normalisation to the data to ensure each feature carried the same value range. This action prevents features with larger ranges from outweighing those with smaller ranges. Min-max normalisation, also called feature scaling, does linear transformation to scale the original data from 0 to 1 [16]. The following formula in Eq. 2 is applied to normalise the data. According to the formula, x is the value to be transformed into x_scaled. x belongs to a numeric feature where x_min and x_max are the minimum and maximum feature values, respectively.

$$x_{scaled} = \frac{x - x_{min}}{x_{max} - x_{min}} \qquad (2)$$

3.3 Creating Target Variable

The Case Fatality Ratio (CFR) was used as a target variable for regression algorithms, but extra steps were required to classify the countries for classification algorithms. We do not know the range for what is considered to be low or high CFR as the number of confirmed cases for certain countries is still increasing. Thus, the data was split according to the quartile values to create distinct classes for classification. As a result, there are four classes in total, which are Q1, Q2, Q3, and Q4 (Fig. 2).

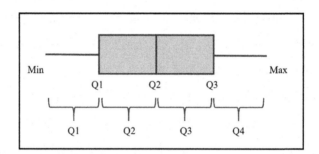

Fig. 2. The CFR is divided into four classes using quartile values: Q1, Q2, Q3 and Q4.

The class labels were decided by following these conditions:

1. "Q1": value is equal to or less than quartile 1 (Q1).
2. "Q2": value is more than quartile 1 (Q1) and less than or equal to quartile 2 (Q2).
3. "Q3": value is more than quartile 2 (Q2) and less than or equal to quartile 3 (Q3).
4. "Q4": value is more than quartile 3 (Q3)

3.4 Feature Selection Methods

We employed feature selection methods, namely filter and wrapper, in this study to determine the important factors related to the mortality rate of COVID-19. The classification result yielded using the feature sets was unsatisfactory, which shall be explained in the next section. Hence, only regression methods were tested for feature selection.

Three filter methods were used in the experiment; Pearson correlation coefficient, F-test, and Mutual Information. Each method has a statistical measure that is used to calculate the relative importance of a feature. Pearson correlation coefficient calculated the correlation coefficient between the factors related to mortality and the target variable, CFR. Similarly, F-test and Mutual Information measured their score for each feature. The factors were then ranked and arranged based on their relative importance in predicting the target variable. From there, we used Random Forest to test the performance using a different top "k" number of variables to select the best number of features. Random Forest Regressor was selected as the model for this test since it performed the best in predicting the CFR compared to the other regression models without feature selection. The full results of the regression models are detailed for comparison in the subsequent section (refer to Table 3). After selecting the best set of features, the parameters of the regression models were tuned and evaluated using the Root Mean Square Error (RMSE) score.

For wrapper methods, we utilised different regression algorithms together with four search methods: Sequential Forward Selection (SFS), Sequential Backward Selection (SBS), Sequential Forward Floating Selection (SFFS), and Sequential Backward Floating Selection (SBFS), to yield feature subsets.

3.5 Supervised Machine Learning Models

After feature selection, we predicted the CFR using classification and regression methods. Six models were tested for each method, and both methods included Random Forest, Decision Tree, K-Nearest Neighbors, Multilayer Perceptron, and Support Vector Machine. In addition, Linear Regression was also used to predict continuous variables for regression problems. Logistic regression, on the other hand, was added for classification problems.

Following model selection, we conducted parameter tuning to find the optimal parameters for each model. To accomplish that goal, we used grid search - an exhaustive search to identify the best set of parameters from a given search space. The search space defines the limits or possible values for the parameters to be tuned. However, this method takes more computation time due to its exhaustive nature, especially for random forest and decision tree algorithms. Thus, we first used randomised search to randomly test 100 different sets of parameters based on the search space for random forest and decision trees. For example, the random forest has continuous parameters such as n_estimators (The number of trees in the forest) and discrete parameters like criterion (the function to measure the quality of a split). The search space defined n_estimators from a range of 200 to 2000 and listed the possible options for criterion, mean squared error and mean absolute error. For randomised search, the search will randomly test possible sets of parameters like "n_estimators = 200 and criterion = mean squared error" or "n_estimators = 500 and criterion = mean absolute error". After randomly selecting 100 different combinations of parameters and testing them, we determined the best set of parameters. After determining the parameters, we were able to narrow the search space to values close to the range of values from the results of the randomised search. The reduced search space was used in grid search to find the best set of parameters.

We used PyTorch's Ax package to perform hyperparameter tuning for multilayer perceptron. Ax assists to find an optimal parameter value in the search space. To find the optimal value, Ax uses Bayesian and bandit optimisations for continuous parameters and discrete parameters, respectively. Both optimisation methods help reduce training time [17]. For the experiments, we set the number of trials to 150 and find the best set of parameters to use. Hence, Ax ran for training 150 times with different sets of parameters and returned the set of parameters that achieved the lowest loss.

3.6   Evaluation

We conducted Five-Fold cross-validation and used Root Mean Square Error and accuracy to evaluate regression and classification algorithms, respectively. RMSE and accuracy were recorded for each iteration.

3.7   Data Visualization

After ensuring that the results are satisfactory, we analysed our findings to gain more insight into the ongoing pandemic. Besides using tables, the results from the experimentation process were also presented using grouped bar charts.

4   Results and Discussion

We summarised and discussed the results of the experiments in this section. We tested classification algorithms first, followed by regression algorithms to observe how well the algorithms performed without feature selection. We then use feature selection to identify if we can achieve similar or better performance with fewer factors.

The classification was conducted using six different machine learning models: Random Forest Classifier (RF), Decision Tree Classifier (DT), K-Nearest Neighbours (KNN), Multilayer Perceptron (MLP), Support Vector Machine (SVM), and Logistic Regression (LR). The results using classification algorithms for all six methods were unsatisfactory, as shown in Table 2. The highest accuracy from the set of results was only around 27%, so we concluded that using classification methods is unsuitable for our problem. Hence, we proceed to test regression methods to determine if we can achieve better results.

Table 2. The accuracy score obtained using the classification algorithms. In general, they performed poorly in predicting CFR.

Algorithm	Accuracy
RF	0.242424
DT	0.248485
KNN	0.266667
MLP	0.254545
SVM	0.272727
LR	0.242424

As for regression, it was conducted using the same six machine learning algorithms. The regression algorithms using the full data set (20 factors) for all six methods yielded better RMSE values, as shown in Table 3.

Table 3. The RMSE scores obtained using the regression algorithms.

Algorithm	RMSE
RF	**0.113612**
DT	0.115016
KNN	0.115164
MLP	**0.113619**
SVM	0.115995
LR	0.135078

The two best performing models (Table 3) are Random Forest Regressor and Multi-layer Perceptron. Both achieved the lowest RMSE score of 0.1136. From these results, it is clear that this problem is better solved using regression algorithms than classification algorithms. Thus, the regression task was continued reducing the data sets using filter methods and wrapper methods (running different search methods). Both Table 4 and Fig. 3 illustrate the RMSE scores of all the algorithms using different feature selection methods.

For Fig. 3, note that the y-axis starts at 0.10 rather than 0. When comparing all of the methods, the feature selection methods reduced the number of features while still retaining good average RMSE scores. In Table 4, we noticed that KNN using the SFS method performed the best overall with the lowest average RMSE score of 0.1058 by including 13 variables (as shown in Table 5). However, Random Forest Regressors using SFS, SBS, SFFS, and SBFS gave an RMSE value of 0.1064 by using just one variable, Household_Debt 2018. Such RMSE value is very closed to the best RMSE score produced by KNN using the SFS method. Though the results are very similar, the number of variables chosen is very different since KNN using the SFS method, selected 12 more features than the aforementioned algorithms.

Table 4. The RMSE scores (with the number of features) of all methods tested. **KNN using the SFS method**, performed the best (the bold number).

Method	RF	DT	KNN	MLP	SVM	LR
Full data set (full features)	0.1136 (20)	0.115016 (20)	0.115164 (20)	0.1136 (20)	0.1160 (20)	0.1351 (20)
Pearson Correlation Coefficient	0.1127 (6)	0.1137 (6)	0.1071 (6)	0.1138 (6)	0.1160 (6)	0.1169 (6)
Mutual Information	0.1141 (13)	0.1470 (13)	0.1134 (13)	0.1146 (13)	0.1160 (13)	0.1210 (13)
F-test	0.1155 (2)	0.1150 (2)	0.1147 (2)	0.1140 (2)	0.1160 (2)	0.1158 (2)
SFS (Wrapper)	0.1064 (1)	0.1187 (1)	**0.1058 (13)**	0.1150 (3)	0.1148 (5)	0.1136 (1)
SBS (Wrapper)	0.1064 (1)	0.1064 (1)	0.1064 (8)	0.1143 (4)	0.1116 (3)	0.1142 (1)
SFFS (Wrapper)	0.1064 (1)	0.1064 (1)	0.1064 (11)	0.1135 (3)	0.1160 (7)	0.1136 (1)
SBFS (Wrapper)	0.1064 (1)	0.1064 (1)	0.1098 (10)	0.1661 (2)	0.1116 (3)	0.1142 (1)

From Table 5, we can identify the relevant features which play a role in predicting the CFR. Household_Debt 2018 is highlighted because it is the most frequently selected factor (16 times) by the feature selection methods compared to other features.

Our results may be explained using the research conducted by [18], which was conducted to study the relation between ethnicity and socioeconomic deprivation with COVID-19 mortality rates in England. Greater COVID-19 mortality is seen within communities experiencing income deprivation or financial hardship. Having a poor financial situation has been linked with increased mortality, and several studies back this claim. It is claimed that poor economic situations have negatively impacted an individual's health [19]. Acquiring high debt may lead individuals to overwork themselves and causes debt. Without sufficient financial security, the borrower may neglect to spend their limited income on health and medicines and may practice unhealthy coping mechanisms. Brzoska and Rasum [20] assessed the effects of unemployment, income, educational level, population density, regional prosperity, number of hospital beds, and the independent effect of indebtedness on mortality. The results of their research show that indebtedness and unemployment are moderately correlated with mortality. This shows how much of an impact debt has on the well-being of an individual.

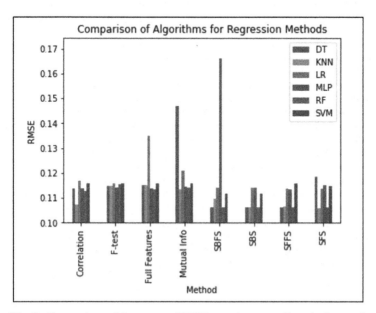

Fig. 3. Comparison of the average RMSE score between all methods tested.

Table 5. The factors (features) selected by the best performing method, KNN using SFS.

	Selected Factors	
CPI 2019	*GDP_Growth 2019*	*Health_Expenditure 2017*
Household_Debt 2018	*International_Tourism 2018*	*Life_Expectancy 2018*
Overweight 2016	*Population_14 2019*	*Population_Density 2018*
Population_Male 2019	*Smoking 2020*	*Total_Population 2019*
Trade 2019		

* represents the important factor that frequently selected by feature selection methods

5   Conclusion and Future Work

In this study, we aim to identify the main demographic and financial factors contributing to the mortality rate of COVID-19 from the data set prepared, which consists of 20 factors and the CFR. We evaluated several supervised learning models (regression and classification) with feature selection methods. Overall, the regression models performed better than the classification models. The best performing algorithm is KNN using the SFS method, which obtained an RMSE value of 0.1058. The algorithm selected 13 factors out of the 20 in total, as shown in Table 5. Out of the 13, Household_Debt 2018 appears to be a significant factor as it is the most frequently selected factor by the feature selection methods. The limitation of the study is the difficulty in collecting data due to unavailability of some factors data, particularly in demographic factors. Our future work may consider to study other factors related to health behaviours, genetics, and environment.

References

1. World Health Organization Declares COVID-19 a 'Pandemic.' Here's What That Means. https://time.com/5791661/who-coronavirus-pandemic-declaration/. Accessed 05 Feb 2021
2. Singhal, T.: A review of coronavirus disease-2019 (COVID-19). Indian J. Pediatr. **87**(4), 281–286 (2020)
3. Dhand, R., Li, J.: Coughs and sneezes: their role in transmission of respiratory viral infections, including SARS-CoV-2. Am. J. Respir. Crit. Care Med. **202**(5), 651–659 (2020)
4. Department of Operational Support. COVID-19 Frequently Asked Questions. https://www.un.org/sites/un2.un.org/files/new_dhmosh_covid-19_faq.pdf. Accessed 28 Jan 2021
5. Hasan, S.S., et al.: Mortality in COVID-19 patients with acute respiratory distress syndrome and corticosteroids use: a systematic review and meta-analysis. Expert Rev. Respir. Med. **14**(11), 1149–1163 (2020)
6. Stojkoski, V., Utkovski, Z., Jolakoski, P., Tevdovski, D., Kocarev, L. The socioeconomic determinants of the coronavirus disease (COVID-19) pandemic. arXiv preprint arXiv:2004.07947 (2020)
7. Goutte, S., Péran, T., Porcher, T.: The role of economic structural factors in determining pandemic mortality rates: evidence from the COVID-19 outbreak in France. Res. Int. Bus. Financ. **54**, 101281 (2020)
8. Prakash, K.B., Imambi, S.S., Ismail, M., Kumar, T.P., Pawan, Y.N.: Analysis, prediction and evaluation of COVID-19 datasets using machine learning algorithms. Int. J. Emerg. Trends Eng. Res. **8**(5), 2199–2204 (2020)
9. Miao, J., Niu, L.: A survey on feature selection. Procedia Comput. Sci. **91**, 919–926 (2016)
10. Ferreira, A.J., Figueiredo, M.A.: Efficient feature selection filters for high-dimensional data. Pattern Recogn. Lett. **33**(13), 1794–1804 (2012)
11. Lotte, F., et al.: A review of classification algorithms for EEG-based brain–computer interfaces: a 10-year update. J. Neural Eng. **15**(3), 031005 (2018)
12. Liu, H., Motoda, H.: Feature Selection for Knowledge Discovery and Data Mining (Vol. 454). Springer, US (2012)
13. Nasteski, V.: An overview of the supervised machine learning methods. Horizons. b, 4, 51–62 (2017)
14. Han, J., Pei, J., Kamber, M.: Data Mining: Concepts and Techniques. Elsevier, Amsterdam (2012)
15. World Health Organization. Estimating mortality from COVID-19. https://www.who.int/news-room/commentaries/detail/estimating-mortality-from-covid-19. Accessed 2020/11/20.
16. Ciaburro, G.: Regression Analysis with R: Design and Develop Statistical Nodes to Identify Unique Relationships Within Data at Scale. Packt Publishing Ltd., Birmingham (2018)

17. Bakshy, E., et al.: AE: a domain-agnostic platform for adaptive experimentation. In: Conference on Neural Information Processing Systems, pp. 1–8 (2018)
18. Rose, T.C., Mason, K., Pennington, A., McHale, P., Taylor-Robinson, D.C., Barr, B.: Inequalities in COVID19 mortality related to ethnicity and socioeconomic deprivation. MedRxiv (2020)
19. Turunen, E., Hiilamo, H.: Health effects of indebtedness: a systematic review. BMC Public Health **14**(1), 1–8 (2014)
20. Brzoska, P., Razum, O.: Indebtedness and mortality: analysis at county and city levels in Germany. Gesundheitswesen (Bundesverband der Arzte des Offentlichen Gesundheitsdienstes (Germany)) **70**(7), 387–392 (2008)

Local Image Analysis of Malaysian Herbs Leaves Using Canny Edge Detection Algorithm

Zuraini Othman[1]([✉]), Sharifah Sakinah Syed Ahmad[1], Fauziah Kasmin[1],
Azizi Abdullah[2], and Nur Hajar Zamah Shari[3]

[1] Fakulti Teknologi Maklumat dan Komunikasi, Universiti Teknikal Malaysia Melaka,
Hang Tuah Jaya, Durian Tunggal, 76100 Melaka, Malaysia
zuraini@utem.edu.my

[2] Center for Artificial Intelligence Technology, Faculty of Information Science and Technology,
Universiti Kebangsaan Malaysia, 43600 Bangi, Selangor Darul Ehsan, Malaysia

[3] Forestry and Environment Division, Forest Research Institute Malaysia (FRIM),
52109 Kepong, Malaysia

Abstract. Machine vision helps a lot with the latest recognition technology. To get the best recognition results, the initial phase of image processing should be done as best as possible. This phase involves the production of an image map of the resulting image using edge detection. The Canny method is often used because of its performance of producing meticulous strong edges but this method is sensitive to changes in image intensity when involving complex images such as image leaves and results in a lot of noise at the edges of the resulting image. This is because in this method the threshold value is selected empirically on the image globally. In this study, local image analysis will be discussed along with its impact on the resulting image edge results. In addition, a set of data from herbal leaves in Malaysia has also been produced by containing ground truth images for each herbal image available. The results from this study found that the locally analysing image approach has outperform the global approach of the conventional Canny method. Findings from this study may help the identification system in the future.

Keywords: Machine vision · Edge detection · Canny method · Local image analysis · Malaysian herbs leaves images

1 Introduction

Nowadays, herbs play a significant role in modern medical treatment as a source of biotechnology [1]. In several countries, such as India, Thailand, and Malaysia, most experts still classify plants using traditional methods based on the expert's knowledge. For example, in India, every region, whether urban or rural, is entirely reliant on plants for survival in the form of food, shelter, clothing, and medicines. Synthetic medicines have grown less economical because of inflation, and their side effects have led to the search for an alternative medical system. Many common disorders and diseases can be treated with Indian medicinal herbs [2].

© Springer Nature Singapore Pte Ltd. 2021
A. Mohamed et al. (Eds.): SCDS 2021, CCIS 1489, pp. 254–263, 2021.
https://doi.org/10.1007/978-981-16-7334-4_19

Most studies in Malaysia focus on herbs that are classified based on the scent, leaf shape, and colour of the leaves. Plant classification continues to be a fascinating topic for scientists. It takes a lot of effort and time to recognise the desired plant among thousands of others. As a result, using a vision system to identify plants is advantageous because the pharmacist and botanist do not have to collect them in the usual manner. As a basis, offering a database with information about these herbs is critical in assisting medicinal practitioners and users. This highlights the importance of creating databases and preservation zones where herb species can be cataloged and information methodically acquired to contribute valuable knowledge to society [3].

The initial phase of the study involving herbal images involved the image processing phase. Here, the image map that has been generated from the edge detection method will be used. This phase's success is critical to ensure the entire identification or classification process becomes accurate [4]. In the image of herbs leaves, the most commonly used edge detection method is the Sobel method [5] and Canny method [1, 3, 6–8]. The Canny approach has been utilised in many previous research because it produces a stronger edge with an exceptional outcome because it delivers the most edge details. Because it is resilient and accurate, the Canny operator unquestionably outperforms all other edge detection approaches. Canny enhanced edge information can aid in improving the quality of the edge image [1, 3, 6–8].

In the study of image, herbs leave images is a type of narrow-dominated image data. This narrow-dominated image data sets or also known as weakly segmented images, a globally defined threshold value will result in many of the actual edges not being detected. These narrowly dominated images have subtle and complex textures. A complex image like this uses the Canny method will give results in an image that detects many noises because the Canny method is sensitive to changes in intensity in the resultant image to produce edges. Such images require important local features in the determination of threshold values so that strong edge images can be produced [9]. Local threshold value feature search studies as conducted by [10] have introduced the fixed division method. This method was used by [11] by making the image division into 3×3 fixed division blocks in the image division and the study of [12] used 8×8 fixed division blocks.

In this paper, a study on local image analysis of Malaysian herbs leaves using canny edge detection algorithm had been discussed. In conjunction with this research, a data set of Malaysian herbal images had been created with a reference edge image. This paper consists of a detailed discussion of the materials and methods will be discussed in Sect. 2. It will be used for results and discussions in Sect. 3, as well as the conclusion of the study carried out will be given in Sect. 4 of this paper.

2 Materials and Methods

2.1 Canny Edge Detection Algorithm

Among the edge detection methods developed so far, Canny edge detection algorithm [13] is said to be a robust method that provides good and reliable detection. This is due to its ability to meet three criteria for edge detection and process simplicity for implementation which makes it one of the most popular algorithms for edge detection [14–16]. The process in this Canny edge detection algorithm consists of 5 following steps:

a) Smooth the image to eliminate noise using a Gaussian filter.
b) Find the intensity gradient on the image.
c) Eliminates false responses to edge detection using non-maximum suppression.
d) Determine the potential edge using two threshold values.
e) Detect edges hysterically by selecting edge detection by suppressing all other weak edges and not connecting with strong edges.

The selection of high and low threshold values can have a significant impact on the edge detection results from all the process involved. If the high threshold value setting is too large, many edge points will not be detected but if the setting is too small, many useless false edges will be detected. Therefore, in this study the local threshold values will be analysed compared to Canny method which uses empirical approach in finding the threshold values.

2.2 Method for Local Image Analysis

In this proposed analysis, original image will be partitioned into 2×2 (4 parts) which labelled as P_1, P_2, P_3 and P_4 or formulate as

$$\text{Local image} = P_t \text{ where } t = 1, 2, 3, 4 \tag{1}$$

Then on each part, threshold value generated by Canny will be used for analysing which are high threshold value (H_t) and low threshold value (L_t).

$$H_t \text{ and } L_t \in P_t \tag{2}$$

Next, value of H_t and L_t on each P_t will be used on generating edge result image. Finally, all results obtained will be compared on the groundtruth image to get the measurement value. Figures 1 shows the process in this local image analysis.

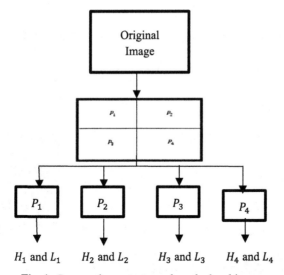

Fig. 1. Proposed process to analyse the local image.

2.3 Image Dataset

In this study the leaves of herbal plants grown in Malaysia will be publish publicly. The ground truth edge images are manually draw. In this data set, there are 180 leaf images among which consist of leaf images of Murraya Koenigii, Curcuma Longa, Citrofortunella Microcarpa, Pandanus Amaryllifolius and others. Pictures of these leaves

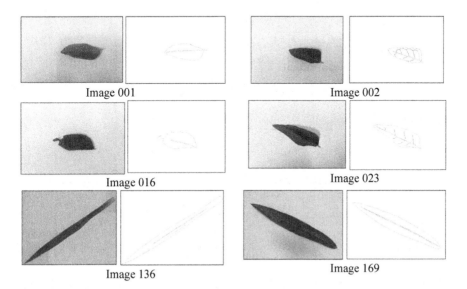

Fig. 2. The example of 6 images from this data set with their respective ground truth edge image

are taken one piece at a time and from different angles. Figures 2 shows the example of 6 images from this data set with their respective groundtruth edge image.

Each folder will represent each image in this data set. Furthermore, each folder has the original leaf image, pre-processing image, reference image, mask image, an edge image from the Canny method and overlay original image with the respective ground truth image (see Figs. 3).

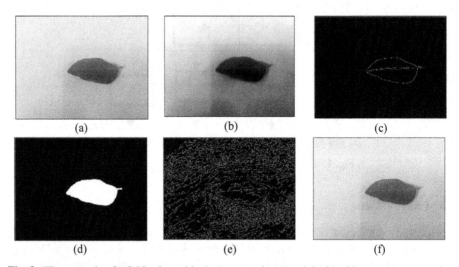

(a)   (b)   (c)

(d)   (e)   (f)

Fig. 3. The example of 1 folder from this dataset contains (a) original leaf image, pre-processing image, reference image, mask image, an edge image from the Canny method and overlay original image with the respective ground truth image

2.4   Performance Measurement

The measurement will be used in the result analysis is

$$\text{F - measure} = \frac{2 \times \text{recall} \times \text{precision}}{\text{recall} + \text{precision}} \tag{3}$$

where

$$\text{recall} = \frac{TP}{TP + FN} \tag{4}$$

and

$$\text{precision} = \frac{TP}{TP + FP} \tag{5}$$

TP, FP and FN denote the true positive, false positive and false negatives values respectively. The higher the value obtained symbolizes the closer the resultant image to the reference image.

3   Results and Discussion

Here, Canny edge detection method will be generated for global image. Then the same process will be used to generate edge for the local 2 × 2 for the same image to get the H_t and L_t for each partition image (P_t). Figures 4 and 5 show the example of edge images generated by using the global image (Canny method) dan the local image from image 001 and 169. In this process, each P_t's high threshold (H_t) and low threshold (L_t) generated will be used to generate new edge image.

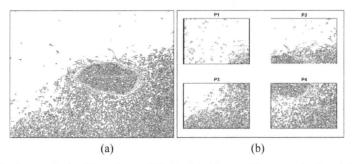

(a)   (b)

Fig. 4. Edge image obtained from image 001 (a) global image (Canny method) and (b) partition image

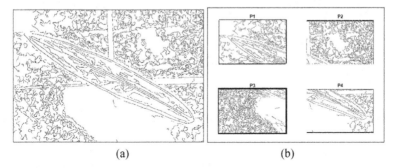

(a)   (b)

Fig. 5. Edge image obtained from image 169 (a) global image (Canny method) and (b) partition image

Table 1 shows their respective H_t and L_t values generated by using Canny method for global image and P_t images from six images. From each image, there are some similar values in partition local image with the global images. For example, in image 001, P_1 and P_2 threshold values are similar with the global value while the values of P_3 and P_4 are different but similar on each other.

Table 1. High threshold H_t and low threshold L_t generated for each image

images	Global image (Canny method)		P_1 image		P_2 image		P_3 image		P_4 image	
	H_t	L_t	H_t	L_t	H_t	L_t	H_t	L_t	H_t	L_t
001	0.0469	0.0188	0.0469	0.0188	0.0469	0.0188	0.0625	0.0250	0.0625	0.0250
002	0.0469	0.0188	0.0469	0.0188	0.0781	0.0312	0.0469	0.0188	0.0781	0.0312
016	0.0625	0.0250	0.0188	0.0469	0.0375	0.0938	0.0625	0.0250	0.0438	0.1094
023	0.0469	0.0188	0.0469	0.0188	0.0938	0.0375	0.0625	0.0250	0.0469	0.0188
136	0.1250	0.0500	0.1250	0.0500	0.1406	0.0563	0.1094	0.0438	0.1562	0.0625
169	0.0625	0.0250	0.0938	0.0375	0.0625	0.0250	0.0781	0.0312	0.0781	0.0312

Next process, each local values threshold from P_t will be used to generated new edge images. Then edge image from the global image and new edge images generated from each P_t will be compared quantitatively with the ground truth image provided by the dataset. This comparison values as shown in Table 2 by using F-measure. The highest F-measure value for each image as highlighted. It shows that the highest value obtained is from the local P_t image for all images used. The respective H_t and L_t values for the highest F-Measure obtained are highlighted as well in Table 1.

Table 2. F-measure values obtained for each image

	F-measure Value				
	Global image (Canny method)	P_1 image	P_2 image	P_3 image	P_4 image
001	0.1708	0.1708	0.1708	0.2035	0.2035
002	0.1893	0.1893	0.2187	0.1893	0.2187
016	0.1617	0.1303	0.2353	0.1617	0.2523
023	0.1038	0.1038	0.1230	0.1226	0.1038
136	0.0777	0.0777	0.0780	0.0742	0.0802
169	0.0520	0.2841	0.0520	0.2551	0.2551

In qualitative comparison, the edge image obtained from each threshold values will be compared with the ground truth image provided. Figure 6 and Fig. 7 show the example of edge images obtained and its ground truth image from image 001 and image 169 respectively. This edge image represents the result from Table 2 qualitatively. Here for Fig. 6, had been found that only two type of edge images generated because there is similar value found in F-measure from Global, P_1 and P_2 images while another type is from P_3 and P_4 images. This result occurs because there is a similarity of threshold values at those places (refer Table 1). Similarly, the comparison of image 169 in Fig. 7. From the results of the edge image obtained, it can be observed that the highest F-measure result that is close to the reference image is from the image of Fig. 6 (c) and Figs. 7 (c) where more complete image boundaries are produced in addition to less noise obtained.

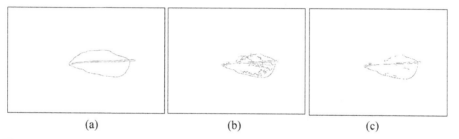

Fig. 6. Comparison edge images obtained from image 001 with the provided (a) Ground truth image and the generated (b) Global (Canny method), P_1 and P_2 images (c) P_3 and P_4 images

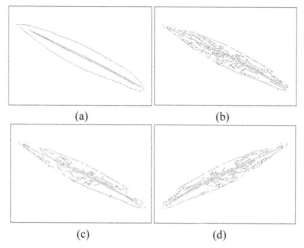

Fig. 7. Comparison edge images obtained from image 169 with the provided (a) Ground truth image and the generated (b) Global image (Canny method) and P_2 images (c) P_1 image (d) P_3 and P_4 images

The threshold value selection factor strongly influences the results of this edge image. It can be observed that the image close to this ground truth image from the higher threshold value generated through the local image compared to the lower threshold value generated through the global value. A lower threshold value will detect a lot of noise because it is in a low area in the image grey level value feature. However, this threshold value should not be too high as it will result in many of the actual edge values not being detected. Figure 8 shows a comparison of F-measure values obtained using the global value approach as used in conventional Canny method with values obtained using the local value approach using the 2×2 method from this study. It is found that the results from the approach using local threshold values are better than using global values from the Canny method.

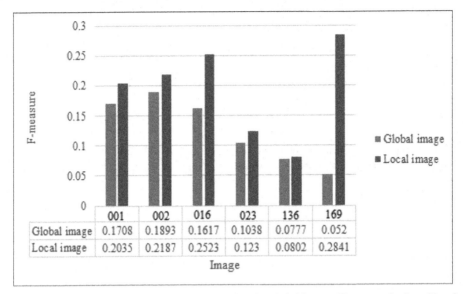

Image	001	002	016	023	136	169
Global image	0.1708	0.1893	0.1617	0.1038	0.0777	0.052
Local image	0.2035	0.2187	0.2523	0.123	0.0802	0.2841

Fig. 8. Comparison histogram for F-measure value for global image (Canny method) and local image approach

4   Conclusion

In this study, a set of data for Malaysian Herbs Leaves images has been developed. In this dataset a ground truth of edge image will be provided for each image respectively. Then, local image analysis was performed for 2×2 by finding the threshold value in the image. Here, the impact of locally selecting this image threshold value was compared with the conventional Canny method which uses threshold values globally. It has been found that this approach can help on finding the best herbs edge image because there are important features that are only obtained by making an assessment on the local value of the image. In addition, the image of the herb itself has a narrow nature that is necessary to undergo detailed image observation. The findings of this study are expected in helping the next phase of the study in machine vision to obtain better recognition. In the future, optimization of the resulting threshold value may be considered to obtain more optimal results.

Acknowledgements. The deepest gratitude and thank you to the Ministry of Education (MOE) and Universiti Teknikal Malaysia Melaka (UTeM) for the financial supports through FRGS. Vote No: FRGS/1/2020/ICT06/UTEM/02/1.

References

1. Halim, S.A., Hadi, N.A., Mat Lazim, N.S. Segmentation and features extraction of Malaysian herbs leaves. J. Phys. Conf. Ser. **1770**(1) (2021). https://doi.org/10.1088/1742-6596/1770/1/012005

2. Roopashree, S., Anitha, J.: Enrich ayurveda knowledge using machine learning techniques. Indian J. Tradit. Knowl. **19**(4), 813–820 (2020)
3. Muneer, A., Fati, S.M.: Efficient and automated herbs classification approach based on shape and texture features using deep learning. IEEE Access **8**, 196747–196764 (2020). https://doi.org/10.1109/ACCESS.2020.3034033
4. Kasmin, F., Othman, Z., Ahmad, S.S.S.: Automatic Road Crack Segmentation Using Thresholding Methods, no. 2, pp. 75–82 (2018)
5. Yusoff, N.M., Halim, I.S.A., Abdullah, N.E.: Real-time hevea leaves diseases identification using sobel edge algorithm on FPGA: a preliminary study. In: 2018 9th IEEE Control System Graduate Research Colloquium, ICSGRC 2018 - Proceeding, no. August, pp. 168–171 (2019). https://doi.org/10.1109/ICSGRC.2018.8657603
6. Mettripun, N.: Thai herb leaves classification based on properties of image regions. In: 2020 59th Annual Conference of the Society of Instrument and Control Engineers of Japan, SICE 2020, pp. 372–377 (2020). https://doi.org/10.23919/sice48898.2020.9240256
7. Shitole, D., Tamboli, F., Motghare, K., Raj, R.K.: Ayurvedic herb detection using image processing. Int. J. Trend Sci. Res. Dev. **Volume-3**(Issue-4), 491–494 (2019). https://doi.org/10.31142/ijtsrd23605
8. Thanikkal, J.G., Dubey, A.K., Thomas, M.T.: Advanced plant leaf classification through image enhancement and canny edge detection. In: 2018 7th International Conference on Reliability, Infocom Technologies and Optimization (Trends and Future Directions) ICRITO 2018, pp. 518–522 (2018). https://doi.org/10.1109/ICRITO.2018.8748587
9. Smeulders, A.W.M., Worring, M., Santini, S., Gupta, A., Jain, R.: Content-based image retrieval at the end of the early years. IEEE Trans. Pattern Anal. Mach. Intell. **22**(12), 1349–1380 (2000). https://doi.org/10.1109/34.895972
10. Abdullah, A., Veltkamp, R.C., Wiering, M.A.: Spatial pyramids and two-layer stacking SVM classifiers for image categorization: a comparative study. In: Proceedings of the International Joint Conference on Neural Networks, pp. 5–12 (2009). https://doi.org/10.1109/IJCNN.2009.5178743
11. Gaur, S.: Adaptive Local Thresholding for Edge Detection Adaptive Local Thresholding for Edge Detection, no. September 2014, pp. 15–18 (2016)
12. Meng, Y., Zhang, Z., Yin, H., Ma, T.: Automatic detection of particle size distribution by image analysis based on local adaptive canny edge detection and modified circular Hough transform. Micron **106**(December 2017), 34–41 (2018). https://doi.org/10.1016/j.micron.2017.12.002
13. Canny, J.: A computational approach to edge detection. IEEE Trans. Pattern Anal. Mach. Intell. **8**(6), 679–98 (1986). http://www.ncbi.nlm.nih.gov/pubmed/21869365
14. Zhao, M., Liu, H., Wan, Y.: An improved canny edge detection algorithm based on DCT. In: 2015 IEEE International Conference on Progress in Informatics and Computing PIC 2015, vol. 2, no. 2, pp. 234–237 (2016). https://doi.org/10.1109/PIC.2015.7489844
15. Abdullah, M.A.M., Dlay, S.S., Woo, W.L., Chambers, J.A.: Robust iris segmentation method based on a new active contour force with a noncircular normalization. IEEE Trans. Syst. Man Cybern. Syst. 1–14 (2016). https://doi.org/10.1109/TSMC.2016.2562500
16. Wang, C., Zhu, Y., Liu, Y., He, R., Sun, Z.: Joint Iris Segmentation and Localization Using Deep Multi-task Learning Framework, pp. 1–13 (2019). http://arxiv.org/abs/1901.11195

Machine and Statistical Learning

Amniotic Fluids Classification Using Combination of Rules-Based and Random Forest Algorithm

Putu Desiana Wulaning Ayu[1,4], Sri Hartati[2(✉)], Aina Musdholifah[2], and Detty S. Nurdiati[3]

[1] Doctoral Program Department of Computer Science and Electronics, Faculty of Mathematics and Natural Science, Universitas Gadjah Mada, Sekip Utara, Yogyakarta 55281, Indonesia
wulaning_ayu@stikom-bali.ac.id
[2] Department of Computer Science and Electronics, Faculty of Mathematics and Natural Science, Universitas Gadjah Mada, Sekip Utara, Yogyakarta 55281, Indonesia
{shartati,aina_m}@ugm.ac.id
[3] Department of Obstetrics and Gynaecology, Faculty of Medicine, Universitas Gadjah Mada, Sekip Utara, Yogyakarta 55281, Indonesia
detty@ugm.ac.id
[4] Department of Information Technology, Faculty Computer and Informatics, Institut Teknologi dan Bisnis STIKOM, Bali, Denpasar, Indonesia

Abstract. One of the studies of fetal anatomy is to measure the volume and echogenicity of the amniotic fluid. This study categorizes amniotic fluid into six, such as Oligohydramnion Clear, Oligodramnion Echogenic, Polygohydramnion Clear, and Polygohydramnion Echogenic, as well as Normal Clear and Normal Echogenic. Meanwhile, the current condition in determining the category of amniotic remains a perception difference among doctors, especially in identifying volume and echogenicity study, which is always conducted manually and visually. Therefore, this research proposed a model for the classification of amniotic fluid by combining the rule-based of the Single Deep Pocket (SDP) method and the Random Forest algorithm. The rule-based used was based on the feature value obtained by extracting the Single Deep Pocket (SDP) feature. Also, the Random Forest algorithm was formed to classify amniotic fluid based on the condition of echogenicity, which includes clear and echogenic based on texture features using First Order Statistical (FOS) and Gray Level Co-occurrence Matrix (GLCM) methods. The average value performance of the proposed model showed an accuracy of 90.52%, a precision of 95.72%, a recall of 75.57%, and an F-measure of 81.51%. Considering this result, the proposed model showed an average increase in accuracy performance of 9.12%, precision of 14.92%, and recall of 0.51% value of the model in previous studies.

Keywords: Amniotic fluid classification · Feature extraction · Rule-based · Random forest

© Springer Nature Singapore Pte Ltd. 2021
A. Mohamed et al. (Eds.): SCDS 2021, CCIS 1489, pp. 267–285, 2021.
https://doi.org/10.1007/978-981-16-7334-4_20

1 Introduction

Amniotic fluid is the liquid that surrounds the fetus during the development process in the womb. Also, it is used to protect the fetus and the umbilical cord in the event of an impact as well as pressure on the uterine wall, respectively [1]. This supports the fetus in movement and promotes its muscle as well as bones development [2]. The amniotic fluid volume continues to increase significantly between the age of 8 to 28 weeks. Furthermore, the volume decreases and stagnates at 35–36 weeks [3]. The amount of volume and condition (echogenic or clear) of the fluid is examined by a doctor using an ultrasound machine. The volume consists of three categories, such as oligohydramnios (lack of fluid), Polygohydramnios (excess of fluid), and Normal. Meanwhile, the condition is divided into two categories, include echogenic and clear. To categorize the volume of amniotic fluid, the doctor indicates the length of the Single Deep Pocket (SDP) by drawing the line at two vertically straight points, which starts from the fluid pocket and intersects with the uterus [4]. Meanwhile, to categorize the condition of the amniotic fluid, it is visually based on the intensity of gray or opaque in the amniotic fluid area. Visual observation was conducted on the opaque or gray patch that spreads into the location of the amniotic fluid, which is similar to the placenta and is categorized as echogenic [5].

The study of amniotic fluid included in the fetal organ image processing in computer science. Meanwhile, previous studies in the field of computer science related to parts of the fetal organ conducted, namely placental organs [6–9], infant aortic [10–12], umbilical cord classification [13, 14], fetal head measurement [15–17], abdominal measurement [18–20], and amniotic fluid segmentation [21–24]. Studies on the placenta focus more on mature placenta staging, where the first step is the segmentation, extraction of texture features, and diagnostic assistance to evaluate the mature placental. Furthermore, a study on the Fetal aortic has focused on measuring the aortic intima-media thickness using segmentation based on anisotropic filtering and level-set methods. Moreover, the umbilical cord organ which focused on segmentation and the amount of coiling were classified into 3, namely, hyper cooling, normal coiling, and hypocoiling, using the assembly multiclassifier method. The first research on amniotic fluid classification was carried out in our previous research was conducted using a machine learning Support Vector Machine (SVM) method. We also used the oversampling method due to the limited availability of data samples, while the experiments were performed using initial image data of 92 b-mode ultrasonography amniotic fluid. The classification stage used the SVM method, which was analyzed on three different kernels: RBF, polynomial, and sigmoid. The results showed that the RBF kernel's proposed feature achieves higher average accuracy, precision, and recall value [25].

This study focuses on an essential part of the fetal measurement object, which includes the amniotic fluid. It proposes the combination rule based of the SDP method with a machine learning algorithm. The rule-based used is based on the feature value from extracting the SDP feature. Meanwhile, it is formed based on the criteria used by obstetricians to measure the amniotic fluid volume. This rule is used to classify amniotic fluid into three classes, namely, Oligohydramnios, Poligohydrmanions, and Normal. Furthermore, a machine learning algorithm is formed to classify amniotic fluid based on the condition of echogenicities, such as clear and echogenic based on texture features using the First Order Statistical (FOS) and Gray Level Co-occurrence Matrix (GLCM) methods. The combination of these two models produces six classes of output, namely Oligohydramnion Clear, Oligohydramnion Echogenic, Poligohydramnion Clear, Poly-gohydramnion Echogenic, Normal Clear, and Normal Echogenic. The novelty of this study lies in the classification separation between echogenicity and volume of amniotic fluid. In this paper, we also compare the Random Forest method with other machine learning methods such as Decision Tree and SVM and also our previous study [25] to determine the performance of the proposed method. Therefore, this study separates the classification model to apply this SDP feature directly to the texture properties of volume and echogenicity to improve the accuracy.

2   Proposed Model

The stages of examination of the amniotic fluid shown in Fig. 1, where the first stage is the patient asked to sleep in a prone position at the examination site. The following process applies gel to the probe to clarify the object's sound and the probe's position placed perpendicular to the uterus (maternal abdomen). Then a screening process is carried out to find the area of amniotic fluid which has the widest area and is under the uterus. The following procedure is to pull the caliper vertically in the widest area, with the upper border of the uterus. There should be no fetal body parts and observe the echogenicity of the amniotic fluid. Then the depth measurement was carried out using the SDP technique to categorize the volume of amniotic fluid and categorize it into six classes. The process in the red box is a process that assisted completed using a machine learning model approach. The proposed model in this study is shown in Fig. 2.

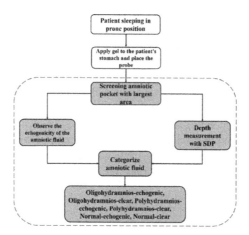

Fig. 1. The process of examining amniotic fluid by a doctor

There are several stages in the process of producing an amniotic fluid classification model in this study. Fig. 2 shows the sequence of the study conducted. Furthermore, the seven main steps in the amniotic fluid classification model include data acquisition, image

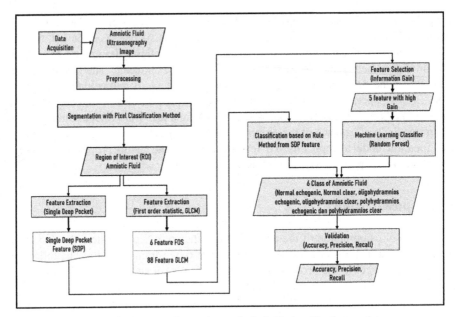

Fig. 2. The proposed stage in amniotic fluid classification model.

preprocessing and segmentation, feature extraction and selection, constructing classification models, as well as validation, and each process is explained in each subsequent section.

2.1 Data Acquisition

The 2D ultrasound image data of amniotic fluid is obtained from Surya Husadha Hospital in Bali. Also, the image data is taken from the Ultrasound Accuvix XG Machine and transducer with a frequency of 3.5 Hz, corresponding to a lateral resolution of 3 mm to 0.2 m, with a gain of 0–8 Hz. The clear and echogenic amniotic fluid data are 53 as well as 42 images. Furthermore, when the data is re-sorted by volume, the classes are normal-clear (31), normal-echogenic (30), oligohydramnios-clear (5), oligohydramnios-echogenic (3), polyhydramnios-clear (17), polyhydramnios-clear echogenic (9), with a total of 95 images. Therefore, the image data used has criteria for a single pregnancy, non-obese pregnant women, and gestational age between 13–37 weeks.

2.2 Image Preprocessing

The image preprocessing stage for a 2D ultrasound of the amniotic cavity begins with the cropping process and continues with the conversion of the image color channel from RGB to Grayscale. Meanwhile, the initial pretreatment process is cropping, which aims to achieve an area of amniotic fluid and eliminate unused information, such as text or descriptions from patient data. The image from the ultrasound machine recording (acquisition process) has a dimension of 800 × 600 pixels, while the cropping process produces a dimension of 427×570 pixels. Furthermore, the conversion of the image into grayscale is conducted to determine the distribution value of gray-level pixels used during the segmentation process.

2.3 Image Segmentation

Based on previous studies, the segmentation of the amniotic fluid area is performed using the pixel classification method [12]. Generally, the segmentation stage of pixel classification begins with the process of collecting the datasets based on local window sampling. Furthermore, the extraction process is performed to obtain local window-based features using the Gray level, First Order Statistical (FOS), and Distance Angle Pixel (DAP) features. This process produces output in the form of a pixel dataset for training and testing. Meanwhile, the training pixel dataset is used as input to train the classification model using the Random Forest method to obtain classes of a pixel, such as amniotic fluid, placenta, fetal body, and uterine.

2.4 Feature Extraction

Single Deep Pocket (SDP)

Single deep pocket (SDP) is a feature adopted by doctors to categorize the volume of amniotic fluid. Conceptually, the SDP method searches for the longest vertical, straight line from the selected amniotic fluid area, which needs not intersect with the fetal body. To determine the volume of amniotic fluid, the caliper drawing is performed vertically by the obstetrician. Meanwhile, this is achieved after determining the area of amniotic fluid that is considered the deepest. The amniotic fluid volume is divided into three categories, namely, Oligohydrmanion when the measurement using a caliper shows the vertical length of less than 2 cm, Normal when the vertical length is between 2 to 8 cm, and Polygohydramnion when the vertical length is above 8 cm. To obtain the SDP feature following the rules from the doctor, this study proposes an algorithm to obtain the longest vertical and straight-line based on the previous studies on SDP feature extraction [25]. This algorithm is shown by the flowchart in Fig. 3.

The Region of Interest (ROI) amniotic fluid is a binary image with a matrix form that is processed based on columns and rows, and the calculation is performed by finding the column in the matrix with the highest value of 1 (white), which represents the amniotic fluid. Meanwhile, the output of this algorithm is the number of pixels in the column with the highest number of 1's and the index of that column. The stages of finding the SDP feature in Fig. 3 begin with initializing the initial value, such as **Max = 0, which** is a variable to store the maximum value (value 1), a **column** which is a variable to store the index of the column that has the highest value 1, **sum[k]** which is a variable to store the number of values 1 of each row and column, and m which is the number of columns.

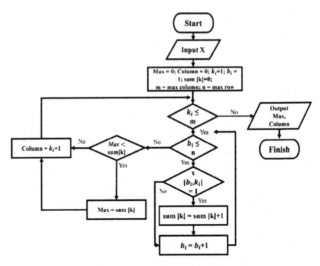

Fig. 3. SDP Flowchart algorithm [2].

Moreover, the iteration in the calculation starts from column 1 and row 1 to row i (b_i), and when the row as well as column have a value of 1, it is added to the **sum[k]** variable, and while at 0, it continues to check to the next row. After iteration, where the row is greater than n (number of rows), the next step is to go to the next column, and the *max* value is set to **sum[k]**. This is similar to iteration in the next column, such as checking until the nth row and the *max* value change to **sum[k]**. When the *max* value is less than the **sum[k]** value from the ith column, the *max* value is updated to the highest number of values 1 in the ith column. The iteration is complete when the column is greater than m, thereby the outputs are the *max,* and the column that has the highest pixel value 1 is the mth column. Meanwhile, the last process is to measure the length of the SDP line in centimeters. The process of calibrating pixels to centimeters (cm) in amniotic fluid ultrasound images uses a reference of a point or stripe in the information area to the right of the ultrasound image. The distance of 2 points or stripes represents a 1-cm (cm) value. Meanwhile, the 1-cm (cm) value in the image represents 28 pixels obtained from the measurement of the stripe length. Therefore, the calibration process from pixels to centimeters becomes $\frac{1}{28}$ = 0.0357 cm. Figure 4, Fig. 5, and Fig. 6 show the results of the proposed SDP feature extraction in the classes of normal, oligohydramnios, and polyhydramnios, respectively.

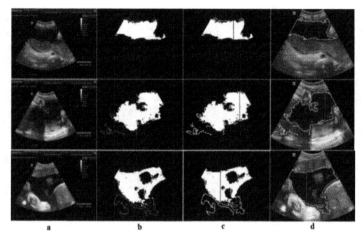

a   b   c   d

Fig. 4. The example of SDP feature output on the normal volume class. a. is an original image, b. segmentation results, c. edge and deepest area detection as well as selection respectively, and d. SDP results on original image.

First Order Statistical (FOS) and Gray Level Co-occurrence Matrix (GLCM) Feature for Echogenicity

GLCM is a statistical method for examining textures and their spatial relationships, which characterizes image texture by calculating how often pairs of pixels with certain values and spatial relationships occur in the image [26]. Meanwhile, to steps to obtain texture features in GLCM are as follows; (1) image transformation into grayscale form, (2) co-occurrence matrix creation, (3) symmetric matrix formation, and (4) normalization, resulting in 22 texture features (autocorrelation, contrast, correlation 1 and 2, cluster prominence and shade, dissimilarity, energy, entropy, homogeneity 1 and 2, maximum probability, sum of squares, average, entropy and variance, difference variance and entropy, information measure of correlation 1 and 2, inverse difference normalized and moment normalized) [27]. The second texture extraction method used is First Order Statistical (FOS) by using statistical calculations based on the original image pixel value to obtain the intensity distribution. Statistical produces several features in the form of mean, skewness, entropy, kurtosis, standard deviation, and variance. Therefore, the feature value used in this study for the FOS method is described in Eqs. (1) to (6).

Mean:

$$\bar{x} = \sum_n \sum_m \sum_{i=0}^{G-1} i_{(n,m)} P_{(n,m)} \tag{1}$$

Skewness:

$$\bar{x}_3 = \sigma^{-3} \sum_n \sum_m \sum_{i=0}^{G-1} \left(i_{(n,m)} - \bar{x}\right)^3 P_{(n,m)} \tag{2}$$

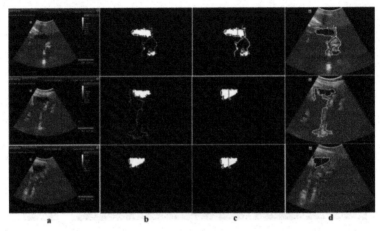

Fig. 5. The example of the SDP feature output on the oligohydramnios volume class. a. an original image, b. segmentation results, c. edge and deepest area detection as well as selection respectively, d. SDP results on the original image.

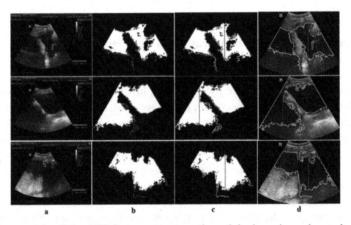

Fig. 6. The example of the SDP feature output on the polyhydramnios volume class. a. is an original image, b. segmentation results, c. edge and deepest area detection as well as selection respectively, d. SDP results on original image.

Entropy:

$$= \sum_n \sum_m \sum_{i=0}^{G-1} P_{(n,m)} log_2 \big[P_{(n,m)} \big] \tag{3}$$

Kurtosis:

$$\overline{x}_4 = \sigma^{-4} \sum_n \sum_m \sum_{i=0}^{G-1} \big(i_{(n,m)} - \overline{x} \big)^4 P_{(n,m)} - 3 \tag{4}$$

Standard deviation:

$$S = \sqrt{\frac{1}{N-1} \sum_n \sum_m \left(x_{(n,m)} - \bar{x}\right)^2} \tag{5}$$

Variance:

$$S = \frac{1}{N-1} \sum_n \sum_m \left(x_{(n,m)} - \bar{x}\right)^2 \tag{6}$$

n and m are the row and column of the window, $h_{(i)}$ Represents the total number of pixels with intensity level (i), N is the total number of pixels with $N = n \times m$, and G is the maximum gray-level.

2.5 Feature Selection

The information gain is a filter-based method of selecting features with the most relevant information for a class. This method detects features that have the most information based on a certain class. To determine the best property, the entropy value, which is a measure of class uncertainty using the probability of certain events or attributes, is first calculated. Meanwhile, the information gain value of a feature is determined using Eqs. 7 and 8.

$$Entrophy(S) = \sum_i^c -P_i log_2 P_i \tag{7}$$

$$Entrophy(x) = \sum_i^c -P_i log_2 P_i \tag{8}$$

Entropy (S) is the value of the class output, and (X) is the entropy value of a feature, c is the total value of a feature, and P_i is the number of samples for the ith class. To calculate the entropy value for each feature against the output class presented in Eq. (9) is used as follows:

$$E(T, X) = \sum_x P(x) Entrophy(x) \tag{9}$$

Where $E(T, X)$ is the entropy value of a feature X to class T, $P(x)$ is the probability of a value in feature X against class T, and entropy (x) is the value in the feature from Eq. (8). After the entropy value of the feature is obtained, the Information Gain value of the feature is calculated by finding the reduction between the target class's entropy and the feature's entropy as in Eq. (10).

$$Information\ Gain = Entrophy(S) - E(T, X) \tag{10}$$

2.6   Constructing Classification Model

The amniotic fluid classification model consists of two models, namely rule-based and machine learning with a Random Forest classifier. The formation of a rule-based model was used to classify amniotic fluid based on the volume measured based on the SDP value from the feature extraction process in the previous stage. The output of the volume classification is divided into three, namely, oligohydramnios, polyhydramnios, and normal. The rules used are as follows:

1. **IF** sdp < 2.0 **THEN** output class == 'oligohydramnios
2. **ELSE IF** sdp < 8.0 output class == 'normal'
3. **ELSE** output class == 'polyhydramnios

This volume classification follows the range of SDP values as in lines 1 to 5 in pseudocode Table 1. Furthermore, the classification method in the second echogenicity model uses a machine-learning algorithm approach with the Random Forest method. The features extracted from this model are using texture and gray-level intensity approaches, and the extraction method uses GLCM and FOS, which are combined and selected to find the features that best contribute and influence the performance of Random Forest. Meanwhile, lines 6 to 9 in the pseudocode Table 1 show the steps used for constructing the random forest mod

The main parameter used is to divide the dataset into 20 sub-datasets using the bootstrap aggregating method, while the last process is to combine the output classes between models one and two to produce six new output classes, which are a combination of echogenicity and volume. Furthermore, lines 10 to 20 in the pseudocode Table 1 show the steps used to combine the output classes of the two models used.

2.7   Validation

The parameters used to measure the performance of the classification model are accuracy, precision, recall, and F-measure, as shown in Eqs. (11)–(14) in Table 2.

TP is True Positive (a positive label that is predicted as an actual label), FP is False Positive (negative label but predicted as a positive label), TN is True Negative (negative data that predicted correctly), and FN is False Negative (a positive label but predicted as a negative label) [25].

3   Experimental Results

3.1   Feature Selection

In this study, the sum of all features with GLCM and FOS is 95 features. Meanwhile, feature selection is performed to get the five features with the highest gain value, and the results are shown in Table 3. Furthermore, the graph of gain values for all features is shown in Fig. 7. These features are input data for the classification process, especially on the echogenicity of the amniotic fluid.

Table 1. Pseudocode amniotic fluid classification

Algorithm. Merge Rule-based and Machine learning algorithm for amniotic fluid classification

---

Input: X: SDP feature value;
      Y: Texture feature vector
      T: Target label

Output: Z: output class echogenicity
          Z_1 : *Normal*
          Z_2 : *Oligo*
          Z_3 : *Poligo*
      V: *output class Volume*
          V_1 : *Clear*
          V_2 : *Echogenic*
      U: *Final class*
          U_1 : *Normal Clear*
          U_2 : *Normal Echogenic*
          U_3 : *Oligohydramnios Clear*
          U_4 : *Oligohydramnios Echogenic*
          U_5 : *Polyhydramnios Clear*
          U_6 : *Polyhydramnios Echogenic*

*// **Constructing Rule-based for volume classification***
1.   if $X < 2.0$
2.   then $Z = Z_2$
3.   else if $X < 8.0$
4.   then $Z = Z_3$
5.   else $Z = Z_1$
*// **Constructing Machine learning Random Forest for echogenicity classification***
6.   iNumBags = 20
7.   str_method = 'classification'
8.   Mdl ← TreeBagger (iNumbags, Y, T, str_method)
9. $V = Predict$ (Mdl, Y)
*// **Merge output class (volume and echogenicity)***
10.   If $Z = Z_1$ && $V = V_1$
11.   Then U ← U_1
12.   Else if $Z = Z_1$ && $V = V_2$
13.   Then U ← U_2
14.   Else if $Z = Z_2$ && $V = V_1$
15.   Then U ← U_3
16.   Else if $Z = Z_2$ && $V = V_2$
17.   Then U ← U_4
18.   Else if $Z = Z_3$ && $V = V_1$
19.   Then U ← U_5
20.   Else U ← U_6

Table 2. Validation parameter

$\text{Accuracy} = \frac{TP + TN}{TP + FP + TN + FN}$	(11)	$\text{Precision} = \frac{TP}{TP + FP}$	(12)
$\text{Recall} = \frac{TP}{TP + FN}$	(13)	$\text{F} - \text{measure} = \frac{\text{Precision. Recall}}{\text{Precision} + \text{Recall}}$	(14)

Table 3. Five features with the highest gain value

No	Feature name	Information Gain Value
1	information_measure_of_correlation1 45°	0.383221589
2	cluster_shade 45°	0.36370615
3	difference_entropy 135°	0.359331636
4	cluster_prominence 45°	0.328765299
5	difference_variance 135°	0.311866701

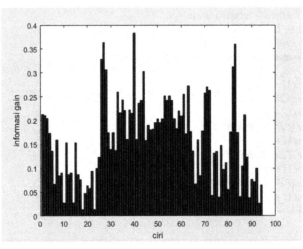

Fig. 7. Graph of gain values on the information gain method to select the echogenicity features of amniotic fluid.

3.2 The Classification Results of Amniotic Fluid Echogenicity with Random Forest Algorithm

In this classification of amniotic fluid echogenicity, the training data is the result of texture feature extraction using the GLCM and FOS methods for a sum of 95 images. This extraction produces 95 features, and the feature selection is carried out using the Information Gain method to select five features with the highest gain value. Meanwhile, studies on the Random Forest method are carried out by observing the correlation between the number of trees formed and the Out of Bag Error (OOB error) that occur at each additional tree/subset sample. This observation is performed to determine the optimal number of trees used to form the Random Forest model. Random forest (RF) is widely applied to solve some segmentation and classification problems in medical images [23, 24, 28–30]. RF classifier is a type of ensemble learning method used to build a final classifier with a set of collections from individual weak classifiers (M) such as the binary tree which on average produces better performance than other machine learning methods [31]. Furthermore, the out-of-bag score is a validation method used in Random Forest by using random samples that are not selected at the tree formation which is a subset of the original training data. This data is used for each tree formed where there is no sample data. Therefore, OOB error is defined as the tree's error rate in classifying the out-of-bag sample. Fig. 8 shows a graph of the relationship between the number of trees and the resulting out-of-bag error. Based on the Fig. 8, the formation of a tree/subset sample with a sum greater than 30 shows a relatively small out-of-bag error value and starts to stabilize.

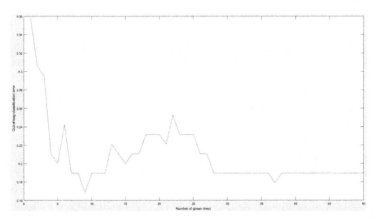

Fig. 8. Out of bag error plot of Random Forest method for amniotic fluid echogenicity

Table 4. The experimental results for the classification of amniotic fluid echogenicity in the random forest method

Total number of tree/subset sample	Performance Parameters			
	Accuracy (%)	Precision (%)	Recall (%)	F-measure (%)
1	88.42	84.44	90.47	87.35
2	92.63	94.87	88.09	91.35
3	93.68	90.9	95.23	93.02
4	95.78	97.5	92.85	95.12
5	97.89	97.61	97.61	97.61
6	96.84	100	92.85	96.29
7	97.89	100	95.23	97.56
8	97.89	100	95.23	97.56
9	98.94	100	97.61	98.79
10	98.94	100	97.61	98.79
11	100	100	100	100
12	100	100	100	100
13	100	100	100	100
14	100	100	100	100
15	100	100	100	100

These results show that the parameter value of the number of grown trees in Random Forest is selected with a value greater than 30. Meanwhile, to measure the performance of this model, the k-fold cross-validation method is used with 5-folds. Furthermore, to determine the accuracy performance of the Random Forest model related to the number of trees used, this study observes the number of trees/subset samples with a correlation to the classification accuracy value. The observation number of trees/subset samples used is 15. The results for the amniotic fluid echogenicity classification on the number of trees/subset samples in the Random Forest method are shown in Table 4.

Based on the results in Table 4, the performance of the Random Forest method in classifying the amniotic fluid echogenicity achieved a satisfactory average performance. In the number of subset samples/trees greater than 10, the Random Forest method achieves the best performance with 100% accuracy, precision, recall, and F-measure. Furthermore, Table 5 shows the confusion matrix for the amniotic fluid echogenicity in the Random Forest method in both classes, clear and echogenic. Therefore, the models correctly predict all data.

3.3 Classification Results of Amniotic Fluid Volume

The classification rules for volume classification based on SDP features are in the form of IF-THEN-ELSE with the rules shown in the Pseudocode in Table 1. Meanwhile, each

Table 5. Confusion matrix of amniotic fluid echogenicity in the random forest method

Random Forest		Predicted	
		Clear	Echogenic
Actual	Clear	**53**	0
	Echogenic	0	**42**

image input produces one SDP feature from the previous feature extraction process. This SDP feature value becomes the reference value to determine the input image class. Furthermore, the performance of the amniotic fluid volume classification results in three classes (normal, oligohydramnios, and polyhydramnios) is measured using the confusion matrix in Table 6. Moreover, this test is performed on 95 images.

Table 6. Confusion matrix for amniotic fluid volume

Random Forest		Predicted		
		Normal	Oligo	Poligo
Actual	Normal	**61**	0	0
	Oligo	4	**4**	0
	Poligo	5	0	**21**

[*]Oligo: oligohydramnios, poligo: polyhydramnios

Based on Table 6, the classification model for this SDP predicts the normal class correctly overall with a sum of 61 data or 100% accuracy. Furthermore, the Oligo class correctly predicts 4 out of 8 data or 50% accuracy, while the Poligo class correctly classifies 21 out of 26 data or 80.7% accuracy.

3.4 Classification Results of Amniotic Fluid Volume and Echogenicity

This section presents the results of the proposed model, which is a combination of two classification models as shown in the pseudocode in Table 1. Meanwhile, the testing scheme on the proposed model uses the k-fold cross-validation method with 5-folds. Furthermore, Table 7 shows the confusion matrix of the proposed model, and the parameters of model performance testing are conducted based on this confusion matrix by finding the accuracy, precision, recall, and F-measure values in each output class. Based on Table 8 shows the performance of the proposed model; it is observed that the average accuracy, precision, recall, and F-measure are 90.52%, 95.72%, 75.57%, and 81.51%, respectively. These results are compared with the previous study using the machine learning method, namely RBF SVM [25], as shown in Table 9.

Table 7. Confusion matrix for the six classes of the amniotic fluid classification

Random Forest		Predicted					
		Normal-Clear	Normal-Echo	Oligo-Clear	Oligo-Echo	Poligo-Clear	Poligo-Echo
Actual	Normal-Clear	**31**	0	0	0	0	0
	Normal-Echo	0	**30**	0	0	0	0
	Oligo-Clear	2	0	**3**	0	0	0
	Oligo-Echo	0	2	0	**1**	0	0
	Poligo-Clear	3	0	0	0	**14**	0
	Poligo-Echo	0	2	0	0	0	**7**

*Oligo: Oligohydramnion; Poligo: Poligohydramnion; Echo: Echogenic

Table 8. The classification result performance of each class on the proposed model

Class	Precision (%)	Recall (%)	F-Measure (%)
Normal-Clear	86.11	100	92.53
Normal-Echogenic	88.23	100	93.75
Oligo-Clear	100	60.0	75.0
Oligo-Echo	100	33.33	50.0
Poligo-Clear	100	82.35	90.32
Poligo-Echo	100	77.77	87.5

Table 9. Comparison of the classification result performance for each class on the proposed model and the our previous model

Class	Rule based + Decision Tree			Rule based + SVM			Proposed model (Rule based + RF)			Previous model (RBF SVM [25])		
	Precision (%)	Recall (%)	F-Measure (%)	Precision (%)	Recall (%)	F-Measure (%)	Precision (%)	Recall (%)	F-Measure (%)	Precision (%)	Recall (%)	F-Measure (%)
Normal-Clear	85.30	93.54	89.23	80.0	90.32	84.85	**86.11**	**100**	**92.53**	64	51	57
Normal-Echogenic	80.55	96.66	87.87	80.0	93.33	86.15	**88.23**	**100**	**93.75**	80	82	81
Oligo-Clear	100	60	75	100	60.0	75	**100**	**60**	**75**	98	100	99
Oligo-Echo	100	33.33	50	100	33.3	50	**100**	**33.33**	**50**	86	89	88
Poligo-Clear	100	70.58	82.75	100	82.35	90.32	**100**	**82.35**	**90.32**	76	85	80
Poligo-Echo	77.77	77.77	77.77	100	77.77	87.5	**100**	**77.77**	**87.5**	-	-	-

Based on the results in Table 9, the proposed model is able to improve the results of precision, recall, and F-measure in four classes, such as Normal-Clear, Normal-Echogenic, Poligo-Clear, and Poligo-Echogenic. Meanwhile, the Oligo-clear and Oligo-echogenic classes show that the previous model has higher precision, recall, and F-Measure values. The average value performance of the proposed model using metode RF + rule-based shows an accuracy of 90.52%, a precision of 95.72%, a recall of 75.57%, and an F-measure of 81.51%. Where the results of the proposed method show a higher performance than the Decision Tree + rule based method with an average accuracy of 83.4%, average precision 82.1%, average recall 83.1% and F-Measure 83.3%. SVM + rule based method shows an average accuracy of 85.26%, precision of 83.3%, recall of 85.1% and F-Measure of 84.4%. Meanwhile, the previous model shows an average accuracy of 81.40%, a precision of 80.80%, recall of 80.8%, and an F-measure of 81.0%.

4 Conclusion

In this study, a model for the classification of amniotic fluid is proposed by combining the rule-based of the single deep pocket (SDP) method and the machine learning algorithm. This rule is based on the feature value from the SDP feature extraction, which is formed according to the specific criteria used by the doctors. Furthermore, this rule is used to classify amniotic fluid into three classes, namely, Oligohydramnios, Poligohydrmanions, and Normal. Meanwhile, a Random Forest algorithm is formed to classify amniotic fluid based on echogenicity conditions, namely clear and echogenic based on texture features using First Order Statistical (FOS) and Gray Level Co-occurrence Matrix (GLCM) methods. This condition has no standard measurement value; therefore, it is only based on the doctor intuition performed in viewing the gray color/textur. Meanwhile, the combination of rule-based and machine learning models shows an average increase in accuracy performance of 9.12%, precision of 14.92%, and recall of 0.51% from the model in the previous study. The proposed model shows an improved performance from the previous model in classifying amniotic fluid. Hence, the model from this study is intended to be an aid for doctors in the initial screening of amniotic fluid.

Acknowledgment. The author would like to thank to the Research Directorate of Universitas Gadjah Mada for funding this research in the RTA (Rekognisi Tugas Akhir) 2021 scheme.

References

1. Ten Broek, C.M.A., Bots, J., Varela-Lasheras, I., Bugiani, M., Galis, F., Van Dongen, S.: Amniotic fluid deficiency and congenital abnormalities both influence fluctuating asymmetry in developing limbs of human deceased fetuses. J. PLoS ONE **8**(11), 1–9 (2013)
2. Tong, X.L., Wang, L., Gao, T.B., Qin, Y.G., Qi, Y.Q., Xu, Y.P.: Potential function of amniotic fluid in fetal development-novel insights by comparing the composition of human amniotic fluid with umbilical cord and maternal serum at mid and late gestation. J. Chinese Med. Assoc. **72**(7), 368–373 (2009)
3. Dallaire, L., Potier, M.: Amniotic fluid, Encyclopedia of Reproduction, vol. 3. Elsevier, pp. 53–97 (2012)

4. Edwards, A.: 3-D ultrasound in obstetrics and gynecology, First Ed., vol. 42, no. 2. Alfred Abuhamad (2004)
5. Karamustafaoglu Balci, B., Goynumer, G.: Incidence of echogenic amniotic fluid at term pregnancy and its association with meconium. Arch. Gynecol. Obstet. **297**(4), 915–918 (2018). https://doi.org/10.1007/s00404-018-4679-7
6. Wanjun, L., Tianfu, W., Dong, N., Siping, C., Baiying, L., Yuan, Y.: Placental maturity evaluation via feature fusion and discriminative learning. Chinese J. Biomed. Eng. **35**(4), 411–418 (2016)
7. Lei, B., et al.: Automatic placental maturity grading via hybrid learning. Neurocomputing **223**(December 2015), 86–102 (2017)
8. Han, M., et al.: Automatic segmentation of human placenta images with u-net. J. IEEE Access **7**, 180083–180092 (2019)
9. Meengeonthong, D., Luewan, S., Sirichotiyakul, S., Tongsong, T.: Reference ranges of placental volume measured by virtual organ computer-aided, vol. 00, no. 00, pp. 1–7 (2017)
10. Tarroni, G., Visentin, S., Cosmi, E., Grisan, E.: Near-automated quantification of prenatal aortic intima-media thickness from ultrasound images. In: Computing in Cardiology (2010), vol. 41, no. January, pp. 313–316 (2014)
11. Linguraru, M.G., Cosmi, E., Veronese, E., Grisan, E., Visentin, S., Tarroni, G.: Estimation of prenatal aorta intima-media thickness from ultrasound examination. Phys. Med. Biol. **59**(21), 6355–6371 (2014)
12. Tarroni G, Visentin S, Cosmi E, Grisan E. Fully-automated identification and segmentation of aortic lumen from fetal ultrasound images. In: *Proceedings of the Annal. Inernational. Conference of the IEEE Engineering in Medicine and Biology Society (EMBC)*, vol. 2015-Novem, no. 2, pp. 153–156 (2015)
13. Pradipta, G.A., Wardoyo, R., Musdholifah, A., Sanjaya, I.N.H.: Improving classifiaction performance of fetal umbilical cord using combination of SMOTE method and multiclassifier voting in imbalanced data and small dataset. Int. J. Intell. Eng. Syst. **13**(5), 441–454 (2020)
14. Pradipta, G.A., Wardoyo, R., Musdholifah, A., Sanjaya, I.N.H.: Radius-SMOTE: a new oversampling technique of minority samples based on radius distance for learning from imbalanced data. IEEE Access **9**, 74763–74777 (2021)
15. Namburete, A.I.L., Stebbing, R.V., Kemp, B., Yaqub, M., Papageorghiou, A.T., Noble, J.A.: Learning-based prediction of gestational age from ultrasound images of the fetal brain. Med. Image Anal. **21**(1), 72–86 (2015)
16. Li, J., et al.: Automatic fetal head circumference measurement in ultrasound using random forest and fast ellipse fitting. IEEE J. Biomed. Heal. Inform. **22**(1), 215–223 (2018)
17. van den Heuvel, T.L.A., Petros, H., Santini, S., de Korte, C.L., van Ginneken, B.: Automated fetal head detection and circumference estimation from free-hand ultrasound sweeps using deep learning in resource-limited countries. Ultrasound Med. Biol. **45**(3), 773–785 (2019)
18. Pradipta, G.A., Ayu, P.D.W.: Fetal weight prediction based on ultrasound image using fuzzy c means clustering and itterative random hough transform. In: Proceedings - 2017 1st International Conference on Informatics and Computational Sciences, ICICoS 2017, 2018, vol. 2018-Janua (2018)
19. Jang, J., Park, Y., Kim, B., Lee, S.M., Kwon, J.Y., Seo, J.K.: Automatic estimation of fetal abdominal circumference from ultrasound images. IEEE J. Biomed. Heal. Inform.**21**94(c), 1–10 (2017)
20. H. Ravishankar, S. M. Prabhu, V. Vaidya, and N. Singhal, "Hybrid approach for automatic segmentation of fetal abdomen from ultrasound images using deep learning," *2016 IEEE 13th Int. Symp. Biomed. Imaging*, pp. 779–782, 2016.
21. Ayu, D.W., Hartati, S., Musdholifah, A.: Amniotic fluid segmentation by pixel classification in b-mode ultrasound image for computer assisted diagnosis. In: Berry, M., Yap, B., Mohamed,

A., Köppen, M. (eds.) Soft Computing in Data Science. SCDS 2019. Communications in Computer and Information Science, vol. 1100. Springer, Singapore (2019). https://doi.org/10.1007/978-981-15-0399-3_5

22. Li, Y., Xu, R., Ohya, J., Iwata, H.: Automatic fetal body and amniotic fluid segmentation from fetal ultrasound images by encoder-decoder network with inner layers. In: Proceedings of the IEEE Engineering in Medicine and Biology Society (EMBC), pp. 1485–1488 (2017)

23. Ayu, P.D.W., Hartati, S., Musdholifah, A., Nurdiati, D.S.: Amniotic fluid segmentation based on pixel classification using local window information and distance angle pixel. Appl. Soft Comput. **107**, 107196 (2021)

24. Ayu, P.D.W., Hartati, S.: Pixel classification based on local gray level rectangle window sampling for amniotic fluid segmentation. Int. J. Intell. Eng. Syst. **14**(1), 420–432 (2021)

25. Ayu, P.D.W., Hartati, S., Musdholifah, A., Nurdiati, D.S.: Amniotic fluid classification based on volume and echogenicity using single deep pocket and texture feature. ICIC Express Lett. **15**(7), 681–691 (2021)

26. Mathwork, T.: Neural Network Toolbox for Use With Matlab. Natick, USA (1999)

27. M. Abdel-Nasser, J. Melendez, A. Moreno, O. A. Omer, and D. Puig, "Breast tumor classification in ultrasound images using texture analysis and super-resolution methods. J. Eng. Appl. Artif. Intell. **59**(August 2016), 84–92 (2017)

28. A. Sarica, A. Cerasa, and A. Quattrone, "Random forest algorithm for the classification of neuroimaging data in Alzheimer's disease: A systematic review," *Front. Aging Neurosci.*, vol. 9, no. OCT, pp. 1–12, 2017.

29. Li, Y., Ho, C.P., Toulemonde, M., Chahal, N., Senior, R., Tang, M.X.: Fully automatic myocardial segmentation of contrast echocardiography sequence using random forests guided by shape model. J. IEEE Trans. Med. Imaging **37**(5), 1081–1091 (2018)

30. Qian, C., Yang, X.: An integrated method for atherosclerotic carotid plaque segmentation in ultrasound image. J. Comput. Methods Programsss Biomed. **153**, 19–32 (2018)

31. Poudel, P., Illanes, A., Ataide, E.J.G., Esmaeili, N., Balakrishnan, S., Friebe, M.: Thyroid ultrasound texture classification using autoregressive features in conjunction with machine learning approaches. IEEE Access **7**(Ml) 79354–79365(2019)

A Modified Inverse Gaussian Poisson Regression with an Exposure Variable to Model Infant Mortality

Selvi Mardalena, Purhadi$^{(\boxtimes)}$, Jerry Dwi Trijoyo Purnomo, and Dedy Dwi Prastyo

Department of Statistics, Faculty of Science and Data Analytics, Institut Teknologi Sepuluh Nopember, Surabaya 60111, Indonesia
purhadi@statistika.its.ac.id

Abstract. Infant mortality has generally been increasing and has become an issue that urgently needs to be addressed. As the number of infant deaths is count data, a Poisson regression model is needed to determine the causal factors. However, the assumption of equidispersion in Poisson regression is rarely satisfied. The overdispersion issue is frequently found in real data. Thus, this research employs mixed Poisson distribution modeling to overcome the overdispersion issue, namely, the inverse Gaussian Poisson regression (IGPR) model. In this study, a simple IGPR model, a modified IGPR model, and the negative binomial regression (NBR) model are compared. The results show that the modified IGPR model and the NBR model with an exposure variable outperform the benchmark, based on the global deviance and Akaike Information Criteria (AIC) value, to model the number of infant deaths in East Nusa Tenggara, Indonesia. The significant predictors that affect the number of infant mortalities are the percentage of complete basic immunization, the percentage of low birth weight (LBW), the percentage of babies under six months who receive exclusive breastfeeding, the percentage of infants who receive vitamin A, and the percentage of births assisted by health workers in the district.

Keywords: Poisson Inverse Gaussian · Negative binomial · Overdispersion · Exposure · Infant mortalities

1 Introduction

According to [1], the condition of residents across provinces in Indonesia today varies considerably. The total fertility rate (TFR) per woman of childbearing age (15–49 years) in some provinces, including East Nusa Tenggara, West Sulawesi, and North Sumatra, is relatively high and is estimated at approximately 2.5. Moreover, the TFR reached a significantly low value of below two in Jakarta, East Java, and Yogyakarta. In 2015, the Inter-Census Population Survey (SUPAS) implemented by the Central Bureau of Statistics discovered that the TFR was 2.28. Additionally, in the 2017 Revision of the World Population Prospects, the United Nations (UN) used the medium-fertility assumption for projecting TFR, where countries have been divided into three levels of fertility: high fertility (more than five births per woman), intermediate fertility (from 2.1 to 5 births per

© Springer Nature Singapore Pte Ltd. 2021
A. Mohamed et al. (Eds.): SCDS 2021, CCIS 1489, pp. 286–300, 2021.
https://doi.org/10.1007/978-981-16-7334-4_21

woman), and low fertility (fewer than 2.1 births per woman) [2]. The UN expected less than one percent of the world's population to live in countries with such high fertility levels in 2025–2030 [3].

Nonetheless, even though the birth rate continues to decrease, Indonesia is still classified as the fourth most populous country on earth. Moreover, owing to its significant influences on the mortality rate, especially infant mortality rates (IMRs), public health is as important as TFR in maintaining the balance of population growth. In addition, increased public health will also increase life expectancy in the long term.

Considering infant mortality as an important indicator of a community's overall physical health, it has become a long-standing issue faced by all countries. Concerning this matter, the Indonesia Basic Health Survey (IBHS) in 2017 discovered that the infant mortality rate (IMR) in Indonesia was 24 deaths per 1,000 live births. These data indicate that 24 of 1000 infants died before their first birthday in 2017. Additionally, East Nusa Tenggara is included in the top ten regions with a high infant and maternal mortality rate in Indonesia. Under this circumstance, Timor Tengah Selatan was perceived as a contributor to the highest number of infant mortalities, which is 114 cases. The number of infant deaths was considered as count data. To determine the factors that cause infant mortality, a research study in the regression model for count data was conducted.

Count data refers to the nonnegative integer value observations starting from zero. Modeling count data aim to estimate the parameters of a probability distribution that best represents the data. In this study, a model of count data is based on a Poisson distribution. The Poisson distribution has a single parameter that defines the mean and variance of the distribution. This feature is referred to as equidispersion, where the mean and variance of the dependent variable are equal. These assumption violations occur frequently. A larger variance than the mean (overdispersion) or vice versa (underdispersion) is sometimes found. Overdispersion is the major problem facing analysts when count data are modeled by Poisson regression. The infringement of these assumptions may result in decision-making errors in hypothesis testing and underestimate the estimated standard error [4].

Several models have been developed to overcome overdispersion, e.g., generalized Poisson regression (GPR) [5–7], negative binomial regression (NBR) [8–10], inverse Gaussian Poisson regression (IGPR) [11–13] and the Poisson-generalized Lindley distribution [14]. Some of these models are derived from a mixed Poisson distribution, which is a blend of the Poisson distribution with other distributions, continuous or discrete. The NBR is derived as a Poisson-gamma mixture model, with the dispersion parameter distributed as gamma shaped. In this study, the IGPR model is used. The IGPR model is based on an inverse Gaussian (IG) mixing distribution discussed in Sect. 2.

Compared to the NBR model, the IGPR can better deal with highly Poisson overdispersed data [15]. In 2016, [13] analyzed motor vehicle crash data using NBR and IGPR models. The results showed that the IGPR model performed better than the NBR model. Hence, the IGPR model is used in this study. [16] first introduced the inverse Gaussian Poisson (IGP) distribution in 1966. Many studies use this model because it has the closed-form likelihood function, and the calculation can be performed effortlessly [17].

In modeling the number of infant mortality cases using inverse Gaussian Poisson regression, an exposure variable is used to adjust the amount of opportunity for infant mortality. The exposure is the amount of time, space, distance, volume, or population size from which the dependent variable is counted. It may also be that the number of individuals at risk from a dependent variable is measured. In this study, the dependent variable was the number of infant deaths in each district of East Nusa Tenggara in 2018. However, the population size among districts will differ that cause the number of deaths over the districts is not comparable. Hence, an exposure variable should be included to allow the number of infant deaths to be comparable and reduce the bias in the model's estimate without including exposure. The dependent variable is the number of infant deaths in 2018 with the observation unit is district; thus, the appropriate exposure variable was the number of live births in each district of East Nusa Tenggara in the same year. Therefore, this research aims to model the number of infant deaths in East Nusa Tenggara Province in 2018 using the IGPR model with the exposure variable. Furthermore, the NBR model is used as a comparison to the IGPR model.

2 Materials and Methods

2.1 An Inverse Gaussian Poisson (IGP) Distribution

Suppose Y is the dependent variable and μ is the mean of Y. The dependent variable Y follows a Poisson distribution with the mean and variance μ, denoted by $Y \sim$ Poisson(μ). For a mixed Poisson distribution, the variance will exceed the conditional mean (overdispersion).

$$\exp\left(\mathbf{x}^T \boldsymbol{\beta} + \boldsymbol{\varepsilon}\right) = \mu \, exp(\varepsilon) = \mu v, \tag{1}$$

V is a positive specific random effect used to incorporate overdispersion. The shape or dispersion parameter depends on the specific distribution of random variable V [13]. Let $f(v)$ be the probability density function for V. Thus, the probability mass function for the mixed Poisson distribution is given by Eq. (2).

$$P(Y = y) = \int e^{-v\mu} \frac{(v\mu)^y}{y!} f(v) dv, \tag{2}$$

The IGP distribution is a mixture of Poisson and Inverse Gaussian distributions. Therefore, the random effects V follow an Inverse Gaussian (IG) distribution, in which the probability density function can be written as Eq. (3).

$$f(v) = \left(2\pi \tau v^3\right)^{-0,5} e^{-(v-1)^2/2\tau v}, \quad v > 0, \tag{3}$$

where $E(V) = 1$ and $Var(V) = \tau$ [11].

The IGP distribution consists of two parameters, μ (mean) as a location parameter and τ (dispersion parameter) as a shape parameter. Let Y follow the IGP distribution and be denoted by $Y \sim$ IGP (μ, τ). The probability density function for Y is formulated as in Eq. (4),

$$f(y; \mu, \tau) = \left(\frac{2z}{\pi}\right)^{1/2} \frac{\mu^y e^{1/\tau} K_s(z)}{(z\tau)^y y!}, \quad y \geq 0 \tag{4}$$

where $s = y - \frac{1}{2}, z = \frac{1}{\tau}\sqrt{2\mu\tau + 1}$ and $K_s(z) = K_{y-\frac{1}{2}}\left(\frac{1}{\tau}\sqrt{2\mu\tau + 1}\right)$ is the third modification of the Bessel function [18]. The expected value and the variance of the IGP distribution are [13].

$$E(Y) = E\{E(Y|\mu v)\} = E(\mu v) = \mu \tag{5}$$

$$Var(Y) = Var\{E(Y|\mu v)\} + E\{Var(Y|\mu v)\} = \mu + \tau\mu^2 \tag{6}$$

where τ is the overdispersion parameter $Var(V)$, which is caused by the presence of heterogeneity or diversity related to the observation unit with a specific character [11].

2.2 Overdispersion Test

Overdispersion on Poisson regression occurs when the variance of the dependent variable is higher than the mean. Overdispersion is caused by a positive correlation or excess variation between dependent probabilities. Overdispersion also arises when there is a violation of the assumption of data distribution; for example, when the data are grouped, it violates the assumption from likelihood independence [19].

The statistical test that can be used to detect overdispersion in data is found on the Applied Econometrics with R (AER) package of R software [20]. This test was developed by [21]. The null hypothesis is that the mean of the dependent variable is equal to its variance (equidispersion). The alternative hypothesis is overdispersion on the dependent variable; thus, $Var(Y_i) = \mu_i + \alpha g(\mu_i)$, where $g(.)$ is a specific function and α is a dispersion symbol. In other words, if the value of $\alpha = 0$, it can be considered equidispersion; otherwise, when $\alpha > 0$, it can be considered overdispersion.

2.3 Modified Inverse Gaussian Poisson Regression (IGPR) with an Exposure Variable

Let y_i be the dependent variable for observation i, where $i = 1, 2,..., n$ and $Y \sim$ IGP (μ, τ) and let $\mathbf{x}$ be a vector of associated p predictors. The regression model for Y that follows an IGP distribution with an associated predictor is called inverse Gaussian Poisson regression (IGPR). $E(Y) = \mu_i = e^{\mathbf{x}_i^T \boldsymbol{\beta}}$ where $\boldsymbol{\beta}$ is the vector of the parameter corresponding to each predictor, and xi is the vector of explanatory variables for the ith observation with dimension $(p + 1)$ x 1. In this study, the IGPR model is modified by adding the exposure variable (q_i) as the weight of each unit observation such that $\mu_i = q_i e^{\mathbf{x}_i^T \boldsymbol{\beta}}$ and the probability density function of Y as in Eq. (7).

$$P(Y_i = y_i | \mathbf{x}_i, \boldsymbol{\beta}, \tau) = \frac{\left(q_i e^{\mathbf{x}_i^T \boldsymbol{\beta}}\right)^{y_i} e^{\frac{1}{\tau}}}{y_i!} \left(\frac{2}{\pi\tau}\right)^{\frac{1}{2}} \left(1 + 2\tau q_i e^{\mathbf{x}_i^T \boldsymbol{\beta}}\right)^{-\frac{\left(y_i - \frac{1}{2}\right)}{2}} K_{s_i}(z_i) \tag{7}$$

Then, the modified IGPR model is shown in the following equation:

$$E(Y_i) = \mu_i = q_i e^{\mathbf{x}_i^T \boldsymbol{\beta}} \tag{8}$$

$$log\left(\frac{\mu_i}{q_i}\right) = e^{\mathbf{x}_i^T \boldsymbol{\beta}}$$

with $\mathbf{x}_i^T = \begin{bmatrix} 1 & x_{1i} & x_{2i} & \cdots & x_{pi} \end{bmatrix}$ is a predictor variable vector with $(p + 1)$ dimension on the i-th observation $(i = 1, 2, \ldots, n$ and $k = 1, 2, \ldots, p)$ and $\boldsymbol{\beta} = \begin{bmatrix} \beta_0 & \beta_1 & \beta_2 & \cdots & \beta_p \end{bmatrix}^T$ vector of regression coefficient with $(p + 1)$ x 1 dimension on the k-th predictor variable.

2.4 Parameter Estimation

The parameters of the IGPR model in Eq. (8) are estimated by the maximum likelihood method. The first step is to determine the likelihood function of the IGP distribution.

$$L(\boldsymbol{\beta}; \tau) = \prod_{i=1}^{n} P(Y_i = y_i | \mathbf{x}_i, \boldsymbol{\beta}, \tau)$$

$$= \prod_{i=1}^{n} \left\{ \frac{\left(q_i e^{\mathbf{x}_i^T \boldsymbol{\beta}}\right)^{y_i} e^{\frac{1}{\tau}}}{y_i!} \left(\frac{2}{\pi \tau}\right)^{\frac{1}{2}} \left(1 + 2\tau q_i e^{\mathbf{x}_i^T \boldsymbol{\beta}}\right)^{-\frac{\left(y_i - \frac{1}{2}\right)}{2}} K_{s_i}(z_i) \right\} \quad (9)$$

The likelihood function is transformed into the form of the logarithm (log).

$$l(\boldsymbol{\beta}; \tau) = log L(\boldsymbol{\beta}; \tau)$$

$$= \sum_{i=1}^{n} y_i log(q_i) + \sum_{i=1}^{n} y_i \mathbf{x}_i^T \boldsymbol{\beta} + \frac{n}{\tau} - \sum_{i=1}^{n} log(y_i!) + \frac{n}{2} log\left(\frac{2}{\pi}\right) - \frac{n}{2} log(\tau)$$

$$- \sum_{i=1}^{n} \left(\frac{2y_i - 1}{4}\right) log\left(1 + 2\tau q_i e^{\mathbf{x}_i^T \boldsymbol{\beta}}\right) + \sum_{i=1}^{n} log\left(K_{s_i}(z_i)\right) \quad (10)$$

There are two primary algorithms to fit the IGPR model with respect to $\boldsymbol{\beta}$ and τ, namely, the Rigby Stasinopoulos (RS) [22] and Cole Green (CG) algorithms [23]. The RS method does not use the cross derivative of the ln likelihood, while the CG algorithm requires information on the first and second cross derivatives of the log-likelihood function with respect to parameters μ and τ. Both RS and CG algorithms used three nested components: the outer iteration, the inner iteration (local scoring algorithm), and the modified backfitting algorithm. Convergence occurs when all three algorithms have converged. The algorithm is implemented in the options method in the function gamlss() within the R package GAMLSS. The combination of both algorithms is also allowed with method = mixed() [24].

In this study, we use the RS algorithm because, empirically, there are no differences among the RS, CG, or mixed algorithm results. Let $\boldsymbol{\theta}$ be the vector of the parameter where $\theta_1 = \mu$ and $\theta_2 = \tau$, thus, $j = 1, 2$. The modified (iterative) dependent variable for fitting the parameter is given by:

$$\mathbf{z}_j = \boldsymbol{\eta}_j + \mathbf{w}_j^{-1} \circ \mathbf{u}_j$$

where z_j, η_j, w_j and u_j are all vectors of length n, e.g., weights vector $\mathbf{w} = \left(w_{j1}, w_{j2}, ..., w_{jn},\right)^T$, $\mathbf{w}_j^{-1} \circ \mathbf{u}_j = \left(w_{j1}^{-1} u_{j1}, w_{j2}^{-1} u_{j2}, ..., w_{jn}^{-1} u_{jn}\right)^T$, is the diagonal matrix of iterative weights, and $\eta_j = g(\theta_j) = X_j \beta_j$ is the predictor vector of the parameter vectors and

$$\mathbf{u}_j = \frac{\partial l}{\partial \eta_j} = \left(\frac{\partial l}{\partial \theta_j}\right)^\circ \left(\frac{\partial \theta_j}{\partial \eta_j}\right)$$

is the score function (the first derivative of the log-likelihood with respect to each parameter corresponding to each predictor). The $\mathbf{w}$ is the iterative weights defined as:

$$\mathbf{w}_j = -\mathbf{f}_j \circ \left(\frac{\partial l}{\partial \theta_j}\right)^\circ \left(\frac{\partial \theta_j}{\partial \eta_j}\right)$$

where there are three different ways to determine $\mathbf{f}_j$ depending on the information available for the specific distribution in the RS algorithms below:

$$\mathbf{f}_j = \begin{cases} E\left(\frac{\partial^2 l}{\partial \theta_j^2}\right) & \text{; if the expectation exists, leading to a Fisher's scoring algorithm,} \\ \frac{\partial^2 l}{\partial \theta_j^2} & \text{; leading to the standard Newton} - \text{Raphson scoring algorithm,} \\ -\left(\frac{\partial l}{\partial \theta_j}\right)^\circ \left(\frac{\partial l}{\partial \theta_j}\right) & \text{; leading to a quasi Newton scoring algorithm.} \end{cases}$$

Let r be the outer iteration index, t be the inner iteration index, and m be the backfitting index. The RS algorithm for the IGPR model is as follows [25]:

Step 1: Start – initialize fitted values $\theta_j^{(1,1)}$. Evaluate the initial value of linear predictors $\eta_j^{(1,1)} = g\left(\theta_j^{(1,1)}\right)$. Note that the initial value of the RS algorithm is only needed for the distribution parameter vectors rather than for the β parameter. The straightforward starting values for the parameter vectors cause the algorithm to have commonly been observed to be stable and rapidly converge [26].

Step 2: Start the outer iteration $r = 1, 2, ...$ until convergence.

a. Start the inner iteration $t = 1, 2, ...$ until convergence.

   (i)   Evaluate the current result.
   (ii) Start the backfitting iteration $m = 1, 2, ...$ until convergence.
   (iii) Fit WLS by regressing the current partial residuals $\varepsilon_0^{(r,t,m)} = z_j^{(r,t)}$ design matrix X with $X_1 = \left[\mathbf{x}_1^T \; \mathbf{x}_2^T \; \cdots \; \mathbf{x}_i^T \; \cdots \; \mathbf{x}_n^T \right]$, using the iterative weights $w_j^{(r,t)}$ to obtain the updated parameter estimates $\beta_j^{(r,t,m+1)}$.
   (iv) End the backfitting iteration on the convergence of $\beta_j^{(r,t)}$ and set $\beta_j^{(r,t,\cdot+1)} = \beta_j^{(r,t,\cdot)}$; otherwise, update m and continue the backfitting iteration.

(v) Calculate the updated $\eta_j^{(r,t+1)}$ and $\theta_j^{(r,t+1)}$.

b. End the inner iteration on the convergence of $\beta_j^{(r,t)}$ and set $\beta_j^{(r,t,\cdot+1)} = \beta_j^{(r,t,\cdot)}$, $\eta_j^{(r+1,1)} = \eta_j^{(r,\cdot)}$, and $\theta_j^{(r+1,1)} = \theta_j^{(r,\cdot)}$; otherwise, update t and continue the inner iteration.

*Note that the design matrix X for $\eta_2 = g(\theta_2) = g(\tau) = \mathbf{X}_k\beta_k$ is $\mathbf{X}_2 = \begin{bmatrix} 1 & 1 & \cdots & 1 \end{bmatrix}^T$.

Step 3: Update the value of j.
Step 4: End the outer iteration if the change in the likelihood is sufficiently small or the global deviance has converged; otherwise, update r and continue the outer iteration.

2.5 Parameter Testing of Inverse Gaussian Poisson Regression

The IGPR parameter testing is calculated using a maximum likelihood ratio test (MLRT), which includes hypothesis testing simultaneously on parameter β as well as partial testing for parameters β and τ. The hypothesis of β parameter testing simultaneously is that all parameters are equal to zero versus at least one $\beta_k \neq 0$ with $k = 1, 2, ..., p$. The test statistic used is in Eq. (11).

$$G = 2\left[log\, L(\hat{\Omega}) - log\, L(\hat{\omega}) \right] \qquad (11)$$

$L(\hat{\omega})$ is a maximum likelihood value for the model under the null hypothesis (does not include the predictor variables). Moreover, $L(\hat{\Omega})$ is a maximum likelihood value for a model under the population that consists of all predictor variables. The likelihood function for each model is shown in Eqs. (12) and (13).

$$L(\hat{\Omega}) = \prod_{i=1}^{n} P(Y_i = y_i | \mathbf{x}_i, \beta, \tau) \qquad (12)$$

$$L(\hat{\omega}) = \prod_{i=1}^{n} P(Y_i = y_i | \beta_0, \tau_\omega) \qquad (13)$$

Substituting Eq. (12) and (13) to Eq. (11), the G statistics for the IGP model is:

$$G = 2\Big\{ \Big[\sum_{i=1}^{n} y_i log(q_i) + \sum_{i=1}^{n} y_i \mathbf{x}_i^T \beta + \frac{n}{\hat{\tau}} - \sum_{i=1}^{n} log(y_i!) - \frac{n}{2}log(\hat{\tau}) - $$
$$\sum_{i=1}^{n} \left(\frac{2y_i - 1}{4} \right) log\left(1 + 2\hat{\tau} q_i e^{\mathbf{x}_i^T \hat{\beta}} \right) + \sum_{i=1}^{n} log(K_{s_i}(z_i)) \Big] - $$
$$\Big[\sum_{i=1}^{n} y_i log(q_i) + \sum_{i=1}^{n} y_i \hat{\beta}_0 + \frac{n}{\hat{\tau}_{\hat{\omega}}} - \sum_{i=1}^{n} log(y_i!) - \frac{n}{2}log(\hat{\tau}_{\hat{\omega}}) - $$
$$\sum_{i=1}^{n} \left(\frac{2y_i - 1}{4} \right) log\left(1 + 2\hat{\tau}_{\hat{\omega}} q_i e^{\hat{\beta}_0} \right) + \sum_{i=1}^{n} log(K_{s_i}(z_{i\hat{\omega}})) \Big] \Big\} \qquad (14)$$

The statistic G is an approximation of the chi-square distribution $\chi^2_{(\alpha,df)}$. The test criterion is to reject H_0 if $G > \chi^2_{(\alpha,df)}$, where df is the degrees of freedom derived from the number of parameters under the population that reduces the number of parameters under the null hypothesis.

2.6 Data Example

This study uses secondary data on the number of infant deaths and the factors that influence it. Data were obtained from the East Nusa Tenggara Health Department. Additionally, the data consist of 22 districts in the year 2018.

The dependent variable used in this study is the number of infant mortality which includes infants who were born alive but who died before the age of one year. Infant mortality is categorized into three types: (1) perinatal mortality, i.e., the mortality that occurs in babies that die before the age of one week, including stillbirth; (2) neonatal mortality, i.e., the mortality that occurs in infants before the age of 28 days; and (3) post-neonatal mortality, the mortality that occurs in infants between the ages of 28 days up until one year [27]. The infant mortality rate (IMR) is the number of infant deaths under one-year-old per 1,000 live births in a given year. Thus, the number of infant deaths includes perinatal, neonatal, and post-neonatal mortality.

The frame of mind regarding the causes of infant and child mortality is described in [28]. Infant and child mortality related socioeconomic factors are divided into five categories: (1) maternal factors, such as education level, parity, and age; (2) environmental factors, such as the condition of water, food, air, and disease-carrying insects; (3) nutritional factors, such as breastfeeding, feeding patterns, lack of calories, protein, vitamins, and others; (4) injury factors, such as accidents; and (5) individual disease control factors, in the form of prevention and treatment.

According to the framework in [35], thus the independent variables used in this study are the antennal care visits by pregnant women, the birth assisted by health workers, the complete neonatal visits, the low birth weight (LBW), babies under six months who get exclusive breastfeeding, the complete basic immunization, the infant received vitamin A, and the infant health services.

3   Results and Discussion

3.1   Results

Knowing the characteristics of the research variables is essential in data analysis. Although descriptive analysis cannot provide definitive estimates, it helps researchers determine the case study for further analysis. An overview of the number of infant deaths in East Nusa Tenggara by district can be seen in Fig. 1. Based on Fig. 1, Timor Tengah Selatan has the highest infant mortality, with 114 cases. Malaka has the lowest infant mortality case number, with as many as 16 cases.

The distribution of the number of infant deaths is presented in Fig. 2. We used two different count data distributions, namely, the inverse Gaussian Poisson distribution and the negative binomial distribution. The Akaike information criterion (AIC) is used to choose between those two distributions. The AIC value of the IGP distribution is lower than that of the NB distribution. Thus, the IGP model is the appropriate model for the number of infant deaths in East Nusa Tenggara in 2018.

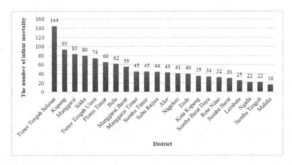

Fig. 1. The number of infant mortalities in East Nusa Tenggara province in 2018

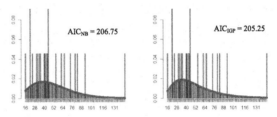

Fig. 2. Fitting a distribution for the number of infant deaths in East Nusa Tenggara

Table 1 shows the mean and standard deviation of predictor variables, which are assumed to influence the number of infant deaths.

Table 1. Predictor variables and their characteristic (unit measurement is a district).

Variables	Natural and environmental factors	Mean (SD[a])	VIF[b]
Y	The number of infant deaths	51.41 (29.68)	–
X_1	The percentage of antenatal care visits by pregnant women	90.08 (11.07)	6.053
X_2	The percentage of births assisted by health workers	89.92 (8.25)	5.418
X_3	The percentage of complete neonatal visits	89.53 (19.74)	1.129
X_4	The percentage of Low Birth Weight (LBW)	10.07 (17.88)	1.924
X_5	The percentage of babies under six months who get exclusive breastfeeding	69.76 (24.44)	1.529
X_6	The percentage of complete basic immunization	70.53 (16.40)	3.642
X_7	The percentage of infants received vitamin A	94.54 (21.76)	3.119
X_8	The percentage of infant health services	95.17 (26.67)	4.017
q	The number of live births	4377 (2124)	–

a: SD = Standard deviation.
b: VIF = Variance Inflation Factor.

The results showed that the mean percentage of babies under six months who received exclusive breastfeeding was 69.76%. This result indicated that only approximately 69% of babies under six months receive exclusive breastfeeding, while others do not. Additionally, the mean percentage of complete basic immunization is 70.53%. These two factors are very important to maintain the baby's immune system.

Breastfeeding is a complete source of nutrients, helping the growth and physical development and brain development of children. Moreover, complete basic immunization for babies is carried out to avoid diseases such as hepatitis, polio, and measles. If the baby's immune system is weak, then the baby is susceptible to disease, and the risk of the baby dying is even greater. Furthermore, the number of live births was used as the weight of observation. It is assumed that a district is different from another district, so data across districts are worth comparing [28].

Because the model uses more than one predictor variable, it is necessary to perform the multicollinearity test. One way to detect multicollinearity is by the variance inflation factor (VIF), a value that describes the increase in the estimated parameter variance between the predictor variables. If the VIF value is more than 10, multicollinearity can occur. If the VIF value is less than 10, it can be said there is no multicollinearity, and regression modeling can be continued [29]. Based on the results in Table 1, the value of VIF for each predictor variable was no more than 10. The conclusion was no multicollinearity among predictor variables such that all of the predictor variables had fulfilled the non-multicollinearity assumption and could be used for regression analysis.

Using the AER package in R software, $\alpha = 10.909$ had a P-value of 0.000153, which is smaller than the significance level of 5%, resulting in the decision to reject H_0. The conclusion is that the variance is not equal to the mean, so the data experience overdispersion.

The IGPR analysis is performed because of the overdispersion results. Compared to the proposed modified IGPR model, we used a simple IGPR model without an exposure variable and the NBR model with an exposure variable. The selection of variables in the models was chosen by a stepwise method. Hence, the models only contain a significant variable at a 5% or 10% significance level. The results of the IGPR models can be seen in Table 2, and the results of the NBR model are presented in Table 3.

The hypothesis test simultaneously for all possible models was performed by the MLRT method. Table 2 and Table 3 show that the test statistic G for the IGPR and NBR models is larger than $\chi^2_{(\alpha, df)}$. Hence, the models are significant or have at least one variable significant in the model. Furthermore, two different significant variables influenced the number of infant deaths in both models. The models can be written as follows:

$$\hat{\mu}_{1i} = \exp(4.758 - 0.009X_{3i}) \tag{15}$$

$$\hat{\mu}_{2i} = q_i \exp(0.191 + 0.037X_{1i} - 0.070X_{2i} - 0.008X_{3i} + 0.009X_{5i} - 0.016X_{7i}) \tag{16}$$

$$\hat{\mu}_{3i} = q_i \exp(0.184 + 0.030X_{1i} - 0.072X_{2i} - 0.007X_{3i} + 0.009X_{5i} - 0.017X_{7i}) \tag{17}$$

Table 2. The IGPR Model Parameters Testing

Model without exposure variable				Model with exposure variable		
Parameter	Estimate	Standard Error	P-value	Estimate	Standard Error	P-value
β_0	4.758	0.485	$7.11 \times 10^{-9}*$	0.191	1.028	0.8549
β_1	–	–	–	0.037	0.012	0.0054*
β_2	–	–	–	−0.070	0.016	0.0005*
β_3	−0.009	0.005	0.0942	−0.008	0.003	0.0213*
β_4	–	–	–	–	–	–
β_5	–	–	–	0.009	0.003	0.0102*
β_6	–	–	–	–	–	–
β_7	–	–	–	−0.016	0.004	0.0009*
β_8	–	–	–	–	–	–
τ	−1.4164	0.357	0.0008*	−2.763	0.442	$1.57 \times 10^{-5}*$
Global deviance	198.48		$7.97 \times 10^{-44}*$	176.63		$1.76 \times 10^{-35}*$
AIC	204.48			190.63		

* Significant at 5% level

Table 3. The NBR Model Parameters Testing

The NBR model with exposure variable			
Parameter	Estimate	Standard Error	P-value
β_0	0.184	1.018	0.8591
β_1	0.038	0.011	0.0046*
β_2	−0.072	0.016	0.0004*
β_3	−0.007	0.003	0.0516
β_4	–	–	–
β_5	0.009	0.003	0.0111*
β_6	–	–	–
β_7	−0.017	0.004	0.0007*
β_8	–	–	–
τ	−2.839	0.411	$5.01 \times 10^{-6}*$
Global deviance	175.22		$3.52 \times 10^{-35}*$
AIC	189.22		

* Significant at 5% level

The model in Eq. (15) refers to a simple IGPR model without an exposure variable, the model in Eq. (16) refers to a modified IGPR model with an exposure variable, and the model in Eq. (17) refers to the NBR model with an exposure variable.

3.2 Discussion

In this section, we will discuss the results of each model in Eqs. (15), (16), and (17). In addition to interpretation, we will also discuss the pattern of relationships between dependent variables and predictor variables and how well these models predict the number of infant deaths in East Nusa Tenggara.

The significant variable that affects the number of infant deaths (Y) by a simple IGPR model in Eq. (15) is the percentage of complete neonatal visits (X3). The percentage of complete neonatal visits (X3) negatively relates to the number of infant deaths. Therefore, improving complete neonatal visits will reduce the number of infant deaths. Neonatal visits are an essential means to reduce infant mortality. Through this facility, monitoring the baby's health condition can prevent various diseases that might threaten the baby's health. However, the field conditions sometimes do not deliver optimal health services, both in service and community care. Hence, cooperation between the government and the community is needed to achieve the objectives of this facility.

The significant variables that affect the number of infant deaths (Y) by a modified IGPR model in Eq. (16) are the percentage of antenatal care visits by pregnant women (X_1), the percentage of births assisted by health workers (X_2), the percentage of complete neonatal visits (X_3), the percentage of babies under six months who receive exclusive breastfeeding (X_5), and the percentage of infants who receive vitamin A (X_7). Variable X_2 has the most considerable effect and an appropriate relationship with variable Y. Variables X_3 and X_7 have negative dependencies with variable Y, while variables X_1 and X_5 have an inappropriate relationship with variable Y. Births assisted by health workers (X_2) can prevent complications during childbirth as well as provide first aid for mothers and babies after delivery to reduce infant mortality. Neonatal visits (X_3) and vitamin A for infants (X_4) also have an important role in monitoring the baby's health to avoid various diseases. A lack of vitamin A can cause preventable blindness in children and increase the risk of illness and death. The intake of vitamin A from daily food is still low, so nutritional supplementation in the form of vitamin A capsules is needed.

Although the percentage of antenatal care visits by pregnant women (X_1) and the percentage of babies under six months who receive exclusive breastfeeding (X_5) have an inappropriate relationship with the number of infant deaths (Y), these two factors must also be considered. Antenatal care has an important role in monitoring the mother's health during pregnancy to prevent complications during childbirth. The goal, of course, is to minimize the risk of maternal and infant mortality. Moreover, exclusive breastfeeding plays a role in maintaining the baby's immune system. Breastfeeding contains colostrum rich in antibodies and it contains high amounts of protein for endurance. Thus, exclusive breastfeeding can reduce the risk of death in infants.

The significant variables that affect the number of infant deaths (Y), as modeled using the NBR model in Eq. (17), are the percentage of antenatal care visits by pregnant women (X_1), the percentage of births assisted by health workers (X_2), the percentage of babies under six months who receive exclusive breastfeeding (X_5), and the percentage of infants who receive vitamin A (X_7). Variables X_2 and X_7 have negative dependencies with variable Y, while variables X_1 and X_5 have an inappropriate relationship with variable Y.

Based on Table 2 and Table 3, it was found that the AIC values of the modified IGPR model and the NBR model were almost the same. Furthermore, the mean square of error (MSE) and the root mean square of error (RMSE) of the modified IGPR model and the NBR model gave the same results, as shown in Table 4.

Table 4 shows that the RMSE values of the modified IGPR model and the NBR model are close to the standard deviation of the number of infant deaths in East Nusa Tenggara, which is reported in Table 1. Thus, the predicted dependents of these empirical results are relatively close to the observations' values.

Table 4. The MSE and RMSE of a simple IGPR model, a modified IGPR model, and the NBR model

	A simple IGPR model	A modified IGPR model	The NBR model with an exposure variable
MSE	790.04	242.02	242.02
RMSE	28.11	15.56	15.56

To support the RMSE results, the comparison between the observed and estimated values for the models is shown in Fig. 3. The fitting values for the number of infant deaths (Y) by a modified IGPR model and the NBR model are better than those of a simple IGPR model.

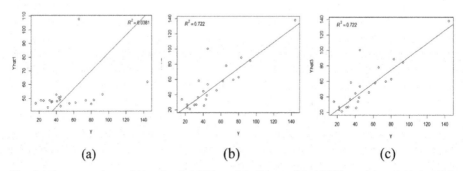

(a)   (b)   (c)

Fig. 3. The comparison of (a) a simple IGPR model, (b) a modified IGPR model, and (c) the NBR model with exposure variable

The R^2 measure is a tool to determine the predictive ability of a linear regression model. However, the R^2 measure is biased when the sample size is small and the number of covariates in the model is large. Hence, the adjusted R^2 is preferable in this study [30]. Figure 2 shows that the modified IGPR model and the NBR model have higher R-squared values than the simple IGPR model. Hence, the modified IGPR and the NBR model have better predictive ability than the simple IGPR model.

4 Conclusion

The modified IGPR model with an exposure variable is a model for count data with overdispersion. The exposure variable allows the dependent variable to be interpreted as a count from the original scale rate. The parameter estimation method employs the maximum likelihood estimation (MLE) approach. Because the MLE solution is not closed form, the iterative method, i.e., the Rigby and Stasinopoulos (RS) algorithm, was applied. The proposed method is applied to real data on the infant mortality rate in East Nusa Tenggara Province in Indonesia.

The descriptive statistics show that the highest number of infant deaths in 2018 occurred in Timor Tengah Selatan, with 144 cases. The statistical analysis of the number of infant deaths experienced overdispersion such that the IGPR model plays a role in overcoming that issue. This study used a simple IGPR model, a modified IGPR model, and the NBR model with an exposure variable. According to this study's results, the modified IGPR model and the NBR model with the exposure variable are better than the simple IGPR model without an exposure variable based on the global deviance and AIC value.

The modified IGPR model and the NBR model give the same results for predictive ability. However, there are differences in the significant factors in the model. The significant factors obtained from the modified IGPR model that affect infant mortality in East Nusa Tenggara Province are the percentage of complete basic immunization, the percentage of low birth weight (LBW), the percentage of babies under six months who receive exclusive breastfeeding, the percentage of infants who receive vitamin A, and the percentage of births assisted by health workers in the district.

References

1. Ministry of Health of Republic of Indonesia, Profil Kesehatan Indonesia 2017. Jakarta, Indonesia: Ministry of Health of Republic of Indonesia (2018)
2. United Nations, Department of Economic and Social Affairs, Population Division, World Population Prospects: the 2017 Revision—Methodology of the United Nations Population Estimates and Projections. World Popul. Prospect. 2017 Revis., p. 48 (2017)
3. United Nations. The End of High Fertility is Near (2017)
4. Hilbe, J.M.: Varieties of Count Data in Modeling Count Data, pp. 1–34. Cambridge University Press, Cambridge (2014)
5. Consul, P.C., Famoye, F.: Generalized poisson regression model. Commun. Stat. – Theor. Methods **21**(1), 89–109 (1992)
6. Harris, T., Yang, Z., Hardin, J.W.: Modeling underdispersed count data with generalized Poisson regression. Stata J. **12**(4), 736–747 (2012)
7. Sellers, K.F., Morris, D.S.: Underdispersion models: Models that are 'under the radar.' Commun. Stat. – Theor. Methods **46**(24), 12075–12086 (2017)
8. Hutchinson, M.K., Holtman, M.C.: Analysis of count data using Poisson regression. Res. Nurs. Heal. **28**(5), 408–418 (2005)
9. Lawless, J.F.: Negative binomial and mixed Poisson regression. Can. J. Stat. **15**(3), 209–225 (1987)
10. Stasinopoulos, D.M., Rigby, R.A.: Generalized additive models for location scale and shape (GAMLSS) in R. J. Stat. Softw. **23**(7), 1–46 (2007)

11. Dean, C., Lawless, J.F., Willmot, G.E.: A mixed Poisson-inverse-Gaussian regression model. Can. J. Stat. **17**(2), 171–181 (1989)
12. Hilbe, J.M.: Poisson Inverse Gaussian Regression in Modeling Count Data, pp. 162–171. Cambridge University Press, Cambridge (2014)
13. Zha, L., Lord, D., Zou, Y.: The Poisson inverse Gaussian (PIG) generalized linear regression model for analyzing motor vehicle crash data. J. Transp. Saf. Secur. **8**(1), 18–35 (2016)
14. Wongrin, W., Bodhisuwan, W.: The Poisson-generalised Lindley distribution and its applications. Songklanakarin J. Sci. Technol. **38**(6), 645–656 (2016)
15. Hilbe, J.M.: Modeling Count Data. Cambridge University Press, Cambridge (2014)
16. Holla, M.S.: On a Poisson-inverse Gaussian distribution. Metr. Int. J. Theor. Appl. Stat. **11**(1), 115–121 (1967)
17. Karlis, D., Xekalaki, E.: Mixed Poisson distributions. Int. Stat. Rev. **73**(1), 35–58 (2005)
18. Willmot, G.E.: The Poisson-inverse gaussian distribution as an alternative to the negative binomial. Scand. Actuar. J. **1987**(3–4), 113–127 (1987)
19. Hilbe, J.M.: Testing Overdispersion in Modeling Count Data, pp. 74–107. Cambridge University Press, Cambridge (2014)
20. Dormann, C.F.: Overdispersion, and how to deal with it in R and JAGS (2016)
21. Cameron, A.C., Trivedi, P.K.: Regression-based tests for overdispersion in the Poisson model. J. Econom. **46**(3), 347–364 (1990)
22. Rigby, R.A., Stasinopoulos, D.M.: Mean and dispersion additive models. Stat. Theor. Comput. A. Smoothing, 215–230 (1996). https://doi.org/10.1007/978-3-642-48425-4_16
23. Cole, T.J., Green, P.J.: Smoothing reference centile curves: the LMS method and penalized likelihood. Stat. Med. **11**(10), 1305–1319 (1992)
24. Stasinopoulos, M.D., Rigby, R.A., Heller, G.Z., Voudouris, V., Bastiani, D.F.: Flexible Regression and Smoothing: Using GAMLSS in R. CRC Press, Boca Raton (2017)
25. Rigby, R.A., Stasinopoulos, D.M.: Generalized additive models for location, scale and shape. J. R. Stat. Soc. Ser. C Appl. Stat. **54**(3), 507–554 (2005)
26. Stasinopoulos, M.D., Rigby, R.A., Heller, G.Z., Voudouris, V., Bastiani, D.F.: Flexible regression and smoothing: Using GAMLSS in R (2017)
27. UNICEF, WHO. The World Bank, and United Nations, "Child Mortality 2015" (2015)
28. Mosley, W.H., Chen, L.C.: An analytical framework for the study of child survival in developing countries. Popul. Dev. Rev., **10**, 25–45 (1984). Supplement: Child Survival: Strategies for Research
29. Montgomery, D.C., Peck, E.A., Vining, G.G.: Introduction to Linear Regression Analysis - Douglas C Elizabeth A. Peck, G. Geoffrey Vining - Google Books. Wiley, Hoboken (2021)
30. Heinzl, H., Mittlböck, M.: Adjusted R2 measures for the inverse Gaussian regression model. Comput. Stat. **17**(4), 525–544 (2002). https://doi.org/10.1007/s001800200125

Poisson and Logistic Regressions for Inhomogeneous Multivariate Point Processes: A Case Study in the Barro Colorado Island Plot

Ahmad Husain[✉] and Achmad Choiruddin

Department of Statistics, Institut Teknologi Sepuluh Nopember, Surabaya, Indonesia
choiruddin@its.ac.id

Abstract. This study aims to extend the estimating equations based on the Poisson and logistic regression likelihoods to model the intensity of a multivariate point process. The proposed approaches result in a framework equivalent to the estimation procedure for generalized linear model. The estimation is different from the existing methods where repetition independently with respect to the number of types of point process is obliged. Our approach does not require repetition and hence could be computationally faster. We implement our method to analyze the distribution of 9-species of trees in the Barro Colorado Island rainforest with respect to 11-environmental variables.

Keywords: Logistic regression · Multivariate point pattern · Poisson regression

1 Introduction

Multivariate point patterns data have been increasing recently in applications. For examples, Baddeley et al. [1] investigated the spatial patterns of several types of spines on a dendrite network, while Jun et al. [2] analyzed the effect of atmospheric variable variations on the formation of rain types in the Pacific ocean by using satellite imagery data. Recently, Hessellund et al. [3] studied the arrangement of the locations of multi-type crime events such as robbery, theft, car robbery, and sexual harassment.

The motivation of this paper comes from a study in ecology where more than 350,000 individual trees from around 300 species are recorded in a 50-hectare region of the tropical forest of Barro Colorado Island (BCI) in central Panama [4]. Regarding the highly multivariate data, typically, the main questions are how the high number of tree species continue to coexists and how to understand their habitats by studying their relation to the environment. To answer such questions, first, the locations of each of the tree species is regarded as a spatial point pattern data generated from spatial point process. Second, analysis using a statistical methodology based on spatial point process has been conducted and

© Springer Nature Singapore Pte Ltd. 2021
A. Mohamed et al. (Eds.): SCDS 2021, CCIS 1489, pp. 301–311, 2021.
https://doi.org/10.1007/978-981-16-7334-4_22

developed [5–9]. For examples, the dependence structure among trees species was studied by [7,9,10] while the interaction between each species of tree and the environmental variables were investigated using estimating functions obtained from the Poisson or logistic likelihoods [11,12] or its regularized versions [8].

When the main concern is to understand the habitats of M number of species of trees by assessing their relation to environmental variables, the intensity of multivariate point process should be modeled and studied. While massive studies for a single species have been conducted [e.g., 5,8,11,13,14], the methodology for multivariate point patterns data is unclear except by only repeating the procedure for single species M times [e.g., 15]. Although there might not be any theoretical issue, computational problems would arise, especially when the number of species is large.

This paper extends the use of the estimating equation based on Poisson and logistic likelihoods for inhomogeneous multivariate point process. The intensity model is restructured to fit multi-species distribution and is described in terms of a linear combination of covariates in the form of a matrix. By such a strategy, the estimation procedure is only conducted once (no repetition is required) and, therefore, could gain computational efficiency.

The remainder of the paper is organized as follows. Section 2 describes the methodology, while Sect. 3 details the data description. Results are presented in Sect. 3.2, and conclusion is provided in Sect. 4.

2   Methodology

2.1   Multivariate Point Processes

The multivariate point pattern data is denoted by $\mathbf{x} = \{\mathbf{x}^{(1)}, ..., \mathbf{x}^{(M)}\}$, representing the collection of locations of events observed in $\mathcal{D}$ which could be identified by its type $\mathbf{x}^{(m)}, m = 1, \cdots, M$. In our study, $\mathbf{x}$ represents the set consisting of all recorded tree locations while $\mathbf{x}^{(m)}$ is the set of tree locations of species m. The underlying process generating such a point pattern is a multivariate point process $\mathbf{X}$ in $\mathbb{R}^2$. The multivariate point process can be written by $\mathbf{X} = \{\mathbf{X}^{(1)}, ..., \mathbf{X}^{(M)}\}$, where $\mathbf{X}^{(m)}$ is a point process of type m [6]. If each type of point process $\mathbf{X}^{(m)}$ is an inhomogeneous Poisson point process (IPP) with intensity function λ_β, then

$$\mathbb{E}[N(\mathbf{X}^{(m)} \cap A] = \int_A \lambda_\beta(u, m)du, \tag{1}$$

where $N(\mathbf{X}^{(m)} \cap A)$ is the number of points process of type m located at $A \subseteq \mathcal{D}$.

The intensity of each process has a log-linear form [5,14,16]

$$\lambda_\beta(u, m) = \exp(\boldsymbol{\beta}_{(m)}^\top \mathbf{Z}(u)), \ u \in \mathcal{D}, \tag{2}$$

where $\mathbf{Z}(u) = (Z_1(u), ..., Z_p(u))^\top$ is a vector of p spatial covariates such as the environmental variables depicted by Fig. 1, and $\boldsymbol{\beta}_{(m)} = (\beta_{m1}, ..., \beta_{mp})^\top$ is the corresponding regression parameters. One way to assess $\boldsymbol{\beta}_{(m)}$ is to model each point process separately and repeat it M times. In this study, we introduce the multivariate intensity function and estimate all the parameters simultaneously. The process $\mathbf{X}$ is an inhomogeneous multivariate Poisson point process (IMPP) with log-intensity:

$$\log(\Lambda_\beta(u)) = \bar{\boldsymbol{\beta}}\mathbf{Z}(u),\tag{3}$$

where each component of (3) is

$$\begin{pmatrix} log(\lambda_\beta(u,1)) \\ log(\lambda_\beta(u,2) \\ \vdots \\ log(\lambda_\beta(u,M)) \end{pmatrix} = \begin{pmatrix} \beta_{11} & \beta_{12} & \dots & \beta_{1p} \\ \beta_{21} & \beta_{22} & \dots & \beta_{2p} \\ \vdots & \vdots & \ddots & \vdots \\ \beta_{M1} & \beta_{M2} & \dots & \beta_{Mp} \end{pmatrix} \begin{pmatrix} Z_1(u) \\ Z_2(u) \\ \vdots \\ Z_p(u) \end{pmatrix}.\tag{4}$$

By constructing (4), the parameters $\bar{\boldsymbol{\beta}} = (\boldsymbol{\beta}_{(1)}^\top, \dots, \boldsymbol{\beta}_{(M)}^\top)$ can be estimated simultaneously using the estimating equations based on Poisson and logistic likelihoods detailed in Sects. 2.2 and 2.3.

2.2 Computational Strategy

Poisson Regression. The log-likelihood function for an IMPP in Poisson regression is given by

$$\ell_{\text{IMPP}}(\boldsymbol{\beta}) = \sum_{m=1}^{M} \sum_{u\in\mathbf{x}^{(m)}} \log(\lambda_\beta(u,m)) - \sum_{m=1}^{M} \int_\mathcal{D} \lambda_\beta(u,m)du,\tag{5}$$

where $\lambda_\beta(u,m)$ is given by (2). The log-likelihood function involves integral which requires numerical approach to evaluate. One could consider the "Berman-Turner" method [17] by approximating the integral term of (5) such that

$$\int_\mathcal{D} \lambda_\beta(u,m)du \approx \sum_{j=1}^{n_m+d_m} \lambda_\beta(u_j,m)w_j^{(m)},\tag{6}$$

where n_m and d_m is respectively the number of data and dummy points for the point pattern of type m and where w is a quadrature weight such that $\sum_j w_j = |\mathcal{D}|$. The log-likelihood function (5) then becomes:

$$\ell_{\text{IMPP}}(\boldsymbol{\beta}) \approx \sum_{m=1}^{M} \left(\sum_{j=1}^{n_m} \log(\lambda_\beta(u_j,m)) - \sum_{j=1}^{n_m+d_m} \lambda_\beta(u_j,m)w_j^{(m)} \right).\tag{7}$$

Suppose $y_j^{(m)} = \mathbf{1}((u_j,m)\in\mathbf{x})/w_j^{(m)}$, where $\mathbf{1}((u_j,m)\in\mathbf{x})$ represents an indicator function whether or not (u_j,m) are data points, Eq. (5) writes

$$\ell_{\text{IMPP}}(\boldsymbol{\beta}) \approx \sum_{m=1}^{M} \left(\sum_{j=1}^{n_m+d_m} w_j^{(m)}(y_j^{(m)}\log(\lambda_\beta(u_j,m) - \lambda_\beta(u_j,m)) \right).\tag{8}$$

Equation (8) is formally equivalent to the weighted Poisson regression with "response variable" y_j. To estimate β, we construct (8) in the form of a weighted least square problem using the second-order Taylor expansion. If the current parameter estimate is $\widetilde{\beta}$, then the quadratic approximation of the log-likelihood function (8) is given by

$$\ell_{\mathrm{PR}}(u;\beta) \approx \frac{1}{2} \sum_{m=1}^{M} \sum_{j=1}^{n_m+d_m} v_j^{(m)} \left(y_j^{(m)*} - \beta_{(m)}^{\top} \mathbf{Z}(u_j) \right)^2, \tag{9}$$

and the working response is

$$y_j^{(m)*} = \log\left(\lambda_{\widetilde{\beta}}(u_j,m)\right) + \frac{y_j^{(m)} - \lambda_{\widetilde{\beta}}(u_j,m)}{\lambda_{\widetilde{\beta}}(u_j,m)}, \tag{10}$$

where

$$v_j^{(m)} = w_j^{(m)} \lambda_{\widetilde{\beta}}(u_j,m). \tag{11}$$

We consider iteratively reweighted least squares (IRLS) to solve (9). See Step 4 in Sect. 2.3 for the details.

Logistic Regression. To have a good approximation on the likelihood function, Berman-Turner device often requires a high number of dummy points [6,8], leading to computationally expensive, especially when the number of data points is already large. One strategy to overcome this issue is to draw random dummy points $\mathbf{d}$ generated from point process $\mathbf{D}$ with a known intensity $\delta(u,m)$. The dummy point process $\mathbf{D}$ could be Poisson, binomial, or stratified binomial point process [6,12] and is independent from $\mathbf{X}$. Here we assume $\delta(u,m) = \delta_m$ and the dummy point pattern is denoted by $\mathbf{d} = \{\mathbf{d}^{(1)}, \mathbf{d}^{(2)}, ..., \mathbf{d}^{(M)}\}, m \in \mathcal{M}$.

By such a strategy, the estimating function results in

$$\ell_{\mathrm{IMPP}}(\beta) = \sum_{(u,m)\in\mathbf{x}} \log\left(\frac{\lambda_\beta(u,m)}{\lambda_\beta(u,m)+\delta_m}\right) + \sum_{(u,m)\in\mathbf{d}} \log\left(\frac{\delta_m}{\lambda_\beta(u,m)+\delta_m}\right),$$

$$\approx \sum_{m=1}^{M} \sum_{j=1}^{n_m+d_m} \left(y_j^{(m)} \log(\lambda_\beta(u_j,m)) - \log\left(\lambda_\beta(u_j,m)+\delta_m\right) \right). \tag{12}$$

Equation (12) is equivalent to the logistic regression log-likelihood function with response variable $y_j^{(m)} = \mathbf{1}\left((u_j,m) \in \mathbf{x}\right)$, an offset term $-\log(\delta_m)$, and success probability

$$p(u_j,m) = \Pr\{y_j^{(m)} = 1\} = \frac{\lambda_\beta(u_j,m)}{\lambda_\beta(u_j,m)+\delta_m}.$$

The procedure for maximizing (12) follows along similar lines to the one for maximizing (8). The resulting least squares problem is

$$\ell_{\mathrm{LR}}(\beta) \approx \sum_{m=1}^{M} \sum_{j=1}^{n_m+d_m} \left(v_j^{(m)} \left(y_j^{(m)*} - \beta_{(m)}^{\top} \mathbf{Z}(u_j) \right)^2 \right), \tag{13}$$

where $y_j^{(m)*}$, the working response, is defined by

$$y_j^{(m)*} = \widetilde{\boldsymbol{\beta}}_{(m)}^\top \mathbf{Z}(u_j) + \frac{1}{v_j^{(m)}} \left(y_j^{(m)} - \frac{\lambda_{\widetilde{\beta}}(u_j, m)}{\lambda_{\widetilde{\beta}}(u_j, m) + \delta_m} \right), \tag{14}$$

and

$$v_j^{(m)} = \widetilde{p}(u_j, m)(1 - \widetilde{p}(u_j, m)); \quad \widetilde{p}(u_j, m) = \frac{\lambda_{\widetilde{\beta}}(u_j, m)}{\lambda_{\widetilde{\beta}}(u_j, m) + \delta_m}. \tag{15}$$

Equations (9) and (13) can be maximized using the **ppm** function of the **spatstat** package [6], where method='mpl' is applied for (9) and method='logi' for (13).

2.3   Algorithm

The algorithm to obtain the estimates of $\bar{\boldsymbol{\beta}} = (\boldsymbol{\beta}_{(1)}^\top, \ldots, \boldsymbol{\beta}_{(M)}^\top)^\top$ is as follows. The estimators are denoted by $\hat{\boldsymbol{\beta}}$.

Step 1: Employ numerical methods to obtain Poisson (8) and logistic (12) likelihoods.

Step 2: Construct the design matrices.

1. Build the response variable:
   (a) Poisson Regression: The response variables are $y_j^{(m)} = \mathbf{1}((u_j, m) \in \mathbf{x})/w_j^{(m)}$ where $\mathbf{1}((u_j, m) \in \mathbf{x})$ is indicator function: 1 if u_j is a data point of type m and 0 is a dummy point, and $w_j^{(m)}$ are the quadrature weights.
   (b) Logistic Regression: The response variables are $y_j^{(m)} = \mathbf{1}((u_j, m) \in \mathbf{x})$.
2. The extracted covariates are

$$\mathbf{Z}^\top(u) = \begin{pmatrix} Z_1(u_1) & Z_2(u_1) & \ldots & Z_p(u_1) \\ Z_1(u_2) & Z_2(u_2) & \ldots & Z_p(u_2) \\ \vdots & \vdots & \ddots & \vdots \\ Z_1(u_{n_m+d_m}) & Z_2(u_{n_m+d_m}) & \ldots & Z_p(u_{n_m+d_m}) \end{pmatrix}. \tag{16}$$

Step 3: Both (8) and (12) are approximated by the Taylor expansion resulting in (9) and (13), where the working response are now written by (10) and (14).

Step 4: Parameter estimation used iteratively re-weighted least squares (IRLS) for Poisson regression and logistic regression.

1. Choose an initial value $\widetilde{\beta}^{(0)}$
2. For $q = 0, 1, 2, \ldots$
   (a) Compute the working responses $y_j^{(m)*}$ and weights $v_j^{(m)}$ for the Poisson regression (see (10) and (11)) and the logistic regression (see (14) and (15)). The extracted covariates are given by (16).

(b) Obtain $\widetilde{\beta}^{(q+1)}$

(c) Check to see whether $\hat{\beta}$ has converged; if yes, then stop

For the model comparison, we perform Akaike information criteria (AIC) developed for spatial point process [14] defined by

$$\text{AIC}(\hat{\beta}) = -2\ell(\hat{\beta}) + 2k, \tag{17}$$

where $\ell(\hat{\beta})$ is the maximum of the Poisson likelihood (8) or logistic regression likelihood (12) and k is the size of $\hat{\beta}$.

3   Application to the BCI Plots Data

3.1   Data Description

BCI data provides 3,000 species consisting of 350,000 trees surveyed in ($\mathcal{D} = 1,000\,\text{m} \times 500\,\text{m}$). We pay attention and analyze the nine species ranging from 3,000–5,500 individual trees previously studied by Waagepetersen et al. [7] and Choiruddin et al. [9]. Based on the diversity that occurs, it is suspected that there is an influence from environmental factors [5]. For example, the distribution of cappfr Fig. 1(a) tends to be similar of topographic wetness index Fig. 1(s). So it is of importance to investigate the effect of each environmental factor on the distribution of each of the species (Sect. 3.2) (Table 1).

Table 1. Name and abundance of species.

Code	Species	Abundance
cappfr	Capparis frondosa	3,112
protpa	Protium panamense	3,119
protte	Protium tenuifolium	3,091
swars1	Swartzia simplexvar. grandiflora	3,189
swars2	Swartzia simplexvar. ochnacea	3185
psycho	Psychotria horizontalis	2,639
hirttr	Hirtella triandra	4,552
tet2pa	Tetragastris panamensis	4,961
gar2in	Garcinia intermedia	5,046

Variables of the environment have been previously measured by environmental data such as soil conditions and the topography of the location presented in grid form. We will consider eleven environmental variables as covariates. The available environmental variables are presented in Fig. 1. Standardization, to reduce the correlation between environmental variables, is carried out for each available covariate (the value of the environmental variable is reduced by the

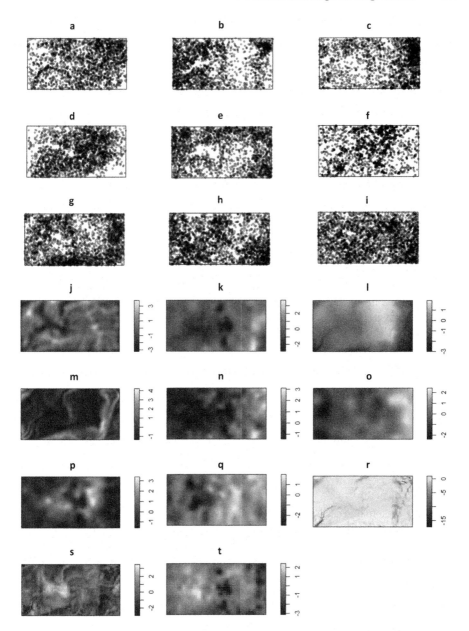

Fig. 1. Species involved and environmental variables in the analysis. (a) Capparis frondosa, (b) Protium panamense, (c) Protium tenuifolium, (d) Swartzia simplexvar. grandiflora, (e) Swartzia simplexvar. ochnacea, (f) Psychotria horizontalis, (g) Hirtella triandra, (h) Tetragastris panamensis, (i) Garcinia intermedia, (j) convergence index with direction to the center cell, (k) copper content (mg/kg of soil) in the surface soil, (l) 5 m resolution elevation model, (m) slope of gradient, (n) potassium content (mg/kg of soli), (o) mineralization needs for nitrogen after a 30-day incubation period, (p) phosphorus content (mg/kg of soil), (q) pH content, (r) incoming mean annual solar radiation, (s) topographic wetness index, (t) Al content in the surface soil.

measure of concentration and divided by the standard deviation of the environmental variable).

Furthermore, Poisson and logistic regression were used to determine the effect of environmental variables on the distribution of species. Both methods have the same scheme in the estimation process, generating dummy points for estimating. Akaike Information Criteria (AIC), to find out the best method, is used as a benchmark. Furthermore, the number of dummy points generated from each method was investigated. It generated will affect the computing time of both methods. These two considerations become the material for evaluating the two methods. The series of analyses are carried out using the ppm function in the spatstat package. The results are explained in Sects. 3.2 and 3.3.

3.2   Model Comparison

Poisson and logistic regressions provide different scenarios of driving the dummy points for parameter estimations described in Sect. 2. The comparison of the two methods based on the Akaike Information Criteria (AIC) value, total dummy points, and computational time is given in Table 2. The computing is conducted using Asus X409JB Intel(R) Core(TM) i5-1035G1 CPU @ (1.00 GHz 1.19 GHz) 20.0 GB RAM 1 TB HDD.

Table 2. AIC, dummy points, and computing time each method.

	Poisson	Logistic
AIC	380,249.6	389,343.2
Dummy points	1,495,208	136,900
Computing time (seconds)	204.12	15.61

The Poisson regression performs slightly better since it produces lower AIC value. However, the logistic regression only requires 16 s for the computation, which is 13 times faster than that of Poisson regression-based approach. The main reason is that the logistic regression uses much less number of dummy points. By that comparison, we in general recommend the logistic-based approach especially when handling a large number of data points.

3.3   Model Interpretation

Figure 2 shows the resulting parameter estimation. The plotted parameters do not involve intercepts. The Poisson and logistic regression approaches obtain similar results with estimates ranging between -0.7 and 0.3. We only interpret the results obtained from the logistic-based technique.

Based on the significance of the effect of environmental variables on the distribution of species (asterisk), forty one possible environmental variables did not significantly affect the distribution of species. According to Fig. 2, we can find out what environmental variables are dominantly affecting most species. As an illustration, the environmental variable (j) topographic wetness index significantly affects seven types of species in the Barro Colorado Islands. So that variable (j) dominantly affects the distribution of the seven available species. Another variable that also significantly affects most species is the variable (l) five meters resolution elevation model. In another comparison, variable (t) Al content in the surface soil dominantly does not significantly affect most species scattered in the Barro Colorado Islands.

By assessing the environmental variable effects, several species share some habitat similarities. For an example, the species *Hirtella triandra* (g) and *Tetragastris panamensis* (h) depend on common environmental factors: copper content in the surface soil (k), 5 m resolution elevation model (l), slope gradient (m), potassium content (n), mineralization needs for nitrogen after a 30-daya incubation period (o), and incoming mean annual solar radiation (r), which could mean a deeper insight that they could be competitive.

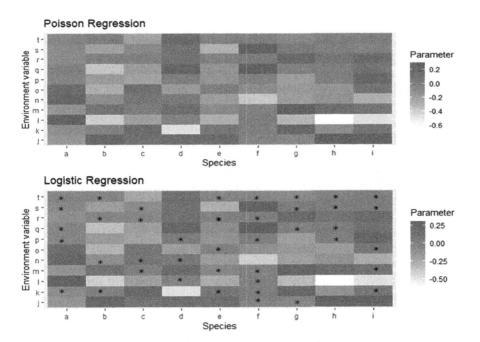

Fig. 2. Parameter estimation of two methods. Asterisk (*) symbolizes do not significant effect of environment variable

After finding an appropriate model and interpreting of each species terms of environmental variable, one can search similarities and dissimilarities in habitat preferences of species by comparing the relations between environmental variable and parameters of covariates each species [9, 15]. Identification using Fig. 2

could lead to a subjective conclusion and only based on the number of significant covariates. Therefore, we applied hierarchical cluster analysis with complete
linkage clustering method to identify similarity habitat preferences.

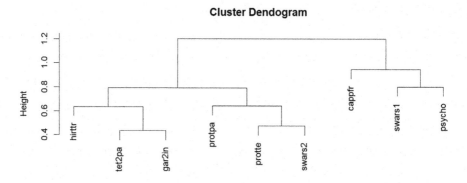

Fig. 3. Dendogram of species cluster based on similarities in habitat preference

By Fig. 3, all of selected species can be categorized into three habitats
groups: (i) `hirttr`, `tet2pa` and `gar2in`; (ii) `protpa`, `protte` and `swars2`; and
(iii) `cappfr`, `swars1` and `psycho`. This is related to the ones previously analyzed by Jalilian [15] and Choiruddin et al., [9] albeit different approaches are
considered.

4   Conclusion

In this paper, we extend the estimating equations based on Poisson and logistic
regression likelihoods for estimating the intensity of a multivariate point process.
The estimation procedure is conducted simultaneously to gain computational
efficiency.

We compare the Poisson and logistic methods to model the distribution
of nine species of trees observed in the BCI rainforest. We find that Poisson
and logistic procedures obtain similar AIC values (Poisson approach produces
a slightly smaller one) but the logistic approach only requires smaller number
of dummy points leading to gain much cheaper computation. By the logistic
approach, the environmental variables such as the topography wetness index
affect the presence of seven tree species in BCI. In addition, species such as
Hirtella triandra (g) and *Tetragastris panamensis* (h) are known to have the
same environmental dominant influence on both.

To determine the important covariates, we decide by the statistical Z-test
[6]. This would not be applicable if a large number of covariates is involved, for
example, when interaction among environmental variables are considered. The
regularization methods are developed for the univariate case [e.g. 8,18] to address
a similar issue. Extending this methodology for multivariate point process would
be an interesting direction for future study.

Acknowledgments. The first author is grateful for the financial support from Lembaga Pengelola Dana Pendidikan (LPDP) under registration no. 20200511301777. We also thank the three reviewers for the helpful comments.

References

1. Baddeley, A., Jammalamadaka, A., Nair, G.: Multitype point process analysis of spines on the dendrite network of a neuron. J. R. Stat. Soc. Ser. C (Appl. Stat.) **63**(5), 673–694 (2014)
2. Jun, M., Schumacher, C., Saravanan, R.: Global multivariate point pattern models for rain type occurrence. Spat. Stat. **31**, 1–19 (2019)
3. Hessellund, K.B., Xu, G., Guan, Y., Waagepetersen, R.: Semiparametric multinomial logistic regression for multivariate point pattern data. J. Am. Stat. Assoc. 1–16 (2021)
4. Hubbel, S., Condit, R., Foster, R.B.: Barro Colorado forest census plot data (2005). http://ctfs.si.edu/datasets/bei
5. Waagepetersen, R.: An estimating function approach to inference for inhomogeneous Neyman-Scott processes. Wiley Int. Biometr. Soc. **63**(1), 252–258 (2007)
6. Baddeley, A., Rubak, E., Turner, R.: Spatial Point Patterns: Methodology and Applications with R. Chapman and Hall/CRC Interdisciplinary Statistics, New York (2015)
7. Waagepetersen, R., Guan, Y., Jalilian, A., Mateu, J.: Analysis of multispecies point patterns by using multivariate log-Gaussian Cox processes. Wiley R. Stat. Soc. **65**(1), 77–96 (2016)
8. Choiruddin, A., Coeurjolly, J.-F., Letue, F.: Convex and non-convex regularization methods for spatial point processes intensity estimation. Electron. J. Stat. **12**, 1210–1255 (2018)
9. Choiruddin, A., Cuevas-Pacheco, F., Coeurjolly, J.-F., Waagepetersen, R.: Regularized estimation for highly multivariate log Gaussian Cox processes. Stat. Comput. JR Stat. Soc. **30**(3), 649–662 (2020)
10. Rajala, T., Murrell, D., Olhede, S.: Detecting multivariate interactions in spatial point patterns with Gibbs models and variable selection. J. R. Stat. Soc.: Ser. C: Appl. Stat. **67**, 1237–1273 (2018)
11. Møller, J., Waagepetersen, R.: Modern statistics for spatial point processes. Cand. J. Stat. **34**, 643–984 (2007)
12. Baddeley, A., Coeurjolly, J., Rubak, E., Waagepetersen, R.: Logistic regression for spatial Gibbs point processes. Biometrika **101**(2), 377–392 (2010)
13. Guan, Y., Shen, Y.: A weighted estimating equation approach for inhomogeneous spatial point processes. Biometrika **97**(4), 867–880 (2010)
14. Choiruddin, A., Coeurjolly, J.-F., Waagepetersen, R.: Information criteria for inhomogeneous spatial point processes. Aust. N. Z. J. Stat. **63**(147) (2021)
15. Jalilian, A.: Modelling and classification of species abundance: a case study in the Barro Colorado Island Plot. J. Appl. Stat. **44**(13), 2401–2409 (2017)
16. Choiruddin, A., Aisah, Trisna, F., Iriawan, N.: Quantifying the effect of geological factors on distribution of earthquake occurrences by inhomogeneous Cox processes. Pure Appl. Geophys. **178**(5), 1579–1592 (2021)
17. Berman, M., Turner, T.R.: Approximating point process likelihoods with GLIM. J. R. Stat. Soc. Ser. C (Appl. Stat.) **41**(1), 31–38 (1992)
18. Choiruddin, A. Coeurjolly, J., Letué, F.: Adaptive lasso and Dantzig selector for spatial point processes intensity estimation. arXiv preprint arXiv:2101.03698 (2021)

Entropy-Based Fuzzy Weighted Logistic Regression for Classifying Imbalanced Data

Ajiwasesa Harumeka[1,2(✉)], Santi Wulan Purnami[1], and Santi Puteri Rahayu[1]

[1] Department of Statistics, Institut Teknologi Sepuluh Nopember, Surabaya, Indonesia
ajiwasesa@bps.go.id
[2] Central Bureau of Statistics-East Java, Surabaya, Indonesia

Abstract. Logistic regression is a popular classification method that has disadvantages when it is applied to large data. Truncated Regularized Iteratively Reweighted Least Square (TR-IRLS) is a method that overcomes this problem. This method is similar to Support Vector Machine (SVM) because both of them have similar loss functions and parameters that can adjust the bias and variance. Both methods were designed with the assumption of balanced data, so that they are not suitable to be applied on imbalanced data. Both methods were developed to overcome problem on imbalanced data. TR-IRLS was developed into Rare Event Weighted Logistic Regression (RE-WLR) and SVM was developed into Fuzzy Support Vector Machine (FSVM). Both RE-WLR and FSVM use weights based on class differences, so that RE-WLR had better performance than TR-IRLS on imbalanced data whereas FSVM was better than SVM. Then, Entropy-based Fuzzy Support Vector Machine (EFSVM) was developed by obtaining weighting values not only based on class differences, but also based on entropy. EFSVM further enhanced minority class interest in imbalanced data than SVM and even FSVM. Therefore, Entropy-based Fuzzy Weighted Logistic Regression (EFWLR) is proposed by adopting the success of Entropy-based Fuzzy Membership (EF) as weight on SVM. This study applied EF as weight on Weighted Logistic Regression for binary classification. Experiments on 20 simulation data and 5 benchmark data with various rarity schemes validated that the EFWLR outperformed TR-IRLS and RE-WLR based on AUC. EFWLR had more efficient AUC than RE-WLR on imbalanced data.

Keywords: Binary classification · Entropy-based Fuzzy · Imbalanced Data · Weighted Logistic Regression

© Springer Nature Singapore Pte Ltd. 2021
A. Mohamed et al. (Eds.): SCDS 2021, CCIS 1489, pp. 312–327, 2021.
https://doi.org/10.1007/978-981-16-7334-4_23

1 Introduction

Logistic regression is a popular and powerful classification method and has several advantages [1], namely: 1) can provide a probability value, 2) can be used in multiclass classification, 3) does not require the assumption of a predictor variable, and 4) can solve the optimization problem without constraints.

Logistic regression does not produce a closed form solution. It takes an iteration method to get the parameter estimator. One of the iteration methods is Iteratively Reweighted Least Square (IRLS). IRLS is an elaboration of the Newton-Raphson iteration method. We need Gradient vector and Hessian matrix, which are the first and second derivatives of the logistic regression likelihood function, respectively. The disadvantage is that when the observations are large, the Hessian matrix will be large too, so that the computational process will take a long time [2]. Then, Truncated Regularized Iteratively Reweighted Least Square (TR-IRLS) was first proposed by Komarek [3]. The TR-IRLS algorithm combines ridge logistic regression, IRLS, and truncated newtons to solve computational problems in large datasets [4].

TR-IRLS is similar with Support Vector Machine (SVM) because they have similar loss functions [5]. In addition, similarity was found in the parameters that can adjust the bias and variance [5]. However, both methods are designed with the assumption of balanced data. In fact, the occurrence of imbalanced data is often encountered.

In binary classification, imbalanced data are events where one class has fewer numbers than another class, such as state failure [6], tornado [1], breast and colon cancers [7], forest covertype [1], underdeveloped village [8], and household poverty [9, 10]. It causes the minority class observation accuracy to be low. Imbalanced data can be handled using two approaches [11], namely the data level approach and the algorithm level approach. The algorithm level approach has an advantage over the data level approach, namely that it does not need to change the data distribution [11].

To accommodate the imbalanced data, TR-IRLS and SVM were developed. Rare Event Weighted Logistic Regression (RE-WLR) was a development of TR-IRLS proposed by Maalouf and Siddiqi [1]. The RE-WLR combines TR-IRLS, weighting, and bias correction. The weight on RE-WLR refers to King and Zeng [12], which can produce more consistent parameters. Bias correction was added to reduce bias due to the addition of ridge regularization [1]. Fuzzy Support Vector Machine (FSVM) was proposed by Lin and Wang [13] by adding fuzzy membership to the SVM objective function. Observations in the minority class were given a higher fuzzy membership value than the observations in the majority class. It aimed to increase the interest of observation in the minority class. So, the accuracy of the minority class could increase [13]. RE-WLR and FSVM have similar value of weight, which is only based on class differences [1, 13]. Then, Entropy-based Fuzzy Support Vector Machine (EFSVM) was proposed by Fan et al. [14] to patch the weaknesses of FSVM. Fuzzy membership on EFSVM was not only based on class differences, but also on the class certainty of each observation. Class certainty was obtained using entropy introduced by Shannon [15]. By adding fuzzy membership based on entropy, the decision surface becomes more flexible [14]. Entropy-based fuzzy membership (EF) has succeeded in increasing the accuracy of the minority class [14].

By adopting the successful of EF to SVM, we propose a classification method that combines TR-IRLS, EF, and bias correction. This method is named Entropy-based Fuzzy Weighted Logistic Regression (EFWLR). EF guarantees the interests of the minority class and at the same time, guarantees the interests of observations that have high class certainty [14]. In addition, truncated newton optimization on the TR-IRLS algorithm makes EFWLR also suitable when it is applied to large data [4]. Bias correction was applied to reduce the effect of bias due to the addition of ridge regularization [1] and imbalanced data [12]. To evaluate the performance of the EFWLR, we tested 20 simulation data and 5 benchmark data. EFWLR performance was compared to TR-IRLS and RE-WLR based on AUC values.

This paper is structured as follows. Section 2 provides an explanation of TR-IRLS. Section 3 provides an explanation of Entropy-based Fuzzy Membership. Section 4 introduces EFWLR algorithm. Section 5 provides a review of the experimental results of the comparison of binary classification performance between TR-IRLS, RE-WLR, and EFWLR. Lastly, concluding remarks are presented in Sect. 6.

2 Truncated Regularized Iteratively Reweighted Least Square (TR-IRLS)

Let y_i is response variabel which has categorical value of 0 as majority class or 1 as minority class. Then, y_i has a Bernoulli distribution with a probability function.

$$P(y_i/\pi_i) = \pi_i^{y_i}(1 - \pi_i)^{1-y_i} \tag{1}$$

π_i is the probability of the i-th occurrence, $i = 1, 2, \ldots, n$. n is the number of mutually exclusive observation. In logistic regression [16], π_i varies according to the equation.

$$\pi_i = \frac{e^{\vec{X}_i^{\mathrm{T}}\vec{\beta}}}{1 + e^{\vec{X}_i^{\mathrm{T}}\vec{\beta}}} \tag{2}$$

$\vec{X}_i$ is a vector of size $p + 1$ which is the predictor variable and $\vec{\beta}$ is a parameter vector of size $p + 1$. p is the number of predictor variables.

Maximum Likelihood (ML) is one of the parameter estimation methods which is often used [16]. So, the ln-likelihood function is

$$ln\left(L\left(\vec{\beta}\right)\right) = \sum_{i=1}^{n} ln\left(\pi_i^{y_i}(1 - \pi_i)^{1-y_i}\right) \tag{3}$$

By substituting Eq. (2) into Eq. (3) and adding ridge regularization $\left(\frac{\lambda}{2}\vec{\beta}^2\right)$, we get ln-likelihood function of regularized logistic regression as follows:

$$Reg.ln\left(L\left(\vec{\beta}\right)\right) = \sum_{i=1}^{n}\left(y_i ln\left(\frac{e^{\vec{X}_i^{\mathrm{T}}\vec{\beta}}}{1 + e^{\vec{X}_i^{\mathrm{T}}\vec{\beta}}}\right) + (1 - y_i)ln\left(\frac{1}{1 + e^{\vec{X}_i^{\mathrm{T}}\vec{\beta}}}\right)\right) - \frac{\lambda}{2}\vec{\beta}^2 \tag{4}$$

where $\vec{\beta}^2 = \sum_{i=1}^{p} \beta_i^2$ and λ is regularization parameter. The objective is to maximize the likelihood function on Eq. (4) to get $\vec{\hat{\beta}}$, which is the estimated value of $\vec{\beta}$. For binary logistic regression, a statistical measure called *Deviance* has the same role as Sum Square Error (SSE) in linear regression [16]. *Deviance* is obtained by the following formula:

$$Deviance = -2ln\left(L\left(\vec{\hat{\beta}}\right)\right) \tag{5}$$

Maximizing Eq. (4) with respect to $\vec{\beta}$ results in a solution that is not closed form. Therefore, numerical method is used [1, 2,17]. One of the numeric iteration methods is Iteratively Reweighted Least Square (IRLS) which uses the Newton-Raphson method [4,17]. This iteration method requires a Gradient vector and a Hessian matrix which are the first and second derivatives of the likelihood function on Eq. (4), respectively. The Gradient vector is:

$$G\left(\vec{\beta}\right) = \frac{dln\left(L\left(\vec{\beta}\right)\right)}{d\vec{\beta}} = X^T\left(\vec{Y} - \vec{\pi}\right) - \lambda\vec{\beta} \tag{6}$$

While the Hessian matrix is:

$$H\left(\vec{\beta}\right) = \frac{dln\left(L\left(\vec{\beta}\right)\right)}{d^2\vec{\beta}} = -X^TVX - \lambda I \tag{7}$$

where $v_i = \pi_i(1 - \pi_i)$, $V = diag(v_1, v_2, \ldots, v_n)$, $i = 1, 2, \ldots, n$, n is the number of observation and I is identity matrix size $(p + 1)X(p + 1)$. Then, the updated Newton-Raphson iteration for $\vec{\beta}$ on $(c + 1)$-iteration is:

$$\vec{\beta}^{(c+1)} = \vec{\beta}^{(c)} + (X^TVX + \lambda I)^{-1}X^T\left(\vec{Y} - \vec{\pi}\right) - \lambda\vec{\beta}^{(c)} \tag{8}$$

Since $\vec{\beta}^{(c)} = (X^TVX + \lambda I)(X^TVX + \lambda I)^{-1}\vec{\beta}^{(c)}$, then

$$\vec{\beta}^{(c+1)} = (X^TVX + \lambda I)(X^TVX + \lambda I)^{-1}\vec{\beta}^{(c)} + (X^TVX + \lambda I)^{-1}X^T\left(\vec{Y} - \vec{\pi}\right) - \lambda\vec{\beta}^{(c)}$$

$$= (X^TVX + \lambda I)^{-1}X^TV\vec{Z}^{(c)} \tag{9}$$

where $z^{(c)} = X\vec{\hat{\beta}}^{(c)} + V^{-1}\left(\vec{Y} - \vec{\pi}\right)$ is adjusted response.

The large unit of observation causes the Hessian matrix to be formed to be large too. This requires a long processing time to compute the inverse of the Hessian matrix [2]. Komarek [3] first introduced Truncated Regularized Iteratively Reweighted Least Square (TR-IRLS) in logistic regression to handle large datasets using Truncated Newtons with linear Conjugate Gradient (CG) to obtain estimates of $\vec{\beta}$. Truncated Newton is a flexible and powerful method for optimizing large data [4, 17].

The TR-IRLS method has two iterations, namely: 1) finding the solution of the Weighted Least Square (WLS) and stopping it when the relative difference of deviance statistics [16] of two consecutive iterations does not exceed the specified limit (ε_1). The WLS sub problem is:

$$(X^T V X + \lambda I)\, \vec{\hat{\beta}}^{\,(c+1)} = X^T V \vec{Z}^{(c)} \tag{10}$$

2) looking for a solution to the Eq. (10) by the CG method [17, 18] and stopping the iteration when the residual value, $\vec{r} = X^T V \vec{Z}^{(c)} - (X^T V X + \lambda I)\, \vec{\hat{\beta}}^{\,(c+1)}$, does not exceed a predetermined limit (ε_2).

3 Entropy-Based Fuzzy Membership (EF)

On the imbalanced data, the minority class is more important than the majority class. For this reason, the observations of the minority class are given a greater weight than that of the observations of the majority class [14]. Based on Fan et al. [14], we guarantee the interest of the minority class observation by giving a weight in the form of the largest fuzzy membership value, namely 1. The majority class observation will be given a weight in the form of a fuzzy membership value according to the class certainty. The class certainty is obtained using entropy.

Entropy is an important measure in information theory that can be used to select an event from several events that have a probability [15]. On two chances of occurrences, entropy is written by the following equation:

$$H_i = -p_{+i} \ln(p_{+i}) - p_{-i} \ln(p_{-i}) \tag{11}$$

p_{+i} and p_{-i} are the probability of the i-th training observation being the minority class and the majority class, respectively. The smaller the entropy value is, the higher the importance of the observation will be. Information of p_{+i} and p_{-i} is based on its k nearest neighbours. Both probablity are calculated using the following equation:

$$p_{+i} = \frac{num_{+i}}{k}$$

$$p_{-i} = \frac{num_{-i}}{k} \tag{12}$$

num_{+i} and num_{-i} are numbers of minority and majority observations in k nearest neighbours, respectively.

Next step is to fuzzify majority observations using their entropy. Let $H_- = \{H_{-1}, H_{-2}, \ldots, H_{-Nmaj}\}$., where $Nmaj$ is the number of majority observations. H_{min} and H_{max} are the minimum and maximum values of H_-, respectively. Then, the entropy of the majority class is divided into m subsets. Each subset is given the following constraints:

$$thrUp_l = H_{min} + \frac{l}{m}(H_{max} - H_{min})$$

$$thrLow_l = H_{min} + \frac{l-1}{m}(H_{max} - H_{min}), l = 1, 2, \ldots, m. \tag{13}$$

$thrUp_l$ and $thrLow_l$ are the upper and lower limits of l-th subset [14]. Finaly, fuzzy membership in each subset is computed by means of:

$$FM_l = 1 - \varphi(l - 1) \tag{14}$$

Where FM_l is the fuzzy membership value of observation on l-th subset and φ is fuzzy membership parameter which has a value of $0 < \varphi \leq \frac{1}{m-1}$. The observations in the same subset have the same fuzzy membership values. This indicates that observations originating in the same subset have the same class certainty [14]. Then, the fuzzy membership of training observation is as follows:

$$w_i = \begin{cases} 1, & \text{if } y_i = 1 \\ \\ FM_l, & \text{if } y_i = 0 \text{ and } \vec{X}_i \epsilon l - thsubset \end{cases} \tag{15}$$

4 Entropy-Based Fuzzy Weighted Logistic Regression (EFWLR)

One of the efforts to overcome the imbalanced data on logistic regression is the provision of weights [12]. The regularized weighted ln-likelihood function [1] becomes:

$$Reg.\ln\left(L_w\left(\vec{\beta}\right)\right) = \sum_{i=1}^{n}\left(w_i\left(y_i ln\left(\frac{e^{\vec{X}_i^T\vec{\beta}}}{1 + e^{\vec{X}_i^T\vec{\beta}}}\right) + (1 - y_i)ln\left(\frac{1}{1 + e^{\vec{X}_i^T\vec{\beta}}}\right)\right)\right)$$
$$-\frac{\lambda}{2}\vec{\beta}^2 \tag{16}$$

Where w_i is fuzzy membership based on entropy which is described in Sect. 3. $\vec{\beta}$ is obtained by using truncated newton with conjugate gradient [1,17], as in TR-IRLS algorithm. The Gradient vector from ln-likelihood on Eq. (16) is:

$$G\left(\vec{\beta}\right) = \frac{dln\left(L\left(\vec{\beta}\right)\right)}{d\vec{\beta}} = X^T W\left(\vec{Y} - \vec{\pi}\right) - \lambda\vec{\beta} \tag{17}$$

Where $W = diag(w_1, w_2, \ldots, w_n)$ and the Hessian Matrix is:

$$H(\vec{\beta}) = \frac{dln(L(\vec{\beta}))}{d^2\vec{\beta}} = -X^TDX - \lambda I \tag{18}$$

Where $D = diag\{v_1w_1, v_2w_2, \ldots, v_nw_n\}$. Then, the WLS sub problem is:

$$(X^TDX + \lambda I)\vec{\beta}^{(c+1)} = X^TD\vec{Z}^{(c)} \tag{19}$$

ML estimators may be biased when n is small [19], the regularization is added [1], or the data are imbalanced [12]. Maalouf and Siddiqi [1] developed the bias vector described by King and Zeng [12] and Mccullagh [19] as follows:

$$B\left(\vec{\hat{\beta}}\right) = (X^TDX + \lambda I)^{-1}X^TD\vec{\xi} \tag{20}$$

Where $\xi_i = 0, 5Q_{ii}((1 + w_1)\widehat{\pi_i} - w_1)$ and w_1 is weight value of minority class, which in this research equals to 1. So, $\xi_i = 0, 5Q_{ii}(2\widehat{\pi_i} - 1)$. Meanwhile, Q_{ii} is the main diagonal of the matrix Q [1] as follows:

$$Q = X(X^TDX + \lambda I)^{-1}X^T \tag{21}$$

The bias vector is optimized by truncated newton with conjugate gradient [1]. WLS sub problem for bias vector is:

$$\left(X^TDX + \lambda I\right)B\left(\vec{\hat{\beta}}\right) = X^TD\vec{\xi} \tag{22}$$

Then, the bias corrected parameter [19] is described as follows:

$$\vec{\hat{\beta}} = \vec{\hat{\beta}} - B\left(\vec{\hat{\beta}}\right) \tag{23}$$

We present the EFWLR algorithm as algorithms 1 to 4. Similar to RE-WLR [1], ε_1 and *maxIRLS* were set to 0.01 and 30, respectively, on algorithm 1. ε_2 and *maxCG*, on algorithm 3 and 4, were set to 0.005 and 200, respectively. m and φ, on Algorithm 2, were set to 10 and 0.05, respectivelly, following Fan et al. [14]. The number of k nearest neighbours was selected from $\{3, 5, 7, 9, 11\}$. The regularization parameter, λ, was selected from $\{10^{-5}, 10^{-4}, 10^{-3}, 10^{-2}, 10^{-1}, 0.5, 0.75, 1, 1.5, 2, 2.5, 3, 4, 5, 10\}$.

Algorithm 1. WLR MLE using IRLS

Input: $X, \vec{Y}, \vec{\beta}^{(0)}, B\left(\vec{\beta}\right)^{(0)}, m, \varphi, k$

Output: $\vec{\beta}$

Begin

$c = 0$

$deltadev = 1$

Get $\vec{w}$ using algorithm 2

  Do While $|deltadev > \varepsilon_1|$ dan $c \leq maxIRLS$

    For i=1 to n

$$\hat{\pi}_i = \frac{\exp\left(\vec{X}_i^T \vec{\beta}^{(c)}\right)}{1 + \exp\left(\vec{X}_i^T \vec{\beta}^{(c)}\right)}$$

$$v_i = \hat{\pi}_i(1 - \hat{\pi}_i)$$

$$z_i = \vec{X}_i^T \vec{\beta}^{(c)} + \frac{(y_i - \hat{\pi}_i)}{v_i}$$

    End For

$$D = diag(v_i w_i)$$

$$A = (X^T D X + \lambda I)$$

$$\vec{b} = X^T D \vec{Z}^{(c)}$$

$$Q = X(X^T D X + \lambda I)^{-1} X^T$$

    For $k = 1$ to n

$$\xi_k = 0{,}5 Q_{kk}(2\hat{\pi}_k - 1)$$

    End For

$$\vec{c} = X^T D \vec{\xi}$$

Get $\vec{\beta}^{(c+1)}$ using algorithm 3

Get $B\left(\vec{\beta}\right)^{(c+1)}$ using algorithm 4

$$deltadev = \frac{deviance^{(c)} - deviance^{(c+1)}}{deviance^{(c+1)}}$$

$c = c + 1$

End While

$$\vec{\beta} = \vec{\beta}^{(c+1)} - B\left(\vec{\beta}\right)^{(c+1)}$$

End

Algorithm 2. Entropy-based Fuzzy Membership to get $\vec{w}$

Input: $X, \vec{Y}, m, \varphi, k$

Output: $\vec{w}$

Begin

  For i = 1 to n

    Get k nearest neighbours of $\vec{X}_i$

  Count num_{+i} and num_{-i} in the k nearest neighbours

    Calculate class probability using equation (12)

    $H_i = -p_{+i}\ln(p_{+i}) - p_{-i}\ln(p_{-i})$

  End For

  Select entropy of majority class, H_-

  $H_{min} = min\{H_-\}$

  $H_{max} = max\{H_-\}$

  For l = 1 to m

$$thrUp = H_{min} + \frac{l}{m}(H_{max} - H_{min})$$

$$thrLow = H_{min} + \frac{l-1}{m}(H_{max} - H_{min})$$

    For j = 1 to n

      If $y_i = majority\ class$

        If $thrLow \le H_i < thrUp$

          $FM = 1 - \varphi(l-1)$

          $w_i = FM$

        End If

      End If

      If $y_i = minority\ class$

        $w_i = 1$

      End If

    End For

  End For

End

Algorithm 3. Linear CG to get $\vec{\beta}$

Input: $A, \vec{b}, \vec{\beta}^{(0)}$

Output: $\vec{\beta}$

Begin

   $t = 0$

   $\vec{r}^{(0)} = \vec{b} - A\vec{\beta}^{(0)}$

   $\vec{d}^{(0)} = \vec{r}^{(0)}$

   Do While $\left\|\vec{r}^{(t)}\right\|^2 > \varepsilon_2$ and $t \leq maxCG$

     $s^{(t)} = \frac{\vec{r}^{T(t)}\vec{r}^{(t)}}{\vec{d}^{T(t)}A\vec{d}^{(t)}}$

     $\vec{\beta}^{(t+1)} = \vec{\beta}^{(t)} + s^{(t)}\vec{d}^{(t)}$

     $\vec{r}^{(t+1)} = \vec{r}^{(t)} - s^{(t)}A\vec{d}^{(t)}$

     $a^{(t)} = \frac{\vec{r}^{T(t+1)}\vec{r}^{(t+1)}}{\vec{r}^{T(t)}\vec{r}^{(t)}}$

     $\vec{d}^{(t+1)} = \vec{r}^{(t+1)} + a^{(t)}\vec{d}^{(t)}$

     $t = t + 1$

   End While

End

Algorithm 4. Linear CG to get $B\left(\vec{\beta}\right)$

Input: $A, \vec{c}, B\left(\vec{\beta}\right)^{(0)}$

Output: $B\left(\vec{\beta}\right)$

Begin

   $t = 0$

   $\vec{r}^{(0)} = \vec{c} - AB\left(\vec{\beta}\right)^{(0)}$

   $\vec{d}^{(0)} = \vec{r}^{(0)}$

   Do While $\left\|\vec{r}^{(t)}\right\|^2 > \varepsilon_2$ and $t \leq maxCG$

     $s^{(t)} = \frac{\vec{r}^{T(t)}\vec{r}^{(t)}}{\vec{d}^{T(t)}A\vec{d}^{(t)}}$

     $B\left(\vec{\beta}\right)^{(t+1)} = B\left(\vec{\beta}\right)^{(t)} + s^{(t)}\vec{d}^{(t)}$

     $\vec{r}^{(t+1)} = \vec{r}^{(t)} - s^{(t)}A\vec{d}^{(t)}$

     $a^{(t)} = \frac{\vec{r}^{T(t+1)}\vec{r}^{(t+1)}}{\vec{r}^{T(t)}\vec{r}^{(t)}}$

     $\vec{d}^{(t+1)} = \vec{r}^{(t+1)} + a^{(t)}\vec{d}^{(t)}$

     $t = t + 1$

   End While

End

5   Experiments and Results

5.1   EFWLR Performance

We compared performance of EFWLR with TR-IRLS and RE-WLR on 20 simulation data and 5 benchmark data described in Table 1 to validate the robustness of EFWLR to imbalanced data. We generated simulation data based on Santosa [10]. The simulation data consisted of 5 schemes for the number of observations, namely: 1) $n = 1,000$, 2) $n = 2,000$, 3) $n = 4,000$, 4) $n = 8,000$, and 5) $n = 16,000$. Each scheme consists of 4 types of minority class percentage (rarity), namely 5%, 10%, 20%, and 40%. They have 2 predictor variables, X_1 and X_2, which were generated from normal distribution with mean equal to 0 and variants equal to 1. Then, β_1 and β_2 are set to 2 and 3, respectively. We give different β_0 values to get 4 types of rarity, namely: 1) $\beta_0 = -6.75$ for 5%, 2) $\beta_0 = -5$ for 10%, 3) $\beta_0 = -3.45$ for 20%, and 4) $\beta_0 = -0.85$ for 40%.

Table 1. Simulation and benchmark datasets

Dataset	Rarity (%)	Observation	Minority	Majority	Predictor
Sim_1000_05	4.6	1,000	46	954	2
Sim_1000_10	10.4	1,000	104	896	2
Sim_1000_20	20.4	1,000	204	796	2
Sim_1000_40	40.4	1,000	404	596	2
Sim_2000_05	4.8	2,000	96	1,904	2
Sim_2000_10	10.25	2,000	205	1,795	2
Sim_2000_20	19.1	2,000	382	1,618	2
Sim_2000_40	40.45	2,000	809	1,191	2
Sim_4000_05	4.7	4,000	189	3,811	2
Sim_4000_10	10.48	4,000	419	3,581	2
Sim_4000_20	19.95	4,000	798	3,202	2
Sim_4000_40	39.95	4,000	1,598	2,402	2
Sim_8000_05	4.76	8,000	381	7,619	2
Sim_8000_10	10.38	8,000	830	7,170	2
Sim_8000_20	20.58	8,000	1,646	6,354	2
Sim_8000_40	40.31	8,000	3,225	4,775	2
Sim_16000_05	4.97	16,000	795	15,250	2
Sim_16000_10	10.32	16,000	1,651	14,349	2
Sim_16000_20	20.22	16,000	3,235	12,765	2
Sim_16000_40	40.83	16,000	6,532	9,468	2

(*continued*)

Table 1. (*continued*)

Dataset	Rarity (%)	Observation	Minority	Majority	Predictor
Ionosphere	35.9	351	126	225	34
Survival	26.47	306	81	225	3
WBCD	37.26	569	212	357	30
Covertype	4.76	10,500	500	10,000	54
Village	2.69	7,721	208	7513	42

The benchmark datasets are ionosphere, survival, Wisconsin Breast Cancer Diagnostik (WBCD), covertype, and village. Ionosphere, survival, and WBCD can be found on the UCI machine learning repository [20]. Covertype which was used by Maalouf and Siddiqi is used for prediction of forest covertype from cartographic variable [1]. Village was used by Sulasih et al. [8]. Village data are used for classifiying underdeveloped villages in East Java, Indonesia.

Performance was described by Area Under ROC Curve (AUC). The AUC is estimated by the trapezoidal method [21, 22]. The 5-fold cross validation [5] with stratification was applied to select the optimal tuning parameters. Specifically in applying RE-WLR, the training data in each fold were divided into several undersampling schemes to get the weighing value. The undersampling scheme was carried out by randomly selecting the majority class observations 1 time, 2 times, 3 times, 5 times, and 8 times the number of minority class observations for simulation data with rarity of 5% and 10%, as well as village and covertype data. Randomly selecting the majority class observations 1 time, 2 times, and 3 times the number of minority class observations for simulation data was carried out with a rarity of 20%. And randomly selecting the majority class observations as many as the minority class observations for simulation data was carried out with 40% rarity, ionosphere, survival, and WBCD.

The problem of imbalanced data is the small minority class accuracy or sensitivity. Based on the simulation data, Fig. 1 shows that the smaller the percentage of the minority class is, the smaller the sensitivity will be, regardless of the number of samples. EFWLR has the highest sensitivity on any rarity type. However, the sensitivity at 5% rarity shows a decreasing pattern with increasing sample size.

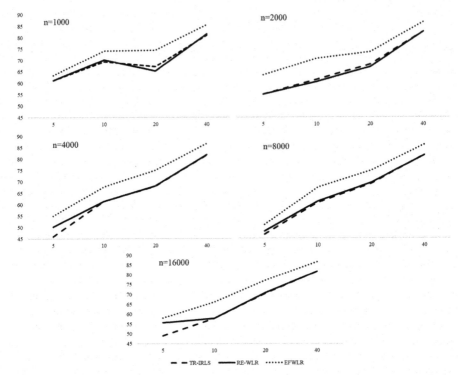

Fig. 1. Sensitivity comparison on simulation data

Table 2 summarizes the result for three methods with the AUC. Bold numbers mean the best AUC among the three methods. EFWLR is dominant when it is applied to simulation data. The more balanced the data are, the more TR-IRLS can compete with RE-WLR and EFWLR. TR-IRLS is better than the other two methods on the ionosphere and WBCD. This is probably due to the fact that the two datasets are relatively balanced.

5.2 Estimator Comparison

We used simulation data with 5% rarity to compare the estimators obtained by the RE-WLR and EFWLR. Each dataset was divided into training and testing sets. On training set, the undersampling scheme was applied. Majority class observations were randomly selected as many as 8 times the number of minority class observations. This was done 100 times of iterations. The estimators were selected from the model with optimal λ. For EFWLR, the parameters k, φ, and m were defined to be equal to 7, 0.05, and 10, respectively.

Table 2. Comparison of stratified 5-fold cross validation AUC

Dataset	TR-IRLS	RE-WLR	EFWLR
Sim_1000_05	80.30	80.40	**81.30**
Sim_1000_10	83.50	83.98	**85.60**
Sim_1000_20	81.04	80.46	**83.90**
Sim_1000_40	85.02	85.35	**86.19**
Sim_2000_05	77.05	76.95	**81.05**
Sim_2000_10	79.59	79.05	**83.86**
Sim_2000_20	81.77	81.37	**83.75**
Sim_2000_40	85.68	85.72	**86.65**
Sim_4000_05	72.57	74.46	**76.72**
Sim_4000_10	79.70	79.76	**82.34**
Sim_4000_20	81.74	81.71	**84.24**
Sim_4000_40	85.77	85.87	**86.76**
Sim_8000_05	72.97	73.56	**74.84**
Sim_8000_10	79.26	79.49	**82.20**
Sim_8000_20	82.16	82.32	**83.91**
Sim_8000_40	85.41	85.41	**86.02**
Sim_16000_05	74.05	77.21	**78.33**
Sim_16000_10	77.83	77.91	**81.44**
Sim_16000_20	83.13	83.26	**85.40**
Sim_16000_40	85.42	85.45	**85.94**
Ionosphere	**85.41**	82.90	84.85
Survival	57.12	54.58	**58.33**
WBCD	**96.54**	95.87	95.93
Covertype	56.28	55.69	**56.73**
Village	71.85	72.63	**76.09**

Table 3 describes estimator comparison of RE-WLR and EFWLR. The numbers in brackets indicate the standard error, while the numbers without brackets indicate the means. RE-WLR produces a more efficient estimator than EFWLR. In addition, the estimator obtained by RE-WLR is also less biased than the one obtained by EFWLR. However, EFWLR performance is better than RE-WLR. The AUC produced by EFWLR is more efficient than the one produced by RE-WLR, indicated by the standard error's AUC of EFWLR that is smaller than that of RE-WLR. Based on this discussion, if we are oriented to minority class accuracy, it is better to use EFWLR. However, if we are oriented towards model interpretation, it is better to use RE-WLR.

Table 3. Estimator and AUC comparison on simulation data with 5% rarity

Data	Method	$\beta 0$	$\beta 1$	β_2	AUC
Sim_1000_05	RE-WLR	−6.597 (0.349)	1.775 (0.157)	2.967 (0.289)	71.39 (5.769)
	EF-WLR	−3.623 (1.266)	1.236 (0.419)	1.970 (0.856)	90.74 (2.326)
Sim_2000_05	RE-WLR	−6.893 (1.109)	1.989 (0.427)	3.159 (0.604)	77.90 (1.970)
	EF-WLR	−5.887 (1.288)	1.963 (0.476)	3.339 (0.786)	87.00 (1.709)
Sim_4000_05	RE-WLR	−6.068 (0.374)	1.860 (0.139)	2.602 (0.186)	70.31 (0.672)
	EF-WLR	−4.619 (0.569)	1.815 (0.219)	2.446 (0.315)	86.26 (0.452)
Sim_8000_05	RE-WLR	−6.090 (0.199)	1.772 (0.085)	2.590 (0.125)	70.62 (0.558)
	EF-WLR	−4.782 (0.412)	1.746 (0.148)	2.573 (0.249)	82.73 (0.297)
Sim_16000_05	RE-WLR	−6.679 (0.089)	1.904 (0.048)	3.013 (0.061)	73.64 (0.597)
	EF-WLR	−5.746 (0.324)	2.008 (0.129)	3.161 (0.196)	82.63 (0.382)

6   Conclusion

The method named EFWLR adopts the success of Entropy-based Fuzzy Membership (EF) which is applied to SVM. By applying EF to the TR-IRLS algorithm, it increases the interest of the minority class and the interest of the majority class observation which has a higher certainty class. Although the EFWLR produces a more biased estimator than the RE-WLR, this method produces a more efficient and higher AUC. It is recommended when we make predictions which are oriented to minority class accuracy, not to model interpretations. If we make model interpretations, RE-WLR is a better method.

More parameters must be tuned on the EFWLR when it is compared to the RE-WLR. However, the EFWLR does not need to apply a sampling scheme to determine the weights. The sampling scheme on RE-WLR can allow for "information loss".

References

1. Maalouf, M., Siddiqi, M.: Weighted logistic regression for large-scale imbalanced and rare events data. Knowl.-Based Syst. **59**, 142–148 (2014)
2. Sulasih, D.E.A., Purnami, S.W., Rahayu, S.P.: The theoretical study of rare event weighted logistic regression for classification of imbalanced data. Int. Conf. Sci. Technol. Humanit. 159–169 (2015)

3. Komarek, P.: Logistic Regression for Data Mining and High-Dimensional Classification. Carnegie Mellon University, Pittsburgh (2004)
4. Komarek, P., Moore, A.W.: Making Logistic Regression a Core Data Mining tool with TR-IRLS, pp. 4–7. Proc. - IEEE Int. Conf. Data Mining, ICDM (2005)
5. James, G., Witten, D., Hastie, T., Tibshirani, R.: An Introduction to Statistical Learning. Springer Science+Business Media, New York (2013).https://doi.org/10.1007/978-1-4614-7138-7
6. King, G., Zeng, L.: Improving forecasts of state failure. World Polit. **53**(4), 623–658 (2001)
7. Ladaya, F., Purnami, S.W., Irhamah, S.W.: Entropy based fuzzy support vector machine untuk klasifikasi microarray imbalanced data. Sepuluh Nopember Inst. Technol. (2018)
8. Sulasih, D.E.A., Purnami, S.W., Rahayu, S.P.: Rare event weighted logistic regression untuk klasifikasi imbalanced data. Sepuluh Nopember Inst. Technol. (2016)
9. Triasmoro, S.P., Ratnasari, V., Rumiati, A.T.: Comparison performance between rare event weighted logistic regression and truncated regularized prior correction on modelling imbalanced welfare classification in Bali. 2018 International Conference on Information and Communications Technology ICOIACT 2018, vol. 2018-Janua, pp. 108–113 (2018)
10. Santosa, B.: Kajian simulasi over sampling K-Tetangga terdekat Pada Regresi Logistik Terboboti dan Penerapannya Untuk Klasifikasi RUmah Tangga Miskin Di Provinsi Daerah Istimewa Yogyakarta. Bogor Agricultural Institute (2018)
11. Ali, A., Shamsuddin, S.M., Ralescu, A.L.: Classification with class imbalance problem: a review. Int. J. Adv. Soft Comput. Appl. **7**(3), 176–204 (2015)
12. King, G., Zeng, L.: Logistic regression in rare events data. J. Stat. Softw. **8**, 137–163 (2003)
13. Lin, C.F., De Wang, S.: Fuzzy support vector machines. IEEE Trans. Neural Networks **13**(2), 464–471 (2002)
14. Fan, Q., Wang, Z., Li, D., Gao, D., Zha, H.: Entropy-based fuzzy support vector machine for imbalanced datasets. Knowl.-Based Syst. **115**, 87–99 (2017)
15. Shannon, C.E.: A mathematical theory of communication. Bell Syst. Tech. J. **27**(4), 623–656 (1948)
16. Hosmer, D.W., Lemeshow, S.: Applied Logistic Regression, Second Edi. Wiley, New Jersey (2000)
17. Rahayu, S.P., Zain, J.M., Embong, A., Juwari, A., Purnami, S.W.: Logistic regression methods with truncated newton method. Procedia Eng., vol. 50, pp. 827–836 (2012)
18. Hestenes, M.R., Stiefel, E.: Methods of conjugate gradients for solving linear systems. J. Res. Natl. Bur. Stand. (1934), **49**(6), 409 (1952)
19. Cordeiro, G.M., McCullagh, P.: Bias correction in generalized linear models. J. R. Stat. Soc. **53**(3), 629–643 (1991)
20. UCI Machine Learning Repository. No Title. https://archive.ics.uci.edu/. Accessed June 15 2021
21. Bekkar, M., Djemaa, H.K., Alitouche, T.A.: Evaluation measures for models assessment over imbalanced data sets. J. Inf. Eng. Appl. **3**(10), 27–38 (2013)
22. Mason, S.J., Graham, N.E.: Areas beneath the relative operating characteristics (ROC) and relative operating levels (ROL) curves: statistical significance and interpretation. Q. J. R. Meteorol. Soc., **128**(584), 2145–2166, PART B (2002)

Quantifying the Impact of Climatic Factors on Dengue Incidence Using Generalized Linear Mixed Model with Spatio-Temporal Bayesian Poisson Random Effects Approach

Nik Nur Fatin Fatihah Sapri[1], Wan Fairos Wan Yaacob[2]([⊠]), and Bee Wah Yap[1,3]

[1] Faculty of Computer and Mathematical Sciences, Universiti Teknologi MARA,
40450 Shah Alam, Selangor, Malaysia
{nikfatinfatihah,yapbeewah}@uitm.edu.my

[2] Faculty of Computer and Mathematical Sciences, Universiti Teknologi MARA Cawangan
Kelantan, Lembah Sireh, 15050 Kota Bharu, Kelantan, Malaysia
wnfairos@uitm.edu.my

[3] Institute for Big Data Analytics and Artificial Intelligence (IBDAAI), Kompleks
Al-Khawarizmi, Universiti Teknologi MARA, 40450 Shah Alam, Selangor, Malaysia

Abstract. Dengue fever is a global life-threatening vector-borne disease which is mainly distributed by the vector *Aedes Aegypt*i mosquito. It is known that the development and survivorship of this vector depends on surrounding climate. The dengue outbreak in Kelantan, Malaysia is alarming. The aim of the study was to compare the fixed effects Negative Binomial GLM and Bayesian Poisson GLMM in prediction of dengue incidence in Kelantan. The data involved daily number of reported dengue cases (1st January 2013–31st December 2017) in ten districts of Kelantan which was collected from Ministry of Health Malaysia. The climate variables, average daily temperature, relative humidity and rainfall (climatic data) were obtained from NASA's Global Climate Change website, while the population data were from Department of Statistics Malaysia. Statistical modeling results revealed that the fixed effects Negative Binomial GLM failed to fit the daily dengue incidence when serious epidemic occurred. The spatio-temporal Bayesian Poisson GLMM model improved the prediction of dengue incidence. Relative humidity at lag 7 days and 21 days and average temperature at lag 21 days were found to be significant contributing factors of dengue incidence in Kelantan. The findings of the study are significant to respective local authorities in providing vital information for early dengue warning systems in a particular area. This is important for authorities to monitor and reduce dengue incidence in endemic areas and to safeguard the community from dengue outbreak.

Keywords: Dengue · Generalized Linear Model · Generalized Linear Mixed Model · Spatio-Temporal Bayesian Poisson Model · Climate

© Springer Nature Singapore Pte Ltd. 2021
A. Mohamed et al. (Eds.): SCDS 2021, CCIS 1489, pp. 328–340, 2021.
https://doi.org/10.1007/978-981-16-7334-4_24

1 Introduction and Motivation

The incidence of dengue has grown dramatically around the world in recent decades. According to Central for Disease Control and Prevention (CDC) [1], dengue has emerged as a worldwide problem since the 1950s. World Health Organization (WHO) [2] estimated that about half of the world's population is now at risk. It is believed that the dengue outbreak is increasing every year. There are approximately 2.5% of worldwide of annual dengue cases are fatal [1, 2]. Transmission of the dengue virus is likely to happen in tropical and sub-tropical regions where-by the transmission is influenced by many factors, including climatic and non-climatic factors [3–7]. Climatic factors are common factors which have always been studied by many researchers in studying the widespread of dengue transmission [8–10]. Climatic factors are believed to influence mosquito biology since Aedes Aegypti mosquito is a climate-sensitive vector.

The Poisson and Negative Binomial fixed effects generalized linear model (GLM) have been extensively used by previous studies either for cross-sectional count data or time series data structure [11–17]. The disadvantage arising from cross-sectional model is that it cannot capture the temporal effect of the data, while, time series model is unable to allow for the spatial effect to be captured in the data. Thus, the spatial and temporal effects are important to be put into consideration as these elements are able to capture the heterogeneity which is due to unobserved confounding factors. In statistical modeling, the panel data models can allow for both spatial and temporal effects to be captured in different area through random effects [18, 19]. To allow for such latent effects and correlation structures, the GLM was extended to generalized linear mixed model (GLMM) by including spatio-temporal structured random effects in the linear predictor using hierarchical Bayesian model (HBM) with the assumption that spatial structured random effects to be a conditional autoregressive (CAR) model. The HBM approach can allow for unobserved confounding factors in the model and temporal correlation within area and spatial effect between areas. This HBM approach can model the overdispersion through spatio-temporal random effects model via Bayesian framework using Markov Chain Monte Carlo (MCMC) to further enhance the prediction of the model.

The aim of the study is to compare the conventional Negative Binomial Generalized Linear Model (GLM) and the Bayesian Poisson Generalized Linear Mixed Model (GLMM) in modeling dengue incidence in Kelantan. This study focused on dengue count issue as it is one of the major public health problems worldwide including Malaysia. It is an epidemic infection disease that has worsened with an increase trend of reported dengue cases in Kelantan. The paper is organized as follows: Sect. 2 provides a brief discussion on some related work on spatio-temporal models. The methodology is explained Sect. 3. Results and discussions are reported in Sect. 4. The paper ends with some concluding remarks and some recommendations for future study in Sect. 5.

2 Related Work

Bayesian modelling approaches have been used in many areas of studies including epidemiological study. In recent years, researchers often implement Bayesian technique approach in epidemiological studies. Recent works by Phanitchat et al. [20] applied a Poisson GLMM with MCMC framework for monthly dengue in Khon Kaen province (Thailand). The random effects model was used in this study to capture the spatio-temporal correlation. They found that the random effect was able to cater for overdispersion problem which commonly occurs in conventional Poisson fixed effects GLM.

Lowe [21] reported that the fixed effects Negative Binomial GLM overestimated the dengue incidence rates in Amazon region. However, the Negative Binomial GLMM with MCMC framework produced better predictions of the dengue incidence rates in a particular area. They concluded that GLM model could not capture spatial variability in dengue incidence rates of microregion of Brazil which tends to have temporal correlation effects within some areas.

Another similar study by Ibarra and Lowe [22] applied the Negative Binomial GLMM with Bayesian MCMC approach in investigating the effect of local climatic factors on dengue incidence in southern Coastal Ecuador of Mexico. They found that the variability in dengue transmission in El Oro province can be captured by the random effects model. Lowe et al. [23] which used similar model found the variability in malaria transmission can be captured after the addition of random effects in the model. Bayesian Poisson GLMM modelling also has been applied by Lekdee and Ingrisawang [18] where the spatial random effects is assumed to be conditional autoregressive (CAR) models with MCMC inference setting for Northern Thailand spatio-temporal dengue data. This spatial random effect helps in capturing spatial dependence among provinces of Northern Thailand. In another spatial and temporal study of dengue which involved 76 provinces of Thailand, in order to account for unobserved heterogeneity, Lowe et al. [24] suggested the random spatial and temporal structures to be considered into the model so that the unknown confounding factors are able to be captured.

Based on these previous studies, the local climatic factors were reported to significantly influence the incidence of dengue in a particular area. Generalized Linear Mixed Models has proven to further improved the prediction of dengue incidence by including the spatial random effects and hierarchical Bayesian method for parameter estimation. GLMM models could capture the temporal correlation effects within some areas and spatial effects between the areas.

3 Methodology

3.1 Fixed Effects Negative Binomial Model

Let us assume the observed number of dengue, y_{it} for a given district, i ($I = 1, 2, ..., N$) and daily time period, t ($t = 1, 2, ... T$), is assumed to follow Negative Binomial family distribution and the linear predictor is modelled by known covariates. The fixed effects Negative Binomial formulation of the dengue model is expressed as follows (Hausman et al. [25]):

$$y_{it} \sim NegBinomial(\lambda_{it})$$
$$\log \lambda_{it} = \log(P_i) + \alpha_i + \beta_0 + \sum_k \beta_k x_{it} + \delta_{t'(t)} + \varepsilon_{it} \tag{1}$$
$$i = 1, 2, ..., N;$$
$$t = 1, 2, ..., T$$

Where λ_{it} is the expected number of dengue for district i in day t, population P_i is an offset, α_i is the fixed effects associated with district i such that $\alpha_i = e^{d_i}$ (where d_i is district specific dummies), β_0 is the overall intercept, β_k is the parameter estimates for $k = 1, 2, .., 15$ of each climatic variables for district i in day t, x_{it} (*Temp, Humid, Rain, Temp7, Humid7, Rain7, Temp14, Humid14, Rain14, Temp21, Humid21, Rain21, Temp28, Humid28, Rain28*) and $\delta_{t'(t)}$ specifies the dummy day at $t'(t)$ (Day 1) is set as the reference level. Note that, district $i = 1, 2, .., 10$ where Bachok serves as the district reference level for Kelantan. The fixed effects Negative Binomial of GLM model was developed using the glm.nb function from MASS package of R open software version 4.1.0. The analysis used the stepwise model selection which were based on Akaike Information Criterion (AIC), Bayesian Information Criterion (BIC) and 'pseudo-R2' (R_D^2).

3.2 Development of Random Effects Poisson Generalized Linear Mixed Model (GLMM) with Bayesian Approach

The Poisson GLMM was developed using Bayesian approach under the proposition of assumption for spatial random effects to be conditional autoregressive structure (CAR). Let us assume the observed number of dengue, y_{it} for a given district, i ($i = 1,2,..., N$) and daily time period, t ($t = 1,2,...T$), is assumed to follow Poisson family distribution and the linear predictor is modelled by known covariates and a vector of random effects. According to Lee et al. [26] and Rushworth et al. [27], the mathematical formulation of mixed effects model can be written as follows:

$$y_{it} \sim Poisson(\lambda_{it})$$
$$\ln(\lambda_{it}) = \log(P_i) + \theta_{it}$$
$$\theta_{it} = \alpha + \beta_k x_{it} + \phi_{it} \tag{2}$$
$$\ln(\lambda_{it}) = \log(P_i) + \alpha + \sum_k \beta_k x_{it} + \phi_{it} + \varepsilon_{it}$$

Where λ_{it} is the expected number of dengue in district i in day t, P_i is the offset variable which refers to population size, α denotes the intercept, β_k is vector of parameters to be estimated for the vector of explanatory variables x_{it} and ϕ_{it} denotes the vector of latent random effects component of spatio-temporal structure. In this Bayesian method, the vector of random effects follows a multivariate first order autoregressive time series process where spatial autocorrelation is modelled to be conditional autoregressive (CAR) prior by Leroux et al. [28].

The Random Effects Poisson GLMM of Bayesian hierarchical models is fitted using ST.CARar function from CARBayesST package of R open software. The Bayesian model were run for 2 200 000 MCMC samples with 200, 000 as burn-in and then thinned by 1000 in order to reduce the autocorrelation of the Markov chain. This generated about 2000 samples for inference. Unlike fixed effects negative binomial GLM, the selection model from Random Effects Poisson GLMM with MCMC framework were based on deviance information criterion (DIC), Watanabe-Akaike Information Criterion (WAIC) and Log Marginal Prediction Likelihood (LMPL).

3.3   The Data and Variables

Kelantan is one out of 14 states of Malaysia which can be found at $5°15'$ North latitude and $102°0'$ East latitude. This state is located at the north-eastern of Peninsular Malaysia. Kota Bharu is the capital city of Kelantan and the royal capital city of Kelantan is Kubang Kerian. The estimated total land area of Kelantan is 15, 040 km^2 and it is populated by 1.89 million of people as of 2019 [29]. Kelantan experiences a tropical climate, with temperatures ranging from 21 °C to 32 °C and received intermittent rain throughout the year [30]. The wet season in Kelantan is during east-coast monsoon season which is happened between November to March every year. During this season, certain districts of Kelantan receives heavy rainfall which can caused a big flood. Kelantan is divided into 10 administrative districts namely, Bachok, Gua Musang, Jeli, Kota Bharu, Kuala Krai, Machang, Pasir Mas, Pasir Puteh, Tanah Merah and Tumpat. In this study, daily dengue cases were collected from Vector Borne and Infectious Diseases Sector (VBIDS) at Ministry of Health (MoH) Malaysia for the ten districts of Kelantan. Meanwhile, the climatic factors such as the daily average of temperature, relative humidity and the amount of rainfall were obtained from climate data online (https://power.larc.nasa.gov/data-access-viewer/) (see Table 1). The data period is between 1st January 2013 to 31st December 2017. Table 2 shows the summary of total dengue incidences and climatic factors for 10 districts in Kelantan. The highest total dengue incidences within the 5 years was in Kota Bahru. The range of climatic factors suggest an ideal condition for *Aedes* mosquitos to breed and disperse which later lead into an outbreak in the community. There were no missing values in the data. In epidemiological studies especially in vector-borne disease study, the outliers are kept because of the importance to analyse in greater detail the evolution of the disease over time [31].

Table 1. Data description

Variables	Measurement Unit	Source
Dengue	Number of daily cases	Vector Borne and Infectious Diseases Sector (Ministry of Health Malaysia)
Average Temperature	°C	NASA climate data online (https://power.larc. nasa.gov/data-access-viewer/)
Rainfall	mm	
Relative Humidity	%	
Population	Millions	Department of Statistics Malaysia

Table 2. Summary of dengue incidences and climatic factors

District	Total Dengue (2013–2017)	Range of Temperature	Range of Relative Humidity	Range of Rainfall
Bachok	1856	24 °C–30 °C	70%–90%	0 mm–250 mm
Gua Musang	225	21 °C–27 °C	70%–95%	0 mm–250 mm
Jeli	171	21.3 °C–28 °C	70%–95%	0 mm–250 mm
Kota Bharu	18, 459	24 °C–31 °C	70%–95%	0 mm–250 mm
Kuala Krai	451	22 °C–29 °C	75%–95%	0 mm–250 mm
Machang	626	22 °C–29 °C	70%–95%	0 mm–500 mm
Pasir Mas	1162	24 °C–30 °C	70%–90%	0 mm–300 mm
Pasir Puteh	480	23.4 °C–29 °C	65%–95%	0 mm–400 mm
Tanah Merah	1269	22 °C–29 °C	70%–95%	0 mm–300 mm
Tumpat	2514	24 °C–30 °C	70%–90%	1 mm–300 mm

4 Results and Findings

The Fixed Effects Negative Binomial GLM and Random Effects Bayesian Poisson Generalized Linear Mixed Models (GLMM) models were developed using all covariates, followed by several different subsets of covariates including the time lag effect for each climatic factor. Table 3 summarizes the model comparison between fixed effects Negative Binomial GLM and Random Effects Bayesian Poisson GLMM. The smaller mean squared error (MSE) for GLMM model revealed that Random Effects Bayesian Poisson GLMM performed better than both fixed effects Negative Binomial GLM in predicting the dengue incidence rates in Kelantan.

Table 3. Model comparison

Parameter	FE NB GLM	RE Bayesian Poisson GLMM		
	Parameter Estimates	Posterior Median	2.5%	97.5%
Intercept	−16.7809*** (0.9407)	−9.58	−10.96	−8.26
Temp_lag21	−	−0.0591	−0.09	−0.03
Temp_lag28	0.0932*** (0.0230)	−	−	−
Humid_lag7	−	−0.0075	−0.01	−0.00
Humid_lag14	−	−	−	−
Humid_lag21	−	−0.0155	−0.02	−0.01
Humid_lag28	0.0132** (0.0044)	−	−	−
Tau	−	0.2322	0.19	0.27
Rho. S (ρ_S)	−	0.9157	0.90	0.93
Rho.T (ρ_T)	−	0.7705	0.74	0.80
Deviance	13000.16	−		
R_D^2	0.3635	−		
AIC	38865.58	−		
BIC	41802.86	−		
DIC	−	33624.88		
WAIC	−	34371.287		
LMPL	−	−17449.166		
Loglikelihood	−	−14564.566		
MSE	2.1258	0.3214		

Note: Only significant variables included in the model

The posterior median and 95% credible intervals for parameter estimates of Random Effects Bayesian Poisson GLMM are shown in Table 4. The Random Effects Bayesian Poisson GLMM revealed that relative humidity of lag 7 days, average temperature of lag 21 days and relative humidity of lag 21 days were found to be statistically significant on the emergence of dengue outbreak in Kelantan. All the three significant climatic factors portrayed a negative relationship towards the emergence of dengue incidence in Kelantan. It was found that the expected risk of dengue incidence in Kelantan would decrease by 0.7%, 1.5% and 5.7% for every 1% increase in the past 7 and 21 days of relative humidity and for every 1 °C increase in the past 21 days of average temperature, respectively. The Geweke diagnostic values reported the model converged as the values fell within ±2.0 as suggested by Lee [32]. The model also revealed the presence of spatial and temporal autocorrelation among microregions (districts) in Kelantan. The spatial and temporal dependence estimated parameters were 0.9157 and 0.7705, respectively. This indicates that high spatial and temporal autocorrelations were present in the data. By looking at the values of dependence parameters, it is noticeable that the heterogeneity amongst the districts of these states is present using GLMM. The model convergence was assessed by inspecting the stationarity of the trace-plots. Figure 1 shows the trace-plot on the left side and density plot on the right side for Kelantan.

Table 4. Posterior Median and 95% Credible Intervals of Dengue rate ratio. (Random Effects Bayesian Poisson GLMM)

Parameter	Reduced Model			
	Posterior Median of Rate Ratio			
	Median	2.50%	97.5%	Geweke Value
Humidity lag 7 days	0.93	0.90	0.98	1.5
Temperature lag 21 days	0.94	0.91	0.97	-0.1
Humidity lag 21 days	0.98	0.97	0.99	-0.3
Rho.s (ρ_S)	0.9157	0.90	0.93	0.5
Rho.T (ρ_T)	0.7705	0.74	0.80	0.6

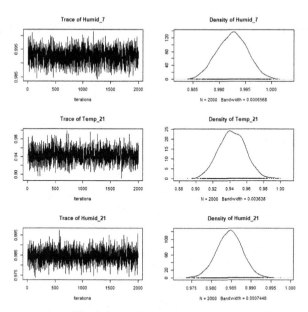

Fig. 1. Trace Plot and Density Plot

Figure 2 compares the temporal series of the daily dengue incidence rates of (a) fixed effects Negative Binomial GLM and (b) Bayesian Poisson GLMM for every district in Kelantan for a daily study period of 1st January 2013 to 31st December 2017. It can be seen that the GLMM with Bayesian framework introduced in the study was able to successfully capture the spatial and temporal variations in the dengue incidence rates even at a small scale of area. For instance, the incidence rates of Bachok, Kota Bharu and Tumpat were best captured by the GLMM compared to fixed effects GLM (underestimation problem) (see Fig. 2). These were also supported by the spatial map of the dengue incidence rates. Note that, only spatial map for the year 2017 was displayed in Fig. 3. Each figure of dengue incidence rates was arranged as (a) observed, (b) fixed effects Negative Binomial GLM, (c) Random Effects Bayesian Poisson GLMM. It can be seen that the fixed effects Negative Binomial GLM predicted the dengue incidence lower than the actual incidences (underestimation problem). Based on the spatial map of every district, it was clear that the dengue incidence rates computed from GLMM successfully captured more variations in dengue incidence rates as compared to the dengue incidence rates computed from fixed effects Negative Binomial GLM. Thus, GLMM was able to correctly capture districts with low to very low incidence rates as well as the districts with extremely high incidence rates. With all of these evidences, it can be justified that Random Effects Bayesian Poisson GLMM was found to be the best panel count model in predicting the daily dengue incidence rates for all districts in Kelantan.

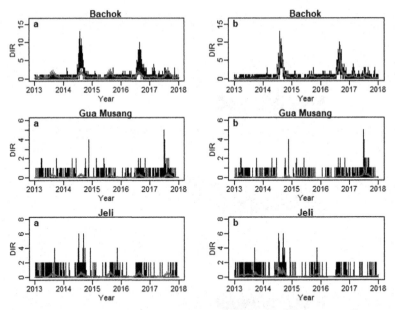

Fig. 2. Temporal Series of Observed DIR and Fitted DIR using (a) Fixed Effects Negative Binomial GLM and (b) Random Effects Bayesian Poisson GLMM

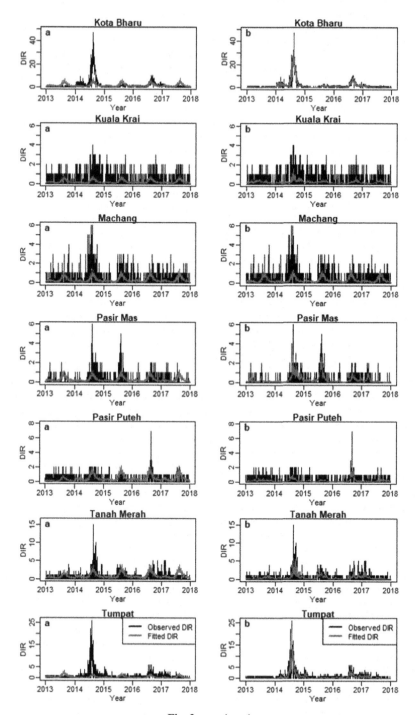

Fig. 2. continued

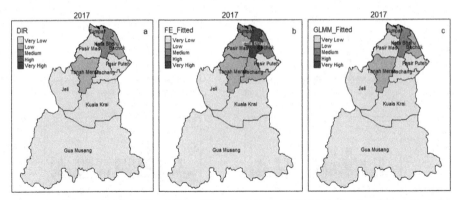

Fig. 3. (a) Observed DIR (b) Fixed Effects Negative Binomial GLM (c) Random Effects Bayesian Poisson GLMM for Kelantan. Categories boundaries are defined by 100, 200, 300, 400 and more than 400 cases per 100 000 population (2017)

5 Conclusion

The spatio-temporal models can capture both spatial and temporal effects. In modeling dengue incident in Kelantan districts, the fixed effects Negative Binomial GLM poorly captured the temporal variability in dengue counts across the districts especially when there are serious outbreaks. The proposed Random Effects Bayesian Poisson GLMM model enables the variability of the heterogeneity factors to be captured by allowing the previous dengue counts trend to be derived in the posterior predictions. The covariates considered are temperature, humidity and rainfall. The results of analysis confirmed that the dengue incidence in Kelantan were significantly associate with temperature and humidity. The Random Effects Bayesian Poisson GLMM introduced in the study had successfully accounted for the unobserved heterogeneity factors and the variations in dengue incidence between spatial regions present in the data which was not captured by the conventional fixed effects Negative Binomial GLM. This study confirmed that the addition of the spatial random effects into the linear predictor improved the over-all model fit as confirmed by more accurate spatial estimates of the dengue incidence rates across districts of Kelantan.

Acknowledgements. We would like to express our appreciation and record our gratitude to the Ministry of Higher Education (MOHE) and Universiti Teknologi MARA for the financial support under the Fundamental Research Grant Scheme (FRGS) (FRGS/1/2016/STG06/UITM/02/2). The authors are also grateful to reviewers for their insightful comments and suggestions.

References

1. Central for Disease Control and Prevention (CDC). Dengue (2021). https://www.cdc.gov/den gue/. Accessed 01 June 2020
2. World Health Organization. Dengue and severe dengue (2021) http://www.who.int/mediac entre/factsheets/fs117/en/. Accessed 01 June 2020

3. Lai, Y.: The climatic factors affecting dengue fever outbreaks in Southern Taiwan: an application of symbolic data analysis. BioMed. Eng. OnLine **17**(148), 49–62 (2018)
4. Alkhaldy, I.: modelling the association of dengue fever cases with temperature and relative humidity in Jeddah, Saudi Arabia—a generalised linear model with break-point analysis. Acta Trop. **168**, 9–15 (2017)
5. Talagala, T.: Distributed lag nonlinear modelling approach to identify relationship between climatic factors and dengue incidence in Colombo district, Sri Lanka. Epidemiol. Biostat. Public Health **12**(4), e11522-1–e11522-8 (2015)
6. Chandren, J.R., Wong, L.P., Abu Bakar, S.: Practices of dengue fever prevention and the associated factors among the orang Asli in Peninsular Malaysia. PLoS Negl. Trop. Dis. **9**(8), e0003954 (2015)
7. Che Him, N., Bailey, T.C., Stephenson, D.B.: Climate variability and dengue incidence in Malaysia. In: 27th International Workshop on Statistical Modelling, vol. 2, pp. 435–440 (2012)
8. Bisht, B., et al.: Influence of environmental factors on dengue fever in Delhi. Int. J. Mosq. Res. **6**(2), 11–18 (2019)
9. Naqvi, S.A.A., et al.: Changing climatic factors favour dengue transmission in Lahore, Pakistan. Environments **6**, 71 (2019)
10. Ruzman, N.S.L.N., Rahman, H.A.: The association between climatic factors and dengue fever: a study in Subang Jaya and Sepang, Selangor, Malaysian. J. Public Health Med. **1**, 140–150 (2017)
11. Ahmed, S.A., Junai, S.S., Sabah, Q., Afaq Ahmed, S.: Analysis of climate structure with Karachi dengue outbreak. J. Basic Appl. Sci. **11**, 544–552 (2015)
12. Atique, S., Syed Abdul, S., Hsu, C.H., Chuang, T.W.: Meteorological influences on dengue transmission in Pakistan. Asian Pac. J. Trop. Med. **9**(10), 954–961 (2016)
13. Choi, Y., et al.: Effects of weather factors on dengue fever incidence and implications for interventions in Cambodia. BMC Public Health **16**, 241 (2016)
14. Chandrakantha, L.: Statistical analysis of climate factors influencing dengue incidences in Colombo, Sri Lanka: poisson and negative binomial regression approach. Int. J. Sci. Res. Publ. **9**(2), 133–144 (2019)
15. Tuladhar, R., Singh, A., Varma, A., Choudhary, D.K.: Climatic factors influencing dengue incidence in an epidemic area of Nepal. BMC Res Notes **12**, 131 (2019)
16. Cheong, Y.L., Burkart, K., Leitao, P.J., Lakes, T.: Assessing weather effects on dengue disease in Malaysia. Int. J. Environ. Res. Public Health **10**, 6319–6334 (2013)
17. Wan Fairos, W.Y., Azaki, W.W., Alias, L.M., Wah, Y.B.: Modelling Dengue Fever (DF) and Dengue Haemorrhagic Fever (DHF) outbreak using poisson and negative binomial model. World Acad. Sci. Eng. Technol. **38**, 903–908 (2010)
18. Lekdee, K., Ingsrisawang, L.: Generalized linear mixed models with spatio random effects for spatio-temporal data: an application to dengue fever mapping. J. Math. Stat. **9**(2), 137–143 (2013)
19. Lowe, R., et al.: Spatio-temporal modelling of climate-sensitive disease risk: towards an early warning system for dengue in Brazil. Comput. Geosci. **37**(3), 371–381 (2011)
20. Phanitchat, T., Zhao, B., Haque, U.: Spatial and temporal patterns of dengue incidence in north-eastern Thailand 2006–2016. BMC Infect. Dis. **19**, 743 (2019). https://doi.org/10.1186/s12879-019-4379-3
21. Lowe, R.: Spatio-temporal modelling of climate sensitive disease risk: towards an early warning system for dengue in Brazil. Doctoral thesis, University of Exeter (2010)
22. Ibarra, A.M.S., Lowe, R.: Climate and non-climate drivers of dengue epidemics in southern coastal Ecuador. Am. J. Trop. Med. Hyg. **88**(5), 971–981 (2013)
23. Lowe, R., Chirombo, J., Tompkins, A.M.: Relative importance of climatic, geographic and socio-economic determinants of Malaria in Malawi. Malar. J. **12**(416), 1–16 (2013)

24. Lowe, R., Cazelles, B., Paul, R., Rodó, X.: Quantifying the added value of climate information in a spatio-temporal dengue model. Stoch. Environ. Res. Risk Assess. **30**(8), 2067–2078 (2015). https://doi.org/10.1007/s00477-015-1053-1

25. Hausman, J., Hall, B., Griliches, Z.: Econometric models for count data with an application to the patents - R&D relationship. Econometrica **52**, 909–938 (1984)

26. Lee, D., Rushworth, A., Napier, G.: Spatio-temporal areal unit modeling in R with conditional autoregressive priors using the CARBayesST package. J. Stat. Softw. **84**(9), 1–39 (2018)

27. Leroux, B.G., Lei, X., Breslow, N.: Estimation of disease rates in small areas: a new mixed model for spatial dependence. In: Halloran, M., Berry, D. (eds.) Statistical Models in Epidemiology, the Environment and Clinical Trials. The IMA Volumes in Mathematics and its Applications, vol. 116, pp. 179–191. Springer, New York (2000). https://doi.org/10.1007/978-1-4612-1284-3_4

28. Rushworth, A., Lee, D., Mitchell, R.: A spatio-temporal model for estimating the long-term effects of air pollution on respiratory hospital admissions in Greater London. Spat. Spatio-Temporal Epidemiol. **10**, 29–38 (2014)

29. Department of Statistics Malaysia (DoSM). Statistics (2021)

30. Malaysian Meteorological Department (MET). https://www.met.gov.my/. Accessed 01 June 2020

31. Cabrera, M.: Spatio-temporal modelling of dengue fever in Zulia State, Venezuela. Doctoral thesis, University of Bath (2013)

32. Lee, D.: A tutorial on spatio-temporal disease risk modelling in R using Markov chain Monte Carlo simulation and the CARBayesST package. Spat. Spatio-Temporal Epidemiol. **34**, 100353 (2020)

The Effect of Cancers Treatment in Quality of Life of the Patient: Meta-analysis Approach

Santi Wulan Purnami$^{(\boxtimes)}$, Prilyandari Dina Saputri, and Bambang Widjanarko Otok

Department of Statistics, Institut Teknologi Sepuluh Nopember, Kampus ITS Keputih Sukolilo, Surabaya 60111, Indonesia
{santi_wp,bambang_wo}@statistika.its.ac.id

Abstract. Cancer treatments are expected to improve the quality of life of the patients. Thus the comparison of the quality of life before and after treatment is important to be present. Several dimensions are used in measuring the quality of life, i.e., global quality of life, physical functioning, role functioning, emotional functioning, cognitive functioning, and social functioning. We applied meta-analysis to identify the comparison based on the included study material from Google Scholar. The duration in measuring the quality of life after treatment is at least two months. The treatments included in this study are chemotherapy, radiotherapy, and surgery. There are 12 suitable study materials from 1150 available articles. The models used are a random-effects model since there exists heterogeneity between studies for all variables. In general, the quality of life of patients after treatment has increased, which is indicated by the negative mean difference estimation for all variables. However, the improvement of the quality of life in the dimensions of role functioning and cognitive functioning is not significant. Meanwhile, in the other four dimensions, the improvement was significant. Thus, cancer treatments are able to improve the patient's quality of life in the long term.

Keywords: Cancer · Meta-analysis · Quality of life · Treatment

1 Introduction

Cancer is one of the foremost causes of death in the world [1]. Many cancer treatments have been developed, including drugs, chemotherapy, radiation, and surgery. In the long term, it is expected that there will be an improvement in the quality of life of cancer patients after they go through the treatment. The quality of life of cancer patients can be measured using the EORTC QLQ-C30 questionnaire [2].

The European Organization for Research and Treatment of Cancer (EORTC) created a multidimensional parameter of quality of life in cancer patients using 30 questions (Quality of Life Questionnaire/QLQ-C30). The dimensions of the quality of life used include global quality of life, physical functioning, role functioning, emotional functioning, cognitive functioning, and social functioning. Quality of life is measured using a scale of 1–100. The higher the scale of a dimension indicates the better quality of life for the corresponding dimension [2].

© Springer Nature Singapore Pte Ltd. 2021
A. Mohamed et al. (Eds.): SCDS 2021, CCIS 1489, pp. 341–350, 2021.
https://doi.org/10.1007/978-981-16-7334-4_25

The previous studies relating to the comparison of the quality of life of cancer patients before and after treatments have been conducted. The different results arise since there are several studies showing the quality of life has improved [3–5], and some others show a deterioration of quality of life [6, 7]. In this study, a meta-analysis has been conducted on the individual research analysis results to integrate a conclusion. The meta-analysis was conducted to determine the difference in the quality of life of cancer patients before and after treatment. The models used are fixed-effects and random-effects models. The selection of the model is completed based on the results of heterogeneity testing and the characteristics of the study materials. Publication bias has been identified using funnel plots and meta-bias testing.

2   Materials and Methods

2.1   Journal Criteria

The inclusion criteria used in this study are described as follows:

- The comparison of quality of life before and after treatment are presented
- The treatment is limited to chemotherapy, radiation, radiotherapy, and surgery
- The type of various cancers are allowed
- The duration of measurement after treatment is at least two months
- The quality of life was measured using the EORTC-QLQ C30 questionnaire.

The exclusion criteria are described as follows:

- The treatment in the form of drugs
- Measurement of post-treatment quality of life is completed in less than two months
- Quality of life measurement using a questionnaire other than EORTC-QLQ C30.

The keywords used in searching the study materials were: "quality of life" AND patient AND "treatment effect" AND cancer AND pre AND post OR before AND after AND EORTC AND "QLQ-C30" AND chemotherapy OR radiotherapy OR surgery.

2.2   Search Process

The included studies used are obtained through Google Scholar. Based on the search for study materials, 1150 articles match the keywords. From 1150 articles, there are 378 articles relevant to the purpose of this study, while the other 772 articles are irrelevant or did not use the EORTC-QLQ C30 questionnaire. From the 378 articles, 313 articles discuss one condition only, specifically when suffering from cancer or after undergoing treatment. Furthermore, from 65 articles, some studies present the differences data only or do not enclose the standard deviation. So there are 33 articles with complete data. However, in the 21 articles, some quality of life measurements after treatment were not presented, and some were less than two months. Thus, the remaining 12 articles met the specified criteria, with three articles discussing more than one treatment. The flowchart of the study search process is shown in Fig. 1.

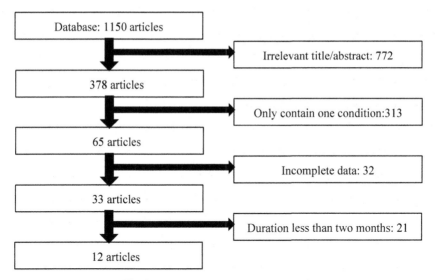

Fig. 1. Search process of Included Study

2.3 Effect Size in Meta-analysis

Meta-analysis is an analysis of individual research results to integrate a conclusion. Effect size is a quantitative index to summarize the results of studies in meta-analysis. In this study, the effect size used is the mean difference to find out whether there is a significant difference in the quality of life of cancer patients before treatment and after treatment. The mean difference can be obtained using the following formula:

$$Y_i = \overline{X}_{1i} - \overline{X}_{2i} \tag{1}$$

where $\overline{X}_1$ and $\overline{X}_2$ are the mean of the two groups. Given S_1 and S_2 are the standard deviation of the two groups, n_1 and n_2 are the sample sizes, the variance for each group can be obtained using the following formula [8]:

$$V_{Y_i} = \frac{S_1^2}{n_1} + \frac{S_2^2}{n_2}. \tag{2}$$

2.4 Fixed-Effects and Random-Effects

In the meta-analysis model, each study will be weighted. In this study, the weight used for the fixed-effects model is the inverse variance, which can be described as follows:

$$W_i = \frac{1}{V_{Y_i}}. \tag{3}$$

where V_{Y_i} is the variance of the ith study. The mean effect size estimate (summary effect) and its variance can be obtained using:

$$M = \frac{\sum_{i=1}^{k} W_i Y_i}{\sum_{i=1}^{k} W_i}, V_M = \frac{1}{\sum_{i=1}^{k} W_i}. \tag{4}$$

From Eq. (4), the standard error of the summary effect is $SE_M = \sqrt{V_M}$. The calculation for the test statistics of the fixed-effects model is

$$Z = \frac{M}{SE_M} \tag{5}$$

In the random-effects model, the parameter τ^2 is used to measure the variability between studies. The estimation of T^2 can be calculated using moment methods (DerSimonian and Laird).

$$T^2 = \frac{Q - df}{C} \tag{6}$$

where,

$$Q = \sum_{i=i}^{k} W_i Y_i^2 - \frac{\left(\sum_{i=1}^{k} W_i Y_i\right)^2}{\sum_{i=1}^{k} W_i}, df = k - 1, C = \sum_{i=i}^{k} W_i - \frac{\sum_{i=1}^{k} W_i^2}{\sum_{i=1}^{k} W_i}$$

The variance of the random-effects model can be obtained using the following formula:

$$V_{Y_i}^* = V_{Y_i} + T^2 \tag{7}$$

The value of $V_{Y_i}^*$ can be used to estimate the weight, mean effect size, variance effect size, and test statistics using similar formulas with the fixed-effects model [8, 9].

3   Result

3.1   The Characteristics of Included Studies

There are 12 studies included in the meta-analysis. We found the different results of the previous studies, which can be found in the forest plot in Fig. 2. Many studies show the improvement of quality of life after treatment, while the others are contrary. The improvement of the quality of life is indicated by the negative mean difference estimated. The characteristics of each study can be found in Table 1.

The included studies are varied in the type of treatment and type of cancer, as shown in Table 1. These conditions lead us to use the random effect model. In the next section, the heterogeneity test is provided to confirm the suitable model. We used the included studies with a duration of more than two months in measuring the quality of life before and after treatment, as we desire the measurement for the quality of life in the long term.

3.2   Effect Size Estimation

We conducted heterogeneity testing to determine the variation between studies. The heterogeneity test has been carried out using the Cochran Q test, which is shown in Table 2.

Table 1. The characteristics of included studies

Study	Location	Study N before/after	Treatment	Type of Cancer	Duration
Perwitasari [10]	Indonesia	38/38	Chemotherapy	Cervical, Ovarian, Nasopharynx	2 months
Wang [6]	China	41/41	Chemotherapy	Breast	2 months
Bang[a] [11]	Korea	46/46	Chemotherapy	Varied	3 cycles
Bang[b] [11]	Korea	49/49	Chemotherapy	Varied	3 cycles
Adamsen [12]	Denmark	134/137	Chemotherapy	Varied	6 months
Shi[a] [3]	Taiwan	57/57	Surgery	Breast	12 months
Shi[b] [3]	Taiwan	83/83	Modified radical mastectomy	Breast	12 months
Bach [13]	Jerman	185/185	Retropubic prostatectomy	Prostate	28 months
Grulke [4]	Jerman	38/38	Haematopoietic stem cell transplantation	Varied	12 months
Meraner [14]	Austria	21/21	Chemotherapy	Ovarian	9 months
Cruzado [15]	Spanyol	81/54	Chemotherapy	Colon	6 months
Schmid [7]	Jerman	32/32	Chemotherapy	Breast	3 months
Hompes [16]	Inggris	92/33	Surgery	Rectal Tumour	12 months
Suwendar[a] [5]	Indonesia	32/32	Cisplatin Chemotherapy	Cervical	3 cycles
Suwendar[b] [5]	Indonesia	11/11	Carboplatin Chemotherapy	Cervical	3 cycles

Table 2. Heterogeneity of Included Study

Variable	I^2	τ^2	Q	p-value
Global Quality of Life	88,3%	110,0384	119,54	<0,0001
Physical Functioning	96,0%	181,6790	350,93	<0,0001
Role Functioning	95,9%	304,8254	338,62	<0,0001
Emotional Functioning	95,7%	231,3584	328,79	<0,0001
Cognitive functioning	96,3%	224,5379	382,36	<0,0001
Social functioning	85,6%	76,8513	97,41	<0,0001

Table 2 shows that all variables comprise high heterogeneity between studies, which is indicated by the I^2 value of more than 76%. The p-value for all variables are less than 0.01, with $\alpha = 0.05$, the null hypothesis is rejected. Thus the study materials used have significant heterogeneity of effect size for all variables.

In the meta-analysis, a forest plot is used to show the mean difference graphically and the weights for each study.

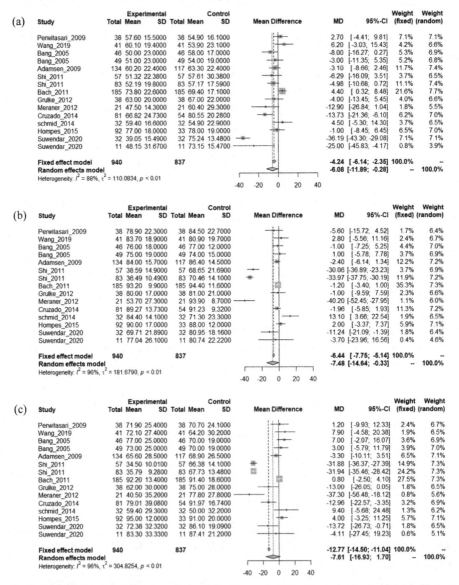

Fig. 2. Forest Plot of (a) global quality of life, (b) physical functioning, (c) role functioning, (d) emotional functioning, (e) cognitive functioning, and (f) social functioning

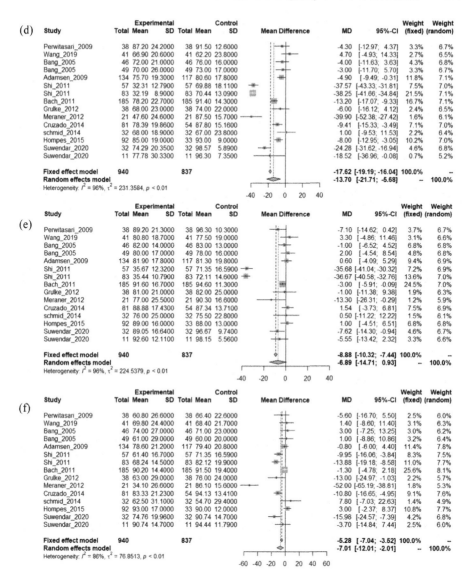

Fig. 2. continued

The forest plot in Fig. 2 shows that there is a difference in weight for each study material. The largest square plot indicates the largest study weight. The largest weights for the fixed-effects and random-effects models are in the same study. However, in the random-effects model, the weights of each study are relatively more homogeneous than the weights in the fixed-effects model. In variable global quality of life, physical functioning, cognitive functioning, and social functioning, the study material with the largest weight is Bach [13]. For the variables of role functioning and emotional functioning, the study from Shi [3] and Bach [13] has the largest weight compared to other studies.

Table 3. Effect Size Estimation

Variable	Fixed-Effects			Random-Effects		
	MD	SE	p-value	MD	SE	p-value
Global Quality of Life	−4,24	0,97	<0,0001	−6,08	2,96	0,0399
Physical Functioning	−6,44	0,67	<0,0001	−7,48	3,65	0,0405
Role Functioning	−12,77	0,88	<0,0001	−7,61	4,75	0,1092
Emotional Functioning	−17,61	0,81	<0,0001	−13,70	4,09	0,0008
Cognitive functioning	−8,88	0,73	<0,0001	−6,89	3,99	0,0843
Social functioning	−5,28	0,90	<0,0001	−7,01	2,55	0,0060

Table 3 shows that all variables show a significant improvement in the quality of life of cancer patients after treatment in the fixed-effects models. While in the random-effects model, the significant variables are the global quality of life, physical functioning, emotional functioning, and social functioning. All models used present an improvement in quality of life, which is indicated by a negative mean difference estimated. In other words, the quality of life after treatment was higher compared to the quality of life before treatment.

3.3   Publication Bias Analysis

We used bias publication analysis to determine the existence of bias in the included studies. Publication bias exists when only significant research is published. So that all selected study results will show a significance that may be higher than the actual significance. We use Begg's funnel plot or Egger's plot to display the presence of publication bias graphically. Statistical tests for publication bias are based on the fact that studies with small sample sizes are more susceptible to publication bias. When estimates from all studies are plotted against their variances, a symmetrical funnel is produced when there is no publication bias. The study with publication bias will result in a skewed asymmetric funnel plot. In this study, identification of publication bias has been completed using funnel plots and meta-bias testing. The funnel plot for all variables can be shown in Fig. 3.

Figure 3 shows that the funnel plots for the global quality of life, physical functioning, and social functioning variables are symmetrical. However, identification based on funnel plots is subjective. Thus, a meta-bias test is carried out with the null hypothesis that there is no publication bias in the included studies. Table 4 shows the results of meta-bias testing for all variables. The null hypothesis failed to be rejected for all variables. Thus, there is no publication bias on all the variables used in the study material.

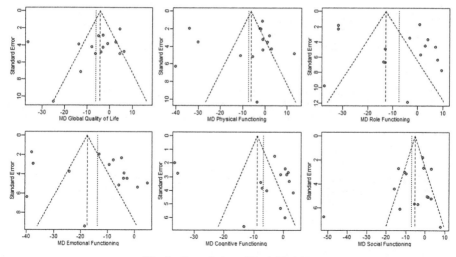

Fig. 3. Funnel plot of Each Variable

Table 4. Testing for Meta Bias

Variabel	bias	SE bias	t-value	p-value
Global Quality of Life	−2,5994	2,2354	−1,16	0,2658
Physical Functioning	−0,9120	2,6374	−0,35	0,7350
Role Functioning	2,6048	2,4147	1,08	0,3003
Emotional Functioning	4,1668	2,9993	1,39	0,1881
Cognitive functioning	2,1934	3,6882	0,59	0,5623
Social functioning	−1,6821	1,6813	−1,00	0,3353

4 Discussion and Conclusion

The models used for all variables are the random-effects model. This selection was based on the heterogeneity test results, which showed high heterogeneity in all variables. In addition, the variation in the type of cancer and treatment also indicates that the random-effects model is more appropriate to be used. In general, patients' quality of life after treatment (more than two months) has improved. It is indicated by the negative estimated mean difference for all variables. However, the improvement in quality of life in role function and cognitive function was not significant. Meanwhile, in the other four aspects, the improvement was significant. The results of the analysis based on the funnel plots show that the funnel tends to be symmetrical. Testing the significance of publication bias also indicates that there is no publication bias in the included study.

References

1. Sung, H., Ferlay, J., Siegel, R.L.: Global cancer statistics 2020: GLOBOCAN estimates of incidence and mortality worldwide for 36 cancers in 185 countries. CA Cancer J. Clin. **71**, 209–249 (2021)
2. Fayers, P., Bottomley, A.: Quality of life assessment and research in the EORTC - the EORTC QLQ-C30. Eur. J. Cancer **38**, S125–S133 (2002)
3. Shi, H.Y., et al.: Two-year quality of life after breast cancer surgery: a comparison of three surgical procedures. Eur. J. Surg. Oncol. **37**, 695–702 (2011)
4. Grulke, N., Albani, C., Bailer, H.: Quality of life in patients before and after haematopoietic stem cell transplantation measured with the European Organization for Research and Treatment of Cancer (EORTC) Quality of Life Core Questionnaire QLQ-C30. Bone Marrow Transplant. **47**, 473–482 (2012)
5. Suwendar, S., Fudholi, A., Andayani, T.M., Sastramihardja, H.S.: Evaluasi Kualitas Hidup Pasien Kanker Serviks yang Mendapat Regimen Kemoterapi Cisplatin-Vinkristin-Bleomisin dan Carboplatin-Paklitaksel. J. Manaj. dan Pelayanan Farm. **10**, 167–175 (2020)
6. Wang, Z., Yin, G., Jia, R.: Impacts of self-care education on adverse events and mental health related quality of life in breast cancer patients under chemotherapy. Complement. Ther. Med. **43**, 165–169 (2019)
7. Schmidt, M.E., et al.: Effects of resistance exercise on fatigue and quality of life in breast cancer patients undergoing adjuvant chemotherapy: a randomized controlled trial. Int. J. Cancer **137**, 471–480 (2014)
8. Borenstein, M., Hedges, L.V., Higgins, J.P.T., Rothstein, H.R.: Introduction to Meta Analysis. Wiley, Hoboken (2009)
9. Chen, D.G., Peace, K.E.: Applied Meta-Analysis with R. CRC Press, Boca Raton (2013)
10. Perwitasari, D.A.: Pengukuran Kualitas Hidup Pasien Kanker Sebelum dan Sesudah Kemoterapi dengan EORTC QLQ-C30 di RSUP Dr. Sardjito Yogyakarta. Maj. Formasi Indones. **20**, 68–72 (2009)
11. Bang, S.M., et al.: Changes in quality of life during palliative chemotherapy for solid cancer. Support. Care Cancer **13**, 515–521 (2005)
12. Adamsen, L., et al.: Effect of a multimodal high intensity exercise intervention in cancer patients undergoing chemotherapy: randomised controlled trial. BMJ **339**, 895–898 (2009)
13. Bach, P., Doring, T., Gesenberg, A., Mohring, C., Mark, G.: Quality of life of patients after retropubic prostatectomy - pre- and postoperative scores of the EORTC QLQ-C30 and QLQ-PR25. Health Qual. Life Outcomes **9**, 93–101 (2011)
14. Meraner, V., et al.: Monitoring physical and psychosocial symptom trajectories in ovarian cancer patients receiving chemotherapy. BMC Cancer **12**, 77–86 (2012)
15. Cruzado, J.A., López-Santiago, S., Martínez-Marín, V., José-Moreno, G., Custodio, A.B., Feliu, J.: Longitudinal study of cognitive dysfunctions induced by adjuvant chemotherapy in colon cancer patients. Support. Care Cancer **22**(7), 1815–1823 (2014). https://doi.org/10.1007/s00520-014-2147-x
16. Hompes, R., et al.: Evaluation of quality of life and function at 1 year after transanal endoscopic microsurgery. Color. Dis. **17**, O54–O61 (2014)

Hybrid of Time Series Regression, Multivariate Generalized Space-Time Autoregressive, and Machine Learning for Forecasting Air Pollution

Hendri Prabowo, Dedy Dwi Prastyo$^{(\boxtimes)}$, and Setiawan

Department of Statistics, Institut Teknologi Sepuluh Nopember, Kampus ITS, Sukolilo,
Surabaya 60111, Indonesia
dedy-dp@statistika.its.ac.id

Abstract. The purpose of this study is to propose a new hybrid of space-time models by combining the time series regression (TSR), multivariate generalized space-time autoregressive (MGSTAR), and machine learning (ML) to forecast air pollution data in the city of Surabaya. The TSR model is used to capture linear patterns of data, especially trends and double seasonal. The MGSTAR model is employed to capture the relationship between locations, and the ML model is used to capture nonlinear patterns from the data. There are three ML models used in this study, namely feed-forward neural network (FFNN), deep learning neural network (DLNN), and long short-term memory (LSTM). So that there are three hybrid models used in this study, namely TSR-MGSTAR-FFNN, TSR-MGSTAR-DLNN, and TSR-MGSTAR-LSTM. The hybrid models will be used to forecast air pollution data consisting of CO, PM_{10}, and NO_2 at three locations in Surabaya simultaneously. Then, the performance of these three-combined hybrid models will be compared with the individual model of TSR and MGSTAR, two-combined hybrid models of MGSTAR-FFNN, MGSTAR-DLNN, MGSTAR-LSTM, and hybrid TSR-MGSTAR models based on the RMSE and sMAPE values in the out-of-sample data. Based on the smallest RMSE and sMAPE values, the analysis results show that the best model for forecasting CO is MGSTAR, forecasting PM_{10} is hybrid TSR-MGSTAR, and forecasting NO_2 is hybrid TSR-MGSTAR-FFNN. In general, the hybrid model has better accuracy than the individual models. This result is in line with the results of the M3 and M4 forecasting competition.

Keywords: Air pollution · Forecast · Hybrid · Machine learning · Space-time

1 Introduction

Air pollution has a destructive impact on human life [1]. In tackling the adverse effects of air pollution, it is possible to forecast the concentration of air pollution as a preventive measure. This air pollution data has several air pollutant parameters such as CO, NO_2, and PM_{10}. Each of these air pollutant variables affects the concentration of other variables [2]. Moreover, it is also influenced by pollutant variables in other locations in the region

© Springer Nature Singapore Pte Ltd. 2021
A. Mohamed et al. (Eds.): SCDS 2021, CCIS 1489, pp. 351–365, 2021.
https://doi.org/10.1007/978-981-16-7334-4_26

[3]. So, the forecasting that will be carried out needs to consider the influence of other variables in the same and other locations.

In time-series analysis, forecasting that considers the location is called space-time data forecasting. Several methods are often used in forecasting space-time data, i.e., Space-Time Autoregressive (STAR) and Generalized Space-Time Autoregressive (GSTAR) [4, 5]. These methods have several disadvantages. They cannot be used to forecast space-time data that has more than one variable in each location. In reality, it is often found that space-time data has more than one variable in each location that influences each other, such as air pollution data. Thus, Suhartono et al. developed the Multivariate Generalized Space-Time Autoregressive (MGSTAR) model [6].

In some cases, the time-series data does not only have linear or nonlinear patterns. It is often found that data with a combination of linear and nonlinear patterns have a location effect, so it is necessary to build a model based on a combination of several models to overcome such issues [7]. Nonlinear models that are often used in the formation of hybrid models for data forecasting are machine learning (ML) models such as feed-forward neural network (FFNN), deep learning neural network (DLNN), and long short-term memory (LSTM) [8–10]. This combined model is called the hybrid model. This hybrid model tends to increase the accuracy of individual models [11].

In this research, a hybrid space-time model is proposed by combining the TSR, MGSTAR, and ML models. The TSR model captures linear patterns from air pollution data, especially trend and double seasonal patterns. The MGSTAR is used to capture the effect of location on air pollution data. The ML model is employed to capture nonlinear patterns from air pollution data. This air pollution data consists of three variables: CO, NO_2, and PM_{10} at three SUF stations (locations) in Surabaya.

Several previous studies have conducted space-time data forecasting using a hybrid model. Suhartono et al. forecast space-time data with exogenous variables using VARX, GSTARX, and hybrid GSTARX-NN [12]. Pusporani et al. used the hybrid MGSTAR-ANN model to forecast air pollution data [13]. Then Laily et al. compared the performance of the individual MGSTAR model with the hybrid MGSTAR-RNN model [14]. Moreover, another study using a hybrid model for forecasting space-time data has also been carried out by Prastyo et al. [15] and Suhartono et al. [16]. In general, several previous studies have shown that hybrid models tend to increase the accuracy of individual models in forecasting space-time data.

The rest of the paper is organized as follows: Sect. 2 reviews the methodology, i.e., TSR, MGSTAR, FFNN, DLNN, LSTM, and hybrid model; Sect. 3 presents the dataset and methodology; Sect. 4 presents the results and discussion; and Sect. 5 presents the conclusion from this study.

2   Methods

2.1   Time Series Regression

Time series regression (TSR) is one of the time series models used to predict data with a linear pattern [17]. In the model of TSR, there are predictor variables that affect the response variable. The predictor variable can be metric or non-metric, for example the dummy variables were employed as predictors [18]. Equation (1) is a multiplicative TSR model with trend and two seasonal elements, the daily seasonal (for half-hour data) and weekly seasonal (seven days).

$$Z_t = \beta_0 + \beta_1 t + \beta_2 H_{1t} M_{1t} + \ldots + \beta_{49} H_{48t} M_{1t} + \ldots + \beta_{336} H_{47t} M_{7t} + \varepsilon_t, \quad (1)$$

where ε_t is an error that satisfies the assumptions of identical, independent, and normally distributed, t is trend component, H_{it} is a dummy variable for half-hour data (i on a day with $i = 1, 2, \ldots, 48$), while M_{mt} is a dummy variable for the day (m in one week with $m = 1, 2, \ldots, 7$).

2.2 Multivariate Generalized Space-Time Autoregressive

The multivariate generalized space-time autoregressive (MGSTAR) model is a development of the GSTAR and VAR models introduced by Suhartono et al. [6]. This model can be used for space-time data forecasting involving more than one variable at each location. For example, the MGSTAR model, involving two variables in three locations with the time and spatial order of one, is shown by Eq. (2).

$$\mathbf{Z}_t = \boldsymbol{\Phi}_0 \mathbf{Z}_{t-1} + \boldsymbol{\Phi}_0^* \mathbf{Z}_{t-1} + \boldsymbol{\Phi}_1 \mathbf{W} \mathbf{Z}_{t-1} + \boldsymbol{\Phi}_1^* \mathbf{W} \mathbf{Z}_{t-1} + \mathbf{a}_t, \quad (2)$$

where $\boldsymbol{\Phi}_0$ is a parameter that describes the lag relationship between the same variable and the same location, $\boldsymbol{\Phi}_0^*$ is a parameter that describes the lag relationship between different variables at the same location, $\boldsymbol{\Phi}_1$ is a parameter that explains the lag relationship between the same variable at different locations, $\boldsymbol{\Phi}_1^*$ is a parameter that describes the lag relationship between different variables at different locations, and $\mathbf{W}$ is the weight of the spatial location.

There are two types of location weights used in this study, i.e., uniform weights and inverse distance weights. Uniform weight assumes that the locations used have homogeneous characteristics and have the same distance between locations indicated by Eq. (3). The inverse weight of the distance is calculated based on the actual distance for each observation location, and then the distance obtained is inversely shown by Eq. (4).

$$\mathbf{W} = \begin{bmatrix} 0 & 0.5 & 0.5 \\ 0.5 & 0 & 0.5 \\ 0.5 & 0.5 & 0 \end{bmatrix}. \quad (3)$$

$$\mathbf{W} = \begin{bmatrix} 0 & 0.497 & 0.503 \\ 0.535 & 0 & 0.465 \\ 0.537 & 0.463 & 0 \end{bmatrix} \quad (4)$$

2.3 Neural Network

One of the machine learning models often used in forecasting time-series data with nonlinear patterns is a neural network (NN). There are many architecture models in NN. The most widely used architecture is the feed-forward neural network (FFNN). In the FFNN structure, there are several layers, namely an input layer, a hidden layer, and an output layer [19]. There is an activation function in the hidden layer and the output layer to capture nonlinear patterns [19]. In the FFNN model, an architecture with more than one hidden layer is often called deep learning neural network (DLNN) [8]. The equation for the FFNN architecture with p input variables and q neurons in the hidden layer is shown in Eq. (5).

$$\hat{Z}_{(t)} = f^o \left[\sum_{j=1}^{q} \left[w_j^o f_j^h \left(\sum_{i=1}^{p} w_{ji}^h X_{i(t)} + b_j^h \right) + b^o \right] \right], \tag{5}$$

where w_j^o, b^o are the parameters of the FFNN model at the output layer; w_{ji}^h, b_j^h are the parameter of the FFNN model in the hidden layer; f^o is the activation function at the output layer; and f_j^h is the activation function in the hidden layer.

2.4   Long-Short Term Memory

Long short-term memory (LSTM) is a development of the recurrent neural network (RNN) method because there are many weaknesses in the RNN method, i.e., long-term lags cannot be accessed in architecture [20]. In addition, LSTM can solve the vanishing gradient problem in RNN [10]. As a result of this vanishing gradient, the information contained in the previous time interval will be reduced or forgotten because of the smaller weight on the information. This results in the loss of information in the previous time interval.

The structure of the LSTM architecture consists of several layers, namely the input layer, the recurrence hidden layer, and the output layer [21]. The difference with FFNN or DLNN is that the input used in LSTM is not only predictor variables but also hidden layer values at the time $(t-1)$. In LSTM, the neurons in the hidden layer are called memory cells. An LSTM cell consists of four gates: the input gate, the input modulation gate, the forget gate, and the output gate, as shown in Fig. 1 [20].

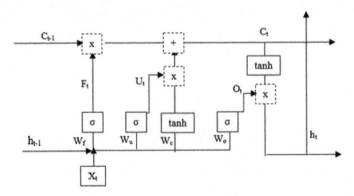

Fig. 1. LSTM cell structure

Figure 1 is the structure of the LSTM cell from which the formulation can be written in Eq. (6) to Eq. (11).

Forget Gate:

$$F_t = \sigma(w_{f,h}h_{t-1} + \sum_{i=1}^{p} w_{f,xi}X_{i(t)} + b_f), \tag{6}$$

Input Gate and Input Modulation Gate:

$$\hat{C}_t = \tanh(w_{c,h}h_{t-1} + \sum_{i=1}^{p} w_{c,xi}X_{i(t)} + b_c), \tag{7}$$

$$U_t = \sigma(w_{u,h}h_{t-1} + \sum_{i=1}^{p} w_{u,xi}X_{i(t)} + b_u), \tag{8}$$

$$C_t = F_t C_{t-1} + U_t \hat{C}_t, \tag{9}$$

Output Gate:

$$O_t = \sigma(w_{o,h}h_{t-1} + \sum_{i=1}^{p} w_{o,xi}X_{i(t)} + b_o), \tag{10}$$

$$h_t = O_t \tanh(C_t), \tag{11}$$

where $w_{f,h}$, $w_{f,xi}$, b_f, $w_{c,h}$, $w_{c,xi}$, b_c, $w_{u,h}$, $w_{u,xi}$, b_u, $w_{o,h}$, $w_{o,xi}$, b_o are the parameters of the LSTM model; h_t is the value of the hidden state; $\sigma(.)$ is the sigmoid activation function; $\tanh(.)$ is the tanh activation function; and p is the number of input variables.

2.5 Hybrid Model

The hybrid model is a combination of two or more models in one function [7, 22]. This hybrid model can help overcome the complex structure of data by capturing linear and nonlinear patterns of time series data simultaneously [23] so that the hybrid model can improve forecast accuracy. The general equation of the hybrid model is shown in Eq. (12).

$$\hat{Z}_t = \hat{L}_t + \hat{N}_t, \tag{12}$$

where L_t is the forecast of the linear component, N_t is the forecast of the nonlinear component, and $\hat{Z}_t$ is the forecast result of the hybrid model.

This research will combine TSR, MGSTAR, and ML models to forecast space-time data involving more than one variable in each location. The hybrid space-time modeling is carried out in two stages. In the first stage, the data will be modeled using the TSR model to capture linear patterns, namely trends and double seasonality in the data. Furthermore, the residuals from the modeling in the first stage will be modeled using the ML model. There are three types of ML models used, i.e., FFNN, DLNN, and LSTM. The input to the ML modeling in the second stage is based on the MGSTAR model. So, it can capture nonlinear patterns and dependencies between locations in the data. The results of the forecasts in the first and second stages will be added together to obtain the results of the hybrid model predictions. Figure 2, Fig. 3, and Fig. 4 are examples of the process of forming the TSR-MGSTAR-FFNN, TSR-MGSTAR-DLNN, and TSR-MGSTAR-LSTM models involving two variables at three locations with the time and spatial order being one.

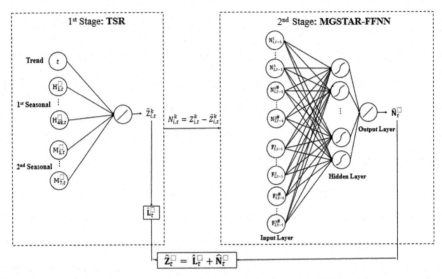

Fig. 2. Flowchart of the Hybrid TSR-MGSTAR-FFNN Model

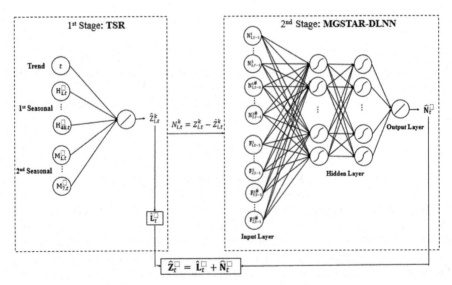

Fig. 3. Flowchart of the Hybrid TSR-MGSTAR-DLNN Model

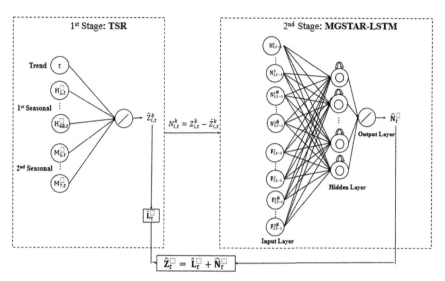

Fig. 4. Flowchart of the Hybrid TSR-MGSTAR-LSTM Model

In general, the difference between each hybrid model is the architecture of the ML model in the second stage. The output of this hybrid model is a vector that produces forecasts of all variables and locations simultaneously at time t. Based on Fig. 2, 3 and Fig. 4, the input used in the first stage is a dummy variable while the input at time t in the second stage is explained as follows:

$$\mathbf{N}^1_{1,t-1} = \begin{bmatrix} N^1_{1,t-1} \\ 0 \\ 0 \\ 0 \\ 0 \\ 0 \end{bmatrix}, \dots, \mathbf{N}^2_{3,t-1} = \begin{bmatrix} 0 \\ 0 \\ 0 \\ 0 \\ 0 \\ N^2_{3,t-1} \end{bmatrix}, \mathbf{N}^{1\#}_{1,t-1} = \begin{bmatrix} 0 \\ 0 \\ 0 \\ N^1_{1,t-1} \\ 0 \\ 0 \end{bmatrix}, \dots, \mathbf{N}^{2\#}_{3,t-1} = \begin{bmatrix} 0 \\ 0 \\ N^2_{3,t-1} \\ 0 \\ 0 \\ 0 \end{bmatrix},$$

$$\mathbf{F}^1_{1,t-1} = \begin{bmatrix} w_{12}N^1_{2,t-1} + w_{13}N^1_{3,t-1} \\ 0 \\ 0 \\ 0 \\ 0 \\ 0 \end{bmatrix}, \dots, \mathbf{F}^2_{3,t-1} = \begin{bmatrix} 0 \\ 0 \\ 0 \\ 0 \\ 0 \\ w_{13}N^2_{1,t-1} + w_{23}N^2_{2,t-1} \end{bmatrix},$$

$$\mathbf{F}^{1\#}_{1,t-1} = \begin{bmatrix} 0 \\ 0 \\ 0 \\ w_{12}N^1_{2,t-1} + w_{13}N^1_{3,t-1} \\ 0 \\ 0 \end{bmatrix}, \dots, \mathbf{F}^{2\#}_{3,t-1} = \begin{bmatrix} 0 \\ 0 \\ w_{13}N^2_{1,t-1} + w_{23}N^2_{2,t-1} \\ 0 \\ 0 \\ 0 \end{bmatrix},$$

where $\mathbf{N}^k_{i,t}$ is the residual of the TSR model in the first stage; $\mathbf{N}^1_{1,t-1}, \mathbf{N}^1_{2,t-1}, \dots, \mathbf{N}^2_{3,t-1}$ show the lag relationship of the same variable at the same location, $\mathbf{N}^{1\#}_{1,t-1}, \mathbf{N}^{1\#}_{2,t-1}, \ldots, \mathbf{N}^{2\#}_{3,t-1}$ explain the lag relationship of different variables at the same location; $\mathbf{F}^{1}_{1,t-1}, \mathbf{F}^{1}_{2,t-1}, \ldots, \mathbf{F}^{2}_{3,t-1}$ show the lag relationship of the same variable at different locations; and $\mathbf{F}^{1\#}_{1,t-1}, \mathbf{F}^{1\#}_{2,t-1}, \ldots, \mathbf{F}^{2\#}_{3,t-1}$ explain the lag relationship of different variables at different locations.

3   Dataset and Methodology

The data used in this study is air pollution data at three SUF stations located in three locations in the city of Surabaya, as shown in Fig. 5. Three air pollution variables are measured from each location, i.e., CO, NO_2, and PM_{10}. So, there are nine observational time series in the study. Air pollution data is recorded every half hour. There is a missing value in the air pollution data in this study, as shown by the time series plot in Fig. 6. Then, imputation is carried out using the median every half hour and every day to overcome the missing value. The data used starts from January 1, 2018, at 00:30 to March 31, 2018, at 00:00. Data from January 1, 2018, to March 24 becomes in-sample data, and data from March 25 to 31, 2018, becomes out-of-sample data. In-sample data is used for training in the modeling, while out-of-sample data is used to validate and select the best model.

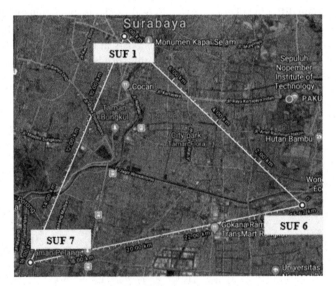

Fig. 5. SUF Station Location

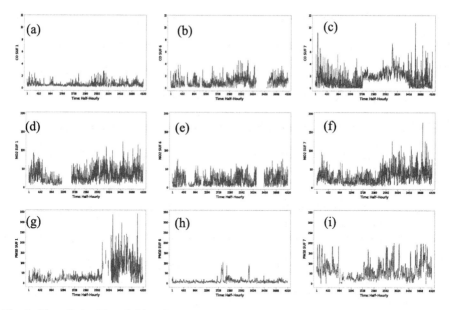

Fig. 6. Time Series Plot of CO at SUF 1 (a), CO at SUF 6 (b), CO at SUF 7 (c), NO_2 at SUF 1 (d), NO_2 at SUF 6 (e), NO_2 at SUF 7 (f), PM_{10} at SUF 1 (g), PM_{10} at SUF 6 (h), and PM_{10} at SUF 7 (i)

The selection of the best model is based on the error of the forecast results. The criteria used in this study are Root Mean Square Error (RMSE) and symmetric Mean Absolute Percentage Error (sMAPE) on out-of-sample data. Equation (13) and Eq. (14) are the formulas of RMSE and sMAPE, respectively [11, 24].

$$RMSE = \sqrt{\frac{1}{L} \sum_{l=1}^{L} (Z_{n+l} - \hat{Z}_n(l))^2},$$

(13)

$$sMAPE = \left(\frac{1}{L} \sum_{l=1}^{L} \frac{2 \left| Z_{n+l} - \hat{Z}_n(l) \right|}{(|Z_{n+l}| + |\hat{Z}_n(l)|)} \right) \times 100\%,$$

(14)

where L is the number of out-of-sample data, Z_{n+l} is out-of-sample actual data, and $\hat{Z}_n(l)$ is out-of-sample forecast data.

4 Result and Discussion

The results of identifying data patterns based on plot intervals in Fig. 7 show that the data tends to have a double seasonal pattern. The concentration of air pollution every day tends to be high at the time of departure, which is around 08:00, and returning from work at 19:00. Moreover, the concentration of air pollution tends to be low on weekends, especially on Sundays.

Modeling on air pollution data in this study will use several individual models and hybrid models. The first individual model used is the TSR model. This TSR modeling uses predictor variables in the form of trend variables and double seasonal dummy variables, namely daily and weekly seasonality. Besides being used as an individual model, the TSR model is also used as a first-stage model in hybrid modeling to capture the trend and double seasonal patterns.

Another individual model used is the space-time model, namely the MGSTAR model. The identification results show that there is an effect of lag 1, lag 48, and lag 336. So several alternative models are tried in this study based on the identification results, i.e., MGSTAR(1_1), MGSTAR($[48]_1$), MGSTAR($[336]_1$), MGSTAR($[1,48]_1$), MGSTAR($[1,336]_1$), MGSTAR($[48,336]_1$), and MGSTAR($[1,48,336]_1$). In this study, two types of spatial weights will be used, i.e., uniform weights and inverse distance weights. The best model obtained from several alternative MGSTAR models is the MGSTAR($[48,336]_1$) model with inverse distance weight based on RMSE and sMAPE values in out-of-sample data. Furthermore, the individual modeling uses the ML model, i.e., MGSTAR-FFNN, MGSTAR-DLNN, and MGSTAR-LSTM, based on the MGSTAR model ($[48,336]_1$) so that the input used in the ML model is based on the MGSTAR($[48,336]_1$) model.

Hybrid space-time modeling is done by modeling the residuals from the TSR model. It is the second stage of hybrid modeling. Then, the forecast results in this second stage will be added to the TSR model forecast results in the first stage and produce the hybrid model forecast results. In the hybrid linear TSR-MGSTAR modeling, the residuals from the TSR model will be modeled using the MGSTAR model. From the identification results, the best MGSTAR model is MGSTAR($[48]_1$) with inverse distance weight. Then, the hybrid nonlinear TSR-MGSTAR-ML model, i.e., TSR-MGSTAR-FFNN, TSR-MGSTAR-DLNN, and TSR-MGSTAR-LSTM, was developed based on the MGSTAR($[48]_1$) model. So, the input used in the MGSTAR-FFNN, MGSTAR-DLNN, and MGSTAR-LSTM models in modeling the residual TSR model is based on the MGSTAR($[48]_1$) model.

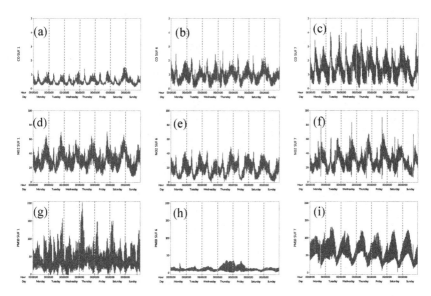

Fig. 7. Interval Plot of CO at SUF 1 (a), CO at SUF 6 (b), CO at SUF 7 (c), NO_2 at SUF 1 (d), NO_2 at SUF 6 (e), NO_2 at SUF 7 (f), PM_{10} at SUF 1 (g), PM_{10} at SUF 6 (h), and PM_{10} at SUF 7 (i)

In the MGSTAR-FFNN modeling, one hidden layer was used, and one to five neurons were tested in the hidden layer. Moreover, two hidden layers are used in the MGSTAR-DLNN model. The activation function used in the MGSTAR-FFNN and MGSTAR-DLNN hidden layer models is the tanh. The tanh activation function is used to capture nonlinear patterns in time series data using the FFNN and DLNN models [25]. Then in the MGSTAR-LSTM model, one recurrent hidden layer was used, and one to five neurons were tried. We tried several different numbers of neurons in the ML model because there was no evidence that a particular number of neurons yielded the best forecast accuracy. The selection of the optimum neuron in each ML model is based on the smallest RMSE and sMAPE values in the out-of-sample data. Figure 8 (a) shows the best architecture for the hybrid TSR-MGSTAR-FFNN model with four neurons in the hidden layer. Figure 8 (b) shows the best architecture for the hybrid TSR-MGSTAR-DLNN model with four neurons in the first hidden layer and five neurons in the second hidden layer. Figure 8 (c) shows the best architecture of the hybrid TSR-MGSTAR-LSTM model with four neurons in the hidden layer.

The comparison between actual data and forecast results for each air pollution variable in out-of-sample data for each method is shown by sMAPE and RMSE in Table 1. Based on Table 1 using RMSE and sMAPE on out-of-sample data, it can be seen that the best model for each parameter of air pollution in the city of Surabaya is different. In the CO variable, the best model obtained is MGSTAR. Then on the NO_2 variable, the best model obtained is hybrid TSR-MGSTAR, and the best model for the PM_{10} air pollution variable is hybrid TSR-MGSTAR-FFNN. The best models obtained for each of these air pollution variables are different.

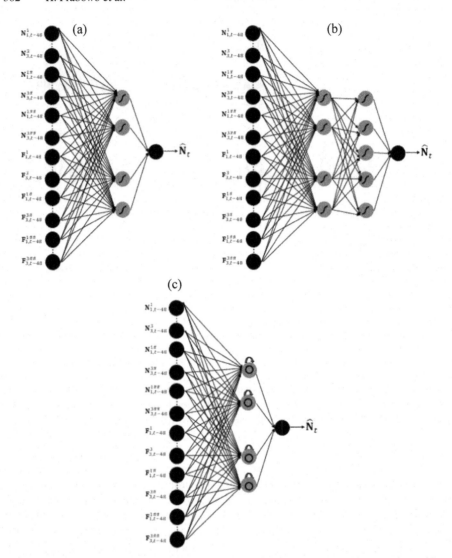

Fig. 8. Best MGSTAR-FFNN Architecture (a), Best MGSTAR-DLNN Architecture (b), and Best MGSTAR-LSTM Architectures (c) in TSR Residual Modeling

Furthermore, the RMSE and sMAPE ratio values to the TSR model as the simplest model is shown in Table 2. It can be seen that the MGSTAR model is only able to reduce the RMSE and sMAPE values from the TSR model in the CO variable. Then, the hybrid TSR-MGSTAR can derive RMSE and sMAPE from the TSR model for all air pollution variables. Meanwhile, the hybrid TSR-MGSTAR-ML model, namely TSR-MGSTAR-FFNN, TSR-MGSTAR-DLNN, and TSR-MGSTAR-LSTM, can reduce RMSE and sMAPE from the TSR model for NO_2 and PM_{10} variables. In general, the hybrid model can improve the accuracy of individual models. It is in line with one of the

M3 and M4 competition conclusions., the hybrid model tends to increase the forecast accuracy of individual models [11, 26]. From the RMSE and sMAPE ratio values, it can also be seen that the hybrid TSR-MGSTAR model is the best model to forecast air quality in the city of Surabaya simultaneously.

Table 1. Average Value of RMSE and sMAPE in Out-of-Sample Data

Method	RMSE			sMAPE		
	CO	NO_2	PM_{10}	CO	NO_2	PM_{10}
TSR	0.796	17.656	37.527	0.618	0.413	0.546
MGSTAR	0.660	18.101	37.607	0.549	0.421	0.555
MGSTAR-FFNN	0.866	19.095	39.019	1.114	0.424	0.565
MGSTAR-DLNN	0.793	18.894	38.223	1.074	0.429	0.558
MGSTAR-LSTM	0.742	19.842	39.090	1.012	0.442	0.576
TSR-MGSTAR	0.787	17.396	37.304	0.604	0.406	0.552
TSR-MGSTAR-FFNN	0.811	17.500	36.857	0.624	0.410	0.535
TSR-MGSTAR-DLNN	0.811	17.517	36.873	0.658	0.410	0.536
TSR-MGSTAR-LSTM	0.876	17.471	36.978	0.672	0.410	0.538

Table 2. RMSE Ratio and sMAPE Ratio Value to TSR Model in Out-of-Sample Data

Method	RMSE Ratio			sMAPE Ratio		
	CO	NO_2	PM_{10}	CO	NO_2	PM_{10}
MGSTAR	0.829	1.025	1.002	0.887	1.019	1.015
MGSTAR-FFNN	1.087	1.082	1.040	1.802	1.028	1.034
MGSTAR-DLNN	0.996	1.070	1.019	1.737	1.039	1.022
MGSTAR-LSTM	0.932	1.124	1.042	1.637	1.069	1.055
TSR-MGSTAR	0.988	0.985	0.994	0.977	0.983	1.011
TSR-MGSTAR-FFNN	1.019	0.991	0.982	1.010	0.992	0.980
TSR-MGSTAR-DLNN	1.019	0.992	0.983	1.065	0.993	0.981
TSR-MGSTAR-LSTM	1.100	0.990	0.985	1.087	0.992	0.985

5 Conclusion

This study employs three hybrid TSR-MGSTAR-ML models, i.e., TSR-MGSTAR-FFNN, TSR-MGSTAR-DLNN, and TSR-MGSTAR-LSTM, to forecast air pollution data, i.e., CO, NO_2, and PM_{10}, at three locations in the city of Surabaya. This three-combined hybrid model is compared with other models such as individual model (TSR and MGSTAR) and two-combined hybrid model (MGSTAR-FFNN, MGSTAR-DLNN, MGSTAR-LSTM, and hybrid TSR-MGSTAR). The identification results show that air pollution data has a relationship between locations and contains a double seasonal pattern. Then, the analysis results show that the MGSTAR model is the best model for PM_{10} forecasting, the hybrid TSR-MGSTAR is the best model for NO_2 forecasting, and the hybrid TSR-MGSTAR-FFNN is the best model for PM_{10} forecasting. In general, hybrid models provide better forecast accuracy than individual models. This result is in line with one of the conclusions of the M3 and M4 competition, namely, the hybrid model can increase the accuracy of individual models [11, 26]. In further research, a parallel hybrid space-time model can be developed that can capture linear and nonlinear patterns from space-time data simultaneously effectively.

Acknowledgements. This research was supported by *Deputi Bidang Penguatan Riset dan Pengembangan, Kementerian Riset dan Teknologi/ Badan Riset dan Inovasi Nasional* under the scheme *Penelitian Dasar*, project no 3/E1/KP.PTNBH/2021 and 799/PKS/ITS/2021. The authors thank to DRPM ITS for the supports and to anonymous referees for their useful suggestions.

References

1. Chen, S., Zhang, D.: Impact of air pollution on labor productivity: evidence from prison factory data. China Econ. Q. Int. **1**, 148–159 (2021)
2. Liu, Y., Zhou, Y., Lu, J.: Exploring the relationship between air pollution and meteorological conditions in china under environmental governance. Sci. Rep. **10**, 14518 (2020)
3. Reames, T.G., Bravo, M.A.: People, place and pollution: investigating relationships between air quality perceptions, health concerns, exposure, and individual and area-level characteristics. Environ. Int. **122**, 244–255 (2019)
4. Pfeifer, P.E., Deutsch, S.J.: A three stage iterative procedure for space-time modeling. Technometrics **22**(1), 35–47 (1980)
5. Borovkova, S., Lopuhaa, R., Ruchjana, B.N.: Generalized STAR model with experimental weights. In: Proceedings of the 17th International Workshop on Statistical Modelling, pp. 143–151 (2002)
6. Suhartono, Nahdliyah, N., Akbar, M.S., Salehah, N.A., Choiruddin, A.: A MGSTAR: an extension of the generalized space-time autoregressive model. J. Phys. Conf. Ser. **1752**, 012015 (2021)
7. Zhang, G.P.: Time series forecasting using a hybrid ARIMA and neural network model. Neurocomputing **50**, 159–175 (2003)
8. LeCun, Y., Bengio, Y., Hinton, G.: Deep learning. Nature **521**, 436–444 (2015)
9. Srivastava, S., Lessmann, S.: A comparative study of LSTM neural networks in forecasting day-ahead global horizontal irradiance with satellite data. Sol. Energy **162**, 232–247 (2018)
10. Abbasimehr, H., Shabani, M., Yousefi, M.: An optimized model using LSTM network for demand forecasting. Comput. Ind. Eng. **143**, 106435 (2020)
11. Makridakis, S., Spiliotis, E., Assimakopoulus, V.: The M4 competition: results, findings, conclusion and way forward. Int. J. Forecast. **34**, 802–808 (2018)
12. Suhartono, Dana, I.M.G.M., Rahayu, S.P.: Hybrid model for forecasting space-time data with calendar variation effects. Telkomnika **17**(1), 118–130 (2019)

13. Pusporani, E., Suhartono, Prastyo, D.D.: Hybrid multivariate generalized space-time autoregressive artificial neural network models to forecast air pollution data at Surabaya. In: AIP Conference Proceedings, vol. 2194, p. 020090 (2019)
14. Laily, V.O.N., Suhartono, Pusporani, E., Atok, R.M.: A novel hybrid Mgstar-Rnn model for forecasting spatio-temporal data. J. Phys. Conf. Ser. **1752**, 012011 (2021)
15. Prastyo, D.D., Nabila, F.S., Suhartono, Lee, M.H., Suhermi, N., Fam, S.F.: VAR and GSTAR-based feature selection in support vector regression for multivariate spatio-temporal forecasting. In: Yap, B., Mohamed, A., Berry, M. (eds.) Soft Computing in Data Science. SCDS 2018. Communications in Computer and Information Science, vol. 937, pp. 46–57. Springer, Singapore (2018). https://doi.org/10.1007/978-981-13-3441-2_4
16. Suhartono, Prastyo, D.D., Kuswanto, H., Lee, M.H.: Comparison between VAR, GSTAR, FFNN-VAR and FFNN-GSTAR models for forecasting oil production. MATEMATIKA **34**(1), 103–111 (2018)
17. Suhartono, Prabowo, H., Fam, S.F.: A hybrid TSR and LSTM for forecasting NO_2 and SO_2 in Surabaya. In: Berry, M., Yap, B., Mohamed, A., Köppen, M. (eds.) Soft Computing in Data Science. SCDS 2019. Communications in Computer and Information Science, vol. 1100, pp. 107–120. Springer, Singapore (2019). https://doi.org/10.1007/978-981-15-0399-3_9
18. Shummway, R.H., Stoffer, D.S.: Time Series Analysis and Its Application with R Examples. Springer, Pittsburg (2006). https://doi.org/10.1007/0-387-36276-2
19. Fausset, L.: Fundamental of Neural Network: Architectures Algorithms and Applications. Prentice-Hall Inc., Hoboken (1994)
20. Hochreiter, S., Schmiduber, J.: Long short-term memory. Neural Comput. **9**(8), 1735–1780 (1997)
21. Ma, X., Tao, Z., Wang, Y., Yu, H., Wang, Y.: Long short-term memory neural network for traffic speed prediction using remote microwave sensor data. Transp. Res. Part C: Emerg. Technol. **54**, 187–197 (2015)
22. Hajirahimi, Z., Khashei, M.: Hybrid structures in time series modeling and forecasting: a review. Eng. Appl. Artif. Intell. **86**, 83–106 (2019)
23. Khashei, M., Bijari, M.: A novel hybridization of artificial neural networks and ARIMA models for time series forecasting. Appl. Soft Comput. **11**, 2664–2675 (2011)
24. Wei, W.W.S.: Time Series Analysis Univariate and Multivariate Methods, 2nd edn. Pearson Education Inc., Boston (2006)
25. Suhartono, Suhermi, N., Prastyo, D.D.: Design of experiment to optimize the architecture of deep learning for nonlinear time series forecasting. Procedia Comput. Sci. **144**, 269–276 (2018)
26. Makridakis, S., Hibbon, M.: The M3-competition result, conclusions and implications. Int. J. Forecast. **16**, 451–676 (2000)

Hybrid Machine Learning for Forecasting and Monitoring Air Pollution in Surabaya

Suhartono[1], Achmad Choiruddin[1(✉)], Hendri Prabowo[1],
and Muhammad Hisyam Lee[2]

[1] Department of Statistics, Institut Teknologi Sepuluh Nopember, Jl. Raya ITS Sukolilo,
Surabaya 60115, Indonesia
choiruddin@its.ac.id
[2] Department of Mathematical Science, Universiti Teknologi Malaysia, 81310 Skudai,
Johor, Malaysia

Abstract. This research aims to propose hybrid machine learnings for forecasting and monitoring air pollution in Surabaya. In particular, we introduce two hybrid machine learnings, i.e. hybrid Time Series Regression – Feedforward Neural Network (TSR-FFNN) and hybrid Time Series Regression – Long Short-Term Memory (TSR-LSTM). TSR is used to capture linear patterns from data, whereas FFNN or LSTM is used to capture non-linear patterns. Fifteen half-hourly series data, i.e. CO, NO_2, O_3, PM_{10}, and SO_2 in three SUF stations at Surabaya, are used as the case study. We compare the forecasting accuracy of these hybrid machine learnings with several individual methods (i.e. TSR, ARIMA, FFNN, and LSTM), and combined methods (i.e. TSR with AR error and TSR with ARMA error). The identification step showed that these air pollution data have double seasonal patterns, i.e. daily and weekly seasonality. The comparison results showed that no superior method that yields the most accurate forecast for all series data. Moreover, the results showed that hybrid methods gave more accurate forecast at 8 series data, whereas the individual methods yielded better results at 7 series data. It supported that methods that are more complex do not always produce better forecasts than simple methods, as shown by the first result of the M3 competition. Additionally, the results of the forecast of air pollution index for monitoring air pollution in Surabaya show that the air quality is in good and moderate air pollution levels for duration of 19.30 to 03.00 and 0.30 to 19.30, respectively.

Keywords: Air pollution · Forecasting · Hybrid · Machine learning · Monitoring

1 Introduction

Air pollution has become one big issue in Surabaya and other big cities in Indonesia [1]. To evaluate air quality, one considers the Air Pollution Index (API). API is computed from the concentrations of PM_{10}, CO, SO_2, NO_2, and O_3 recorded every half hour each day. It is a major importance to provide a good air quality forecasting in Surabaya to monitor and control air pollution.

© Springer Nature Singapore Pte Ltd. 2021
A. Mohamed et al. (Eds.): SCDS 2021, CCIS 1489, pp. 366–380, 2021.
https://doi.org/10.1007/978-981-16-7334-4_27

Statistical methodologies based on time series regression (TSR) [e.g. 2, 3] and machine learning (ML) [4, 5] are commonly used for the analysis of time series data such as air quality. TSR is useful if the data has linear patterns while machine learning is advantageous for non-linear time series data. Studies considering the hybrid method, which is a combination of two distinct methods [6], demonstrate major benefits. Zhou et al. [7] forecast $PM_{2.5}$, PM_{10} and NO_x at five cities in Taiwan using shallow multi-output long short-term memory (SM-LSTM) and deep multi-output long short term-memory (DM-LSTM) and show that DM-LSTM has a better prediction accuracy. Hao et al. [8] consider hybrid Elman neural network (ENN) to forecast $PM_{2.5}$, PM_{10}, SO_2, NO_2, CO and O_3 in Jinan, Shanghai and Harbin (China). Cheng et al. [9] forecast $PM_{2.5}$ in 5 cities in China by comparing autoregressive integrated moving average (ARIMA), support vector machine (SVM), artificial neural network (ANN) and hybrid-GARCH methods for $PM_{2.5}$ forecasting in 6 cities in China and find that hybrid-GARCH performs best. It is worth emphasizing that all the studies mentioned earlier develop hybrid method for only capturing non-linear patterns. Therefore, it is of interest to propose a new method involving both linear and non-linear patterns for the analysis of air quality in Surabaya. In this study, we focus on hybrid TSR-FFNN and hybrid TSR-LSTM to forecast the concentrations of CO, NO_2, O_3, PM_{10}, and SO_2 to determine API at three SUF stations in Surabaya. TSR method is used to capture linear patterns from data, particularly the deterministic trend and seasonal patterns. Whereas, FFNN or LSTM is used to capture non-linear patterns from data. Moreover, this hybrid method is proposed due to air pollution data is frequently consisted of linear and non-linear patterns.

The rest of the article is organized as follows. Section 2 reviews time series regression and machine learning. We detail in Sect. 3 the dataset and methodology. We discuss our results in Sect. 4 and provide conclusion in Sect. 5.

2 Statistical Methods and Machine Learning

2.1 Statistical Methods

We provide brief review of time series regression and Autoregressive Integrated Moving Average (ARIMA) in this section.

Time Series Regression (TSR). TSR is one of the time series models which plays the same role as the linear regression model [3]. In this study, two types of predictors will be considered, i.e. 1 dummy variable to capture trend patterns and 336 dummy variables to capture double seasonal patterns. The TSR model is illustrated in Fig. 1.

Autoregressive Integrated Moving Average (ARIMA). Some time series data are stationary and some are not stationary in the mean and variance. To do an analysis using ARIMA, the data must be stationary in the mean and variance. ARIMA (p, d, q) is an ARMA model (p, q) that experiences differencing as much as d in order to be stationary. The ARIMA model can be used for seasonal and non-seasonal data. In general, ARIMA (p, d, q) for non-seasonal data can be written as follows [10]:

$$\phi_p(B)(1 - B)^d Y_t = \theta_0 + \theta_q(B)a_t. \tag{1}$$

If there are seasonal and non-seasonal patterns in time series, then ARIMA multiplicative model, i.e. ARIMA $(p, d, q)(P, D, Q)^S$ can be applied. The general form of ARIMA $(p, d, q)(P, D, Q)^S$ is:

$$\phi_p(B)(1 - B)^d \Phi_P\left(B^S\right)(1 - B^S)^D Y_t = \theta_q(B)\Phi_Q\left(B^S\right)a_t, \tag{2}$$

where p and q are the non-seasonal order of AR and MA, P and Q are the seasonal order of AR and MA. ARIMA modeling is done by implementing Box Jenkins procedure. This procedure consists of identification, parameters estimation, diagnostic check, and forecasting steps [11].

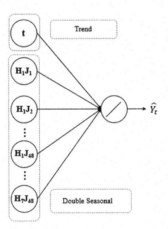

Fig. 1. Scheme for TSR model development

Time Series Regression with ARIMA Error. Error in time series regression models that not fulfill white noise condition can be modeled using AR and ARMA [12] and can be illustrated at Fig. 2.

In general, TSR models with AR errors and TSR with ARMA errors is similar. The different is only in the TSR residual model, that is, the TSR with ARMA error involves the MA component.

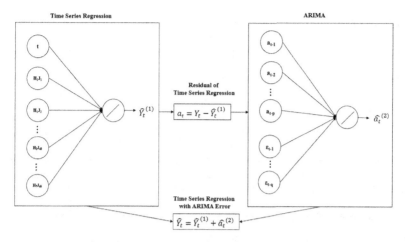

Fig. 2. Scheme to develop TSR with ARIMA error model

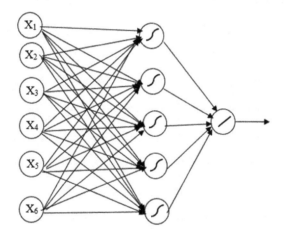

Fig. 3. Architecture of FFNN with 6 neurons in input layer, 5 neurons in hidden layer and 1 neuron in output layer

2.2 Machine Learning

Feed Forward Neural Network. Feed forward neural network (FFNN) is a non-linear model that can be used to model and predict time series data [13]. In the FFNN structure consist several layers, namely the input layer, one or more hidden layers and the output layer which can be seen in Fig. 3. Each neuron will receive information only from neurons in the previous layer where the input neurons come from the output weights on the previous layer [14]. In FFNN the process will begin when neurons receive input which is grouped into the input layer then the information is directed past the hidden layer then sequence until it reaches the output layer [15]. This is the general equations of FFNN.

$$\hat{Y}_{(t)} = f^o \left[\sum_{j=1}^{q} \left[w_j^o f_j^h \left(\sum_{i=1}^{p} w_{ji}^h X_{i(t)} + b_j^h \right) + b^o \right] \right], \tag{3}$$

where b^o, b_j^h ($j = 1, ..., q$), w_{ji}^h ($j = 1,...,q$; $i = 1, ..., p$), w_j^o ($j = 1, ..., q$) is the parameter of the FFNN model, f^o is the activation function in output layer, f_j^h ($j = 1, ..., q$) is the activation function in hidden layer, p is the number of input variables and q is the number of neurons in the hidden layer. In this paper, sigmoid is used as activation function.

Long Short-Term Memory. Long short-term memory is an extension to improve the recurrent neural network (RNN) method since the RNN method cannot access the long-term lag in the architecture [16, 17]. The structure of LSTM consists of several layers, namely the input layer, recurrent hidden layer and output layer [18]. The difference with FFNN is the use of recurrent hidden layers or memory blocks in hidden layers. This hidden layer of the LSTM network contains a memory cell and a pair of adaptive, multiplicative gating units that direct input and output to all cells in the block. An LSTM cell consists of four gates, i.e. input gate, input modulation gate, forget gate and output gate [16], shown in Fig. 4.

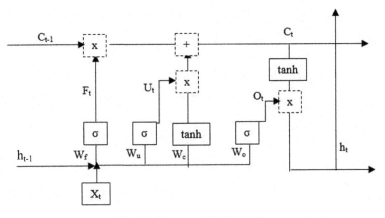

Fig. 4. Structure of LSTM cell

Based on Fig. 4, the equation in LSTM is obtained as follows:
Forget gate:

$$F_t = \sigma(W_{fh}h_{t-1} + W_{fx}X_t + b_f), \tag{4}$$

Input gate:

$$\hat{C}_t = tanh(W_{ch}h_{t-1} + W_{cx}X_t + b_c), \tag{5}$$

$$U_t = \sigma(W_{uh}h_{t-1} + W_{ux}X_t + b_u), \tag{6}$$

Memory cell unit:

$$C_t = F_tC_{t-1} + U_t\widehat{C}_t, \tag{7}$$

Output gate:

$$O_t = \sigma(W_{oh}h_{t-1} + W_{ox}X_t + b_o), \tag{8}$$

$$h_t = O_t tanh(C_t), \tag{9}$$

where W_{fh}, W_{fx}, b_f, W_{ch}, W_{cx}, b_c, W_{uh}, W_{ux}, b_u, W_{oh}, W_{ox}, b_o are the parameters of LSTM model, $\sigma(.)$ is a sigmoid activation function and $tanh(.)$ is a tanh activation function.

3 Dataset and Methodology

3.1 Dataset

We consider air quality data (CO, NO_2, O_3, PM_{10}, and SO_2) recorded every half hour during 2018 at three SUF stations in Surabaya. Figure 5 shows time series plot of PM_{10}. We highlight that missing data are observed. To handle this issue, we replace the missing values by the median of PM_{10} at the same data period (half hour and day). For the other variables with the same issue (CO, NO_2, O_3, and SO_2), the procedure follows along similar lines.

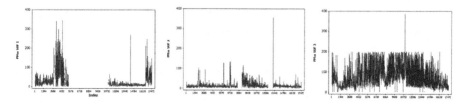

Fig. 5. Time series plot of PM_{10} at the SUF 1 (left), SUF 2 (middle) and SUF 3 (right).

3.2 The Proposed Hybrid Machine Learning

It has been shown [7–9] that hybrid methods are advantageous for forecasting. However, for the air pollution data, it tends to have both linear and non-linear patterns. To handle such issue, we propose in this study the two hybrid methods, namely TSR-FFNN and TSR-LSTM. This combination is expected to overcome complex structures of data [19]. To build a hybrid model, two-step procedure is required. First, we model the data using TSR and obtain the residuals. Second, machine learnings (FFNN and LSTM) are employed to model the residuals. In general, forecasting of hybrid models can be written as follows:

$$Y_t = L_t + N_t + \varepsilon_t, \tag{10}$$

where L_t shows the linear component represented by TSR and N_t shows the non-linear component represented by the machine learning i.e. FFNN or LSTM. In more detail, Hybrid modeling starts with air pollution data modeling using TSR at step 1. Residuals are then obtained by:

$$a_t = Y_t - \hat{L}_t. \tag{11}$$

In the step 2, residuals from the TSR model (see Eq. 11) are modeled using the machine learnings i.e. FFNN or LSTM. The residual equation of the TSR model is

$$a_t = f(a_{t-1}, a_{t-2}, ..., a_{t-k}) + \varepsilon_t, \tag{12}$$

where f is a nonlinear function obtained from the neural network and a_t or $\hat{N}_t$ is the result of forecasting at time-t. The forecast of hybrid model [6] is

$$\hat{Y}_t = \hat{L}_t + \hat{N}_t. \tag{13}$$

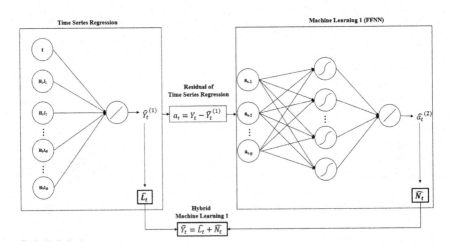

Fig. 6. Scheme for building hybrid TSR-FFNN

The scheme for building hybrid machine learning illustrated at Fig. 6 (hybrid TSR-FFNN) and Fig. 7 (hybrid TSR-LSTM). In general, there is no significant difference between the hybrid TSR-FFNN and hybrid TSR-LSTM models. It only differs at step 2, where different machine learning method is used.

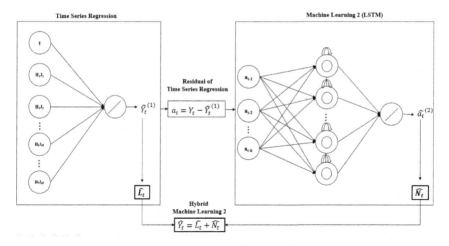

Fig. 7. Scheme for building hybrid TSR-LSTM

3.3 Model Evaluation

To evaluate our methods compared to the others, we consider Root Mean Square Error Prediction (RMSEP) and symmetric Mean Absolute Percentage Error Prediction (sMAPEP) [10, 20], criteria defined by:

$$RMSEP = \sqrt{\frac{1}{L} \sum_{l=1}^{L} (Y_{n+l} - \hat{Y}_n(l))^2}, \qquad (14)$$

$$sMAPEP = \left(\frac{1}{L} \sum_{l=1}^{L} \frac{2\left|Y_{n+l} - \hat{Y}_n(l)\right|}{(|Y_{n+l}| + \left|\hat{Y}_n(l)\right|)} \right) \times 100\%, \qquad (15)$$

where Y_{n+l} is the actual value at out-of-sample data, $\hat{Y}_n(l)$ is the forecast value at out-of-sample data and L is the size of out-of-sample data.

4 Result

The observations include five indicators, which are concentrations of CO, NO_2, O_3, PM_{10} and SO_2, measured at three SUF stations in Surabaya (SUF 1, SUF 2 and SUF 3 as shown in Fig. 8) at the period of January 1 - December 20, 2018. These indicators are considered as a basis to obtain API value. To do so, we first compute API value by:

$$I = \frac{I_a - I_b}{X_a - X_b}(X_x - X_b) + I_b, \qquad (16)$$

where I is the resulting API value, I_a is the upper limit of API, I_b is the lower limit of API, X_a is an ambient upper limit, X_b is an ambient lower limit, and X_x is an ambient

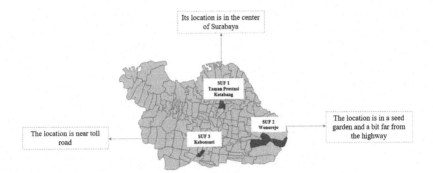

Fig. 8. SUF station map in Surabaya

measurement result. To get I_a, I_b, X_a and X_b, we consider Table 1. Second, we select one among five indicators at three SUF station with the highest concentration. Third, we categorize the resulting API value. The procedure to obtain API values is illustrated by Fig. 9.

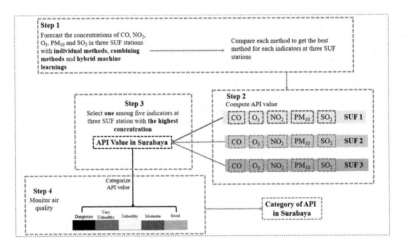

Fig. 9. The procedure to obtain API values

To evaluate our methods, we separate the data into in-sample and out-of-sample data. In-sample data starts from January 1 2018 at 00.30 to December 20, 2018 at 19.00, while the rest is set as out-of-sample data. In the analysis, we compare our two proposed methods (i.e. hybrid TSR-FFNN and hybrid TSR-LSTM) with TSR, ARIMA, FFNN, LSTM, TSR with AR error and TSR with ARMA error. We describe the air quality in Surabaya in 2018 in Sect. 4.1 and demonstrate the performance of our method in Sects. 4.2–4.3.

Table 1. API limit

API	PM_{10} ($\mu g/m^3$)	SO_2 ($\mu g/m^3$)	CO ($\mu g/m^3$)	O_3 ($\mu g/m^3$)	NO_2 ($\mu g/m^3$)
50	50	80	5	120	
100	150	365	10	235	
200	350	800	17	400	1130
300	420	1600	34	800	2260
400	500	2100	46	1000	3000
500	600	2620	57.5	1200	3750

4.1 Characteristics of Air Quality Indicators in Surabaya

Figure 10 depicts the interval plot of each of the five air quality indicators at three SUF stations. Each point of the plot represents the 95% interval value of given indicator given time and day. For example, in the top-left, the first plot shows the 95% interval of the concentration of CO at SUF 1 on Monday at 00:00 occurring in all year is between 0.4–0.8. This information is important to catch seasonal pattern. The general message shows double seasonal patterns, i.e. daily and weekly seasonal.

Let us focus on the first row (concentration of CO). In general, the highest concentration is reached at around 08.00 every day except Sunday. The CO concentration is highly related to the vehicle activities on street, meaning that there are many vehicle activities on Monday-Saturday at around 08.00, but not on Sunday. The general message applies similarly for the other indicators but O_3. Concentration of O_3 reaches the highest at around 12.00. This is because the high intensity of solar radiation at noon increases the concentration of O_3 in the air. The lowest O_3 concentration occurs at early morning.

4.2 Forecasting Air Quality Parameters in Surabaya

In this paper, we compare our hybrid machine learnings i.e. hybrid TSR-FFNN and hybrid TSR-LSTM with several methods i.e. TSR, ARIMA, FFNN, LSTM, TSR with AR error and TSR with ARMA error to predict five air quality parameters at three SUF stations in Surabaya. The identification step shows that the data tend to have small trend and double seasonal patterns, i.e. daily and weekly seasonality. Therefore, we propose TSR model with two main predictors, i.e. 1 trend and 336 seasonal dummy variables. The dummy variables for these double seasonal are dummy for every half hour in a day and dummy for every day in one week. The model evaluation showed that the residual of TSR model did not fulfill white noise condition. Hence, other methods can be used to modelling these residuals.

ARIMA model is applied based on the Box Jenkins procedure. The identification step is conducted to identify stationarity of data. The data are not stationary in the mean and variance so that transformation and differencing are applied on the data. Then, the best ARIMA model for each variable is obtained, i.e. double seasonal and subset ARIMA models. These best ARIMA models satisfy assumption about white noise residuals.

Machine learning modeling is applied using the ARIMA model, particularly about the determination of the inputs. The significant AR lags of the ARIMA model are used as inputs of both FFNN and LSTM. FFNN and LSTM with one hidden layer and number of neurons from 1 to 5 are applied to the data. The sigmoid activation function is used in the hidden layer of FFNN. LSTM uses memory block in hidden layer.

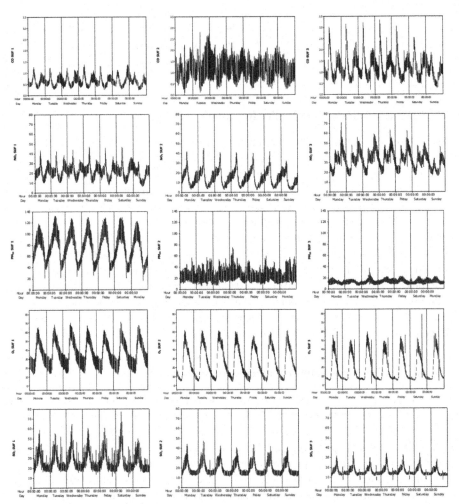

Fig. 10. Interval plot (95%) of each air quality indicators at 3 SUF stations in Surabaya at given half hour and day. The column indicates the station code (First column: SUF 1, second column: SUF 2, and third column: SUF 3). The row indicates the five air quality indicators; from top to bottom: Concentrations of CO, NO_2, PM_{10}, O_3, and SO_2.

The residuals of TSR model are not white noise so other methods can be applied in residual. This residual is modeled with AR. Some AR models were tried based on lag 1, 48 and 336. The AR models tried were $AR(1)$, $AR(1)^{48}$, $AR(1)^{336}$, $AR(1)(1)^{48}$,

$AR(1)(1)^{336}$ and $AR(1)(1)^{48}(1)^{336}$. The results of TSR with AR error have residuals that do not fulfill white noise condition, so the residuals of the TSR model are also modeled with ARMA model.

Two hybrid machine learning models that be proposed in this research, i.e. hybrid TSR-FFNN and hybrid TSR-LSTM. Machine learning algorithm (i.e. FFNN and LSTM) is used to capture non-linear patterns that cannot be captured by TSR model. TSR model only capture linear pattern particularly trend and seasonal pattern. The inputs of machine learning models are based on the AR lags of ARMA model.

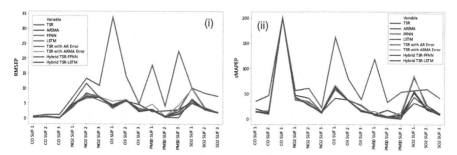

Fig. 11. Comparison of RMSEP (i) and sMAPEP (ii)

The best model will be chosen based on the smallest RMSEP and sMAPEP values in the out-of-sample data. Comparing all the eight methods (See Fig. 11), the time series regression (TSR) performs worst. The ARIMA is used to predict O_3 SUF 1, O_3 SUF 2 and PM_{10} SUF 1. FFNN is used to predict CO SUF 1. LSTM is used to predict CO SUF 2, NO_2 SUF 1 and SO_2 SUF 1. TSR with AR error is used to predict NO_2 SUF 3. TSR with ARMA error is used to predict SO_2 SUF 3. Hybrid TSR-FFNN is used to predict NO_2 SUF 2, PM_{10} SUF 2 and PM_{10} SUF 3. Hybrid TSR-LSTM is used to predict CO SUF 3, O_3 SUF 3 and SO_2 SUF 3.

In general we find two findings regarding the comparison between hybrid and individual methods. First, when hybrid machine learning produces more accurate forecast than individual methods, both linear and non-linear methods, it is in line with the third result of M3 competition and the main result of M4 competition, i.e. hybrid machine learning methods tend to give more accurate forecast than individual methods [21–25]. Second, when individual method produces more accurate forecast than hybrid machine learning methods, it is in line with the first result of M3 competition i.e. more complex methods do not necessary give better forecast than simpler ones. This second result is not in line with the result from Cheng et al. [9], Wang et al. [26] and Wu et al. [27].

4.3 Monitoring Air Quality Parameters in Surabaya

After forecasting each air quality parameter at three SUF stations in Surabaya, the calculation of the air pollution index (API) value is then performed. This API is used to see and monitoring air quality in an area. There are several studies that not only forecast air quality but also monitor air quality [28–30]. The API is calculated by Eq. (16).

In determining the API value in an area, the API value of all air quality parameters at all SUF stations is first calculated, then the highest API value is sought. The highest API value is the API value determined in the region.

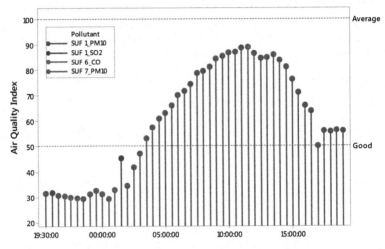

Fig. 12. API value prediction in Surabaya

Figure 12 shows that the forecast value of the API in Surabaya at 19.30 on 20 December 2018 to 03.00 on 21 December 2018 is still in good condition. This means that air quality in the city of Surabaya at that time was still healthy. During this period, the dominant air quality parameters varied, namely PM_{10} in SUF 1, PM_{10} in SUF 7, CO in SUF 6 and SO_2 in SUF 1. After 03.00 until 19.00 the value of the API in the City of Surabaya was in the moderate category, this is because Surabaya citizen have started activities. This means that air quality in the city of Surabaya in this period worsened but still in the safe category because the API value was still less than 100. The dominant air quality parameter was PM_{10} in SUF 7. The highest API forecast value occurred at 11.30.

5   Conclusion

In this study, hybrid machine learning methods i.e. hybrid TSR-FFNN and hybrid TSR-LSTM are proposed for forecasting and monitoring air quality in Surabaya. Based on the RMSEP and sMAPEP criteria in the out-of-sample data, the best method for forecasting five air quality parameters for three SUF stations is obtained. There were no dominant methods of the 8 methods used. This shows that a more complex method does not always produce a better forecasting than a simpler method, this is in line with the first results of the M3 competition i.e. more complex methods did not necessarily yield better forecast that simpler ones [21]. In addition, this also shows that each different air quality parameter in different SUF stations has different data patterns, some linear, non-linear or both linear and non-linear. The results of monitoring the air quality of the city of Surabaya based on

API values showed that from 19.30 to 03.00 the air quality in Surabaya was still good. At 03.00 to 19.30, air quality is moderate. The most dominant air quality parameter is PM_{10} at SUF 7. Involving meteorological factors such as rainfall and humidity could be an interesting future study. In addition, development of new hybrid machine learning models or considering the multivariate structure [31] could be a direction for further research, for example, model that incorporates both spatial and temporal dependence [32, 33].

Acknowledgments. The first author has passed away while the research is being completed. We devote the article as a tribute to the contribution by Dr. Suhartono in the study and in Statistics.

References

1. Santoso, E.B., Kusuma, S.H., Siswanto, V.K.: Air space modeling for living arrangement in Surabaya. Procedia. Soc. Behav. Sci. **227**, 747–753 (2016)
2. Hanke, J.E., Wichern, D.W.: Bussines Forecasting, Eight Pearson Pratice Hall, Hoboken (2005)
3. Shummway, R.H., Stoffer, D.S.: Time Series Analysis and Its Application with R Examples. Springer, Pittsburg (2006). https://doi.org/10.1007/0-387-36276-2
4. Suhartono, Saputri, P.D., Amalia, F.F., Prastyo, D.D., Ulama, B.S.S.: Model selection in feedforward neural networks for forecasting inflow and outflow in Indonesia. In: Mohamed, A., Berry, M., Yap, B. (eds.) Soft Computing in Data Science. SCDS 2017. Communications in Computer and Information Science, vol. 788, pp. 95–105. Springer, Singapore (2017). https://doi.org/10.1007/978-981-10-7242-0_8
5. Srivastava, S., Lessmann, S.: A comparative study of LSTM neural networks in forecasting day-ahead global horizontal irradiance with satellite data. Sol. Energy **162**, 232–247 (2018)
6. Suhartono, Hikmawati, F., Setyowati, E., Salehah, N.A., Choiruddin, A.: A novel hybrid GSTARX-RNN model for forecasting space-time data with calendar variation effect. J. Phys. Conf. Ser. **1463**, 012037 (2020)
7. Zhou, Y., Chang, F., Chang, L., Kao, I., Wang, Y.: Explore a deep learning multi-output neural network for regional multi-step-ahead air quality forecasts. J. Clean. Prod. **209**, 134–145 (2019)
8. Hao, Y., Tian, C.: The study and application of a novel hybrid system for air quality early-warning. Appl. Soft Comput. **74**, 729–746 (2019)
9. Cheng, Y., Zhang, H., Liu, Z., Chen, L., Wang, P.: Hybrid algorithm for short-term forecasting of $PM_{2.5}$ in China. Atmos. Environ. **200**, 264–279 (2019)
10. Wei, W.W.: Time Series Analysis Univariate and Multivariate Methods, 2nd edn. Pearson Education Inc., Boston (2006)
11. Box, G.E., Jenkins, G.M., Reinsel, G.C., Ljung, G.M.: Time Series Analysis: Forecasting and Control. Wiley, Hoboken (2015)
12. Hyndman, R.J., Athanasopoulos, G.: Forecasting: Principles and Practice, 2nd edn. OTexts, Melbourne (2018)
13. Tealab, A.: Time series forecasting using artificial neural networks methodologies: a systematic review. Future Comput. Inform. J. **3**(2), 334–340 (2018)
14. Chong, E., Zak, S.H.: An Introduction to Optimize. Wiley, Toronto (2001)
15. Suhartono: New procedures for model selection in feedforward neural networks. Jurnal Ilmu Dasar **9**, 104–113 (2008)

16. Hochreiter, S., Schmiduber, J.: Long short-term memory. Neural Comput. **9**(8), 1735–1780 (1997)

17. Azzouni, A., Pujjole, G.: A long-short term memory recurrent neural network framework for network traffic matrix prediction. arXiv preprint arXiv:1705.05690 (2017)

18. Ma, X., Tao, Z., Wang, Y., Yu, H., Wang, Y.: Long short-term memory neural network for traffic speed prediction using remote microwave sensor data. Transp. Res. Part C: Emerg. Technol. **54**, 187–197 (2015)

19. Zheng, F., Zhong, S.: Time series forecasting using a hybrid rbf neural network and AR model based on binomial smoothing. World Acad. Sci. Eng. Technol. **75**, 1471–1475 (2011)

20. Makridakis, S., Spiliotis, E., Assimakopoulus, V.: The M4 competition: results, findings, conclusion and way forward. Int. J. Forecast. **34**, 802–808 (2018)

21. Makridakis, S., Hibbon, M.: The M3-competition result, conclusions and implications. Int. J. Forecast. **16**, 451–676 (2000)

22. Chen, S., Wang, J., Zhang, H.: A hybrid PSO-SVM model based on clustering algorithm for short-term atmospheric pollutant concentration forecasting. Technol. Forecast. Soc. Change **146**, 41–54 (2019)

23. Saxena, H., Aponte, O., McConky, K.: A Hybrid machine learning model for forecasting a billing period's peak electric load days. Int. J. Forecast. **35**(4), 1288–1303 (2019)

24. Suhartono, Lee, M.H.: A hybrid approach based on Winter's model and weighted fuzzy time series for forecasting trend and seasonal data. J. Math. Stat. **7**(3), 177–183 (2011)

25. Wang, J., Bai, L., Wang, S., Wang, C.: Research and application of the hybrid forecasting model based on secondary denoising and multi-objective optimization for air pollution early warning system. J. Clean. Prod. **234**, 54–70 (2019)

26. Wang, P., Liu, Y., Qin, Z., Zhang, G.: A novel hybrid forecasting model For PM_{10} and SO_2 daily concentrations. Sci. Total Environ. **505**, 1202–1212 (2015)

27. Wu, Q., Lin, H.: Daily urban air quality index forecasting based on variational mode decomposition, sample entropy and LSTM neural network. Sustain. Cities Soc. **50**, 101657 (2019)

28. Yang, Z., Wang, J.: A new air quality monitoring and early warning system: air quality assessment and air pollutant concentration prediction. Environ. Res. **158**, 105–117 (2017)

29. Li, H., Wang, J., Li, R., Lu, H.: Novel analysis-forecast system based on multi-objective optimization for air quality index. J. Clean. Prod. **208**, 1365–1383 (2019)

30. Wu, Q., Lin, H.: A novel optimal-hybrid model for daily air quality index prediction considering air pollutant factors. Sci. Total Environ. **683**, 808–821 (2019)

31. Suhartono, Nahdliyah, N., Akbar, M.S., Salehah, N.A., Choiruddin, A.: A MGSTAR: an extension of the generalized space-time autoregressive model. J. Phys. Conf. Ser. **1752**, 012015 (2021)

32. Choiruddin, A., Cuevas-Pacheco, F., Coeurjolly, J.-F., Waagepetersen, R.: Regularized estimation for highly multivariate log Gaussian Cox processes. Stat. Comput. **30**(3), 649–662 (2019). https://doi.org/10.1007/s11222-019-09911-y

33. Choiruddin, A., Aisah, Trisnisa, F., Iriawan, N.: Quantifying the effect of geological factors on distribution of earthquake occurrences by inhomogeneous cox processes. Pure Appl. Geophys. **178**(5), 1579–1592 (2021)

Survival Analysis of Diabetes Mellitus Patients Using Semiparametric Approach

Jerry Dwi Trijoyo Purnomo$^{(\boxtimes)}$, Santi Wulan Purnami, Febry Hilmi Anshori, and Albertus Kurnia Lantika

Institut Teknologi Sepuluh Nopember, Surabaya, Indonesia

Abstract. The disease that attacks the human body and often gets special attention is diabetes. Diabetes mellitus is a non-communicable disease that is most commonly suffered by the world's population. Diabetes is a condition that interferes with the body's ability to process glucose in the blood, otherwise known as blood sugar. So, most patients have survival of only a few months. Therefore, research was conducted on the survival of people with diabetes mellitus and factors that affect it during the event. The method used in this study was the cox regression model. The results obtained from this study are three variables that significantly affect the survival of diabetes mellitus patients, namely Genetics, Age, and Diet. Then the variables Genetic, Age, and Diet became part of *Cox Proportional Hazard* (PH) modeling.

Keywords: *Cox Proportional Hazard* · Diabetes mellitus · Survival

1 Introduction

In 2000, the World Health Organization (WHO) stated that 57 million deaths occur annually due to non-communicable diseases from the world's mortality statistics. It is estimated that about 3.2 million people per year of the world's population die from Diabetes mellitus. Furthermore, in 2003 WHO estimated 194 million people, or 5.1% of the world's 3.8 billion people aged 20–79 years old, have Diabetes Mellitus, and by 2025 will increase to 333 million people. Who predicts that in Indonesia, there will be an increase from 8.4 million diabetics in 2000 to around 21.3 million diabetics by 2030. These will make Indonesia ranked 4th in the world after the United States, China, and India in the prevalence of diabetes [1].

Previous studies analyzed the survival patterns of diabetes mellitus patients at Nekemte Hospital, Ethiopia using the Cox Proportional Hazard Regression. Taking alcohol, smoking cigarette, overweight, high blood pressure, and positive family history of diabetics, have higher death rate [2]. In this study, we analyzed what factors most affect the survival of inpatient Diabetes mellitus patients at RAA Soewondo Pati Hospital using the Cox Regression method; Thus, information can be obtained about the factors that have the most significant effect on the survival of Diabetes Mellitus patients at RAA Soewondo Pati Hospital.

© Springer Nature Singapore Pte Ltd. 2021
A. Mohamed et al. (Eds.): SCDS 2021, CCIS 1489, pp. 381–394, 2021.
https://doi.org/10.1007/978-981-16-7334-4_28

This research is expected to provide benefits for medical personnel in improving diabetes mellitus by seeking early detection in diabetic patients and improving counseling on diabetes, especially prevention methods. People can have healthy habits, especially in diet and exercise. People who have diabetes mellitus need to check themselves for early detection of diabetes mellitus actively.

2   Literature Review

2.1   Survival Analysis

Survival analysis is a statistical method where the variable that is observed is the time until the occurrence of an event, commonly called survival time [3]. Survival method should take into account the main problem in the analysis called censorship. Censorship occurs when some information is available from a person's survival, but it is unknown the exact time of their survival. There are three main reasons why censoring may happen: 1) The study ends if the study has ended, but the patient has not yet experienced a failure event; 2) Loss to follow up if a patient does not continue treatment or a hospital move during the study; and 3) Patient withdraws from the study, that is, if a patient dies of other causes. There are three censored data types, right censoring (when a patient leaves the study before an event occurs or an event occurs after end of study), left censoring (the failure happened before study begin) and interval censoring (almost the same as left censoring, however, the failure occurred within some given time period).

2.2   Survival Function and Hazard Function

Survival analysis contains two basic quantities that are often used, namely survival function and hazard function. The survival function, $S(t)$, is defined as an individual's chances of survival with a survival time of up to $t(t > 0)$, while the hazard $h(t)$ function is the speed at which an individual experience an event in time intervals t to $t + \Delta t$ provided that the individual survives for a set time. The survival function $S(t)$ can be stated as follows [3–14].

$$S(t) = P(T \geq t)$$

Based on the definition of the cumulative distribution function, the survival function can also be expressed by,

$$S(t) = 1 - P(T \geq t)$$
$$= 1 - F(t) \tag{1}$$

The hazard $h(t)$ function can be stated as follows.

$$h(t) = \lim_{\Delta t} \frac{P(t \leq T < t + \Delta t | T \geq t)}{\Delta t} \tag{2}$$

If the opportunity distribution function when T is a random variable representing survival time, then the opportunity density function can be expressed in Eq. (3) and for the cumulative function expressed in the Eq. (4).

$$f(t) = \lim_{\Delta t} \frac{P(t \leq T \prec t + \Delta t)}{\Delta t} \tag{3}$$

$$F(t) = P(T \leq t) = \int_0^t f(u)du \tag{4}$$

Kaplan-Meier analysis is used to estimate survival functions [4]. The Kaplan-Meier method is based on the individual's survival time. It assumes that the sensor data is independently based on survival time, i.e., the reason censored observation is not related to the cause of failure time. Kaplan-Meier analysis is used to interpret survival functions. Kaplan-Meier Survival Curve illustrates the relationship between the estimation of survival function at t time and survival time. The estimation of survival function is obtained from the following equation:

$$\hat{S}(t_{(j)}) = \hat{S}(t_{(j-1)}) * \hat{P}r[T > t_{(j)} | T \geq t_{(j)}] \tag{5}$$

Kaplan-Meier method particularly continued with the Log Rank test. Log-Rank test is a nonparametric statistical test and suitable for use when the data is not symmetrical, i.e., data is tilted to the right. In addition, Log-Rank trials are widely used in clinical trials to see the efficiency of a new treatment compared to the old treatment when measured is the time until an event occurs. Log-Rank test used to compare Kaplan-Meier in different groups [3].

The following are the hypotheses used in Log-Rank testing.

H_0: There is no difference in the survival curve between different groups.
H_1: There are differences in the survival curve between different groups.

Test Statistics

$$\chi^2 = \sum_{i=1}^{G} \frac{(O_i - E_i)^2}{E_i} \tag{6}$$

where,

$$O_i - E_i = \sum_0^h (m_{if} - e_{if}) \tag{7}$$

$$e_{if} = \left(\frac{n_{if}}{\sum_{i=1}^G n_{if}} \right) \left(\sum_{i=1}^G m_{if} \right) \tag{8}$$

The null hypothesis is rejected if $\chi^2 > \chi^2_{(\alpha(i-1))}$.

2.3 Proportional Hazard Assumption

A condition is said to meet the assumption of proportional hazard if the situation has a constant hazard ratio value against the time [4]. On finding out if a situation achieves proportional hazard assumptions or not can be seen through the following two approaches.

Graphical Approach. Two types of graphs can be used to test proportional hazard assumptions. The most widely used chart approaches in survival analysis are the $\ln[-\ln S(t)]$ chart and the observed versus expected survival curve chart that can be described in Fig. 1.

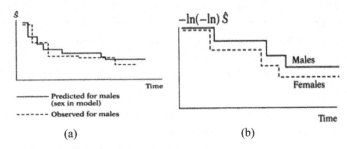

(a)   (b)

Fig. 1. Kaplan-Meier curve

Illustrated that there are variables sex, males and females. Furthermore, an analysis will be conducted on whether sex variables meet proportional hazard assumptions or not. Based on Fig. 1 (a), it appears that proportional hazard assumptions are met because the lines representing males data are aligned with the lines representing females data. Similarly, when viewed from the observed versus expected graph in Fig. 1 (b), sex variables meet proportional hazard assumptions due to the observed survival curve and expected to be close to each other [4].

Goodness of Fit. The second approach used in the analysis of proportional hazard assumptions is the attachment to the goodness of fit (GOF) test. This test is done because GOF uses a statistical approach and generates a p-value. There are several steps taken in this GOF test [15, 16].

1. Regress survival time with its predictors to obtain Schoenfeld residual values.
2. Look for correlations between Schoenfeld residual variables and survival time (sorted from small to large).
3. Conducted correlation testing between Schoenfeld's residual and survival time, sorted from large to small. The hypotheses used in this correlation test are as follows.

Hypothesis:

$H_0: \rho = 0$

$H_1: \rho \neq 0$

The decision to reject H_0 when the p-value value is more than a significant level (α) is used which means there is a correlation between Schoenfeld's residual and survival time that has been sorted from large to small. Therefore, proportional hazard assumptions can be met when the correlation test is insignificant [4].

2.4 Cox Proportional Hazard Models

In the survival analysis, we can see the pattern of survival time with predictor variables that are considered to affect survival time by using assessment with regression modeling. The regression model in survival analysis that is often used is cox proportional hazard regression model. Cox regression is known as Cox proportional hazard regression, where the variable relationship (Y) and (X) in Cox proportional hazard regression have hazard function in exponential form at any given time. Cox proportional hazard regression aims to determine the effect of multiple variables on survival data together [3, 17]. Here is a writable model of Cox proportional hazard.

$$h(t) = h_0(t) \exp\left[\sum_{y=1}^{p} \beta_y x_{ijy}\right] \tag{9}$$

In Cox Proportional Hazard modeling there is an assumption that must be fulfilled, namely Proportional Hazard assumption. The way to find out the fulfillment of Proportional Hazard assumptions is to use Goodness of Fit (GOF) testing.

2.5 Cox Proportional Hazard Parameter Estimation

The estimated parameter in Cox Proportional Hazard modeling is by maximizing the partial likelihood function or commonly called Maximum Partial Likelihood Estimation (MPLE). Suppose there are n individuals with r individuals experiencing events, so $n - r$ is the number of individuals who are censored and it is assumed that only one individual is experiencing the event at a given time. Survival time is sorted from r individuals experiencing notified event $t_{(1)} < t_{(2)} < \ldots < t_{(r)}$. The set of individuals experiencing the event before the $t_{(l)}$ time is notified as $R(t_{(l)})$ so that the partial likelihood function of the Cox Proportional Hazard model can be formulated as follows.

$$L(\beta) = \prod_{l=1}^{r} \frac{\exp(\beta^t x_{(l)})}{\sum f \in R(t_{(l)}) \exp(\beta^t x_f)} \tag{10}$$

$x_{(l)}$ is a vector variable of an individual that fails at the lth time with a time of $t_{(l)}$. Notation $R(t_{(l)})$ is all individuals who are at risk of failure at the time of- l. Once the partial likelihood function is obtained, the next step is to maximize the first derivative of the $lnL(\beta)$ function. Due to the implicit estimation of parameters obtained, the numeric iteration method, the Newton-Raphson method, is used [10].

If $g(\beta)$ is a $p \times 1$ sized vector that is the first derivative of the $lnL(\beta)$ function against the β parameter. $H(\beta)$ is a $p \times p$ hessian matrix containing the second derivative of the $lnL(\beta)$ function, then the estimated parameter in iteration to $(l + 1)$ is as follows.

$$\beta^{(l+1)} = \beta^l - H^{-1}(\beta^{(l)})g(\beta^l) \tag{11}$$

As a prefix $\beta^{(0)}$ Iteration will stop if, $||\beta(l + 1) - \beta l|| \leq \varepsilon$ where ε is a very small number.

3 Data Example

This study used secondary data sources related to medical record data of Diabetes Mellitus patients at RAA Soewondo Hospital, Pati with data on 65 observations and 7 variables. The response variable used in this study are as follows (Table 1).

Table 1. Research dependent variables

Variabel	Variable name	Description	Scale
Time	Survival Time	When Diabetes Mellitus patients undergo treatment until declared dead/moved during the study	Ratio
Status	Patient Status	1: Diabetes Mellitus Patient Disease	Nominal
		0: Diabetes Mellitus patients do not disease	

Meanwhile, the predictor variables used are five factors that are suspected to affect a person's survival with Diabetes Mellitus as follows (Table 2).

Table 2. Independent variable research

Variables	Code		
	0	1	2
Genetics (X1)	No hereditary diabetes	There are descendants of diabetes	–
Age (X2)	≤49 Years	≥50 Years	–
Diet (X3)	Yes, regularly	Sometimes	No diet
Sport (X4)	Yes, regularly	Sometimes	No sports
Weight (X5)	Nominal scale (units of kilograms)		

4 Analysis and Discussions

4.1 Descriptive Statistics

Identify data characteristics using descriptive statistics. Identification of diabetes mellitus patients is carried out for each factor on continuous variables and discrete variables.

Patient Characteristics Based on Continuous Variables. The characteristics of weight variable data are identified using descriptive statistics shown in the table as follows.

Table 3. Continuous variables descriptive statistics

Variable	Mean	Minimum	Maximum
Time	6.511	1,7	20
Weight (X5)	62,51	48	75

Based on Table 3, it is known that the average survival time of diabetes mellitus patients is 6.5 years, where the survival time is the shortest for 1.7 years and the longest survival time for 20 years. In addition, the average condition of the patient as a whole has a bodyweight of 62.5 kg.

Patient Characteristics Based on Discrete Variables. The characteristics of genetic, age, diet, and exercise variable data are identified using descriptive statistics shown in the table as follows.

Table 4. Discrete variables descriptive statistics

Variables	Frequency
Status	Deceased: 15
	Censored: 50
Genetic (X1)	No hereditary diabetes: 30
	There are descendants of diabetes: 35
Age (X2)	≤49 Years: 14
	≥50 Years: 51
Diet (X3)	Yes, regularly: 35
	Sometimes: 14
	No diet: 16
Sport (X4)	Yes, regularly: 9
	Sometimes: 19
	No sports: 37

According to Table 4, it is known that the frequency of patients who died as many as 15 people with the type of event experienced is dying. Factors that affect patients dying are categorical data such as genetics, age, diet, and exercise.

4.2   Kaplan-Meier Curve and Log Rank Test

Kaplan Meier curve is a curve to know the survival characteristics of diabetes mellitus patients based on the factors that affect it. In this study, the Kaplan-Meier curve focuses more on categorical data types, namely genetic variables, age, diet, and exercise. The Log Rank test is used to determine the difference in survival curve between groups of each factor.

Genetic Factors (X1).

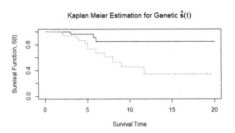

Fig. 2. Kaplan Meier genetic curve

Figure 2 shows that the survival function on genetic no diabetic offspring (black line) is higher than having diabetic offspring (red line). It means that there is a tendency of patients with diabetes mellitus who do not have diabetes offspring to have a higher chance of surviving longer than patients who have diabetes. Then, to support the guess on the Kaplan Meier curve, a log-rank test will be conducted.

Table 5. Log-rank test based on genetics

Log-rank	df	P-value
6.1	1	0.01

Table 5 gives the result that in genetic variables obtained p-value (0.01) < α (0.05), then the decision to reject H_0, so it can be concluded that there is a significant difference in survival time between diabetes mellitus patients who do not have diabetes and diabetes mellitus patients who have diabetes descent with a significance level of 5%.

Age Factor (X2).

Figure 3 shows that the graph of survival function at the age of ≥50 years (red line) is higher than the age of ≤49 years (black line). It means there is a tendency of diabetes mellitus patients who have a ≥50 years old to have a higher chance of surviving longer than patients who have the age of ≤49 years. Then, to support the allegations on the Kaplan-Meier curve, a log-rank test will be conducted.

Table 6 gives the result that in the variable age obtained p-value (0.000) < α (0.05), then the decision is to reject null hypothesis. Hence, it can be concluded that there is a

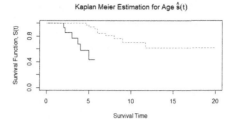

Fig. 3. Kaplan Meier age curve

Table 6. Log rank test by age

Log-rank	df	P-value
21.2	1	0.000

significant difference in survival time between diabetes mellitus patients with the age of ≥50 years and diabetes mellitus patients with the age of ≤49 years with a level of significance 5%.

Dietary Factors (X3).

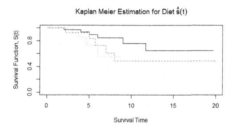

Fig. 4. Kaplan Meier curve diet

Figure 4 shows that the survival function chart has a regular diet (black line) higher compared to having a diet sometimes (red line) and not having a diet (green line). Hence, there is a tendency of patients with diabetes mellitus who experience a regular diet to have a higher chance of surviving longer than patients who experience a diet sometimes and do not diet. Then, to support the guess on the Kaplan-Meier curve, a log-rank test will be conducted. Furthermore, a log-rank test is conducted to strengthen the analysis of the Kaplan Meier curve as follows.

Table 7 gives the result that in the diet variable obtained p-value $(0.2) > \alpha$ (0.05), then the decision failed to reject H_0, so it can be concluded that there is no significant difference in survival time between patients with diabetes mellitus who experience a regular diet and diabetes mellitus patients who experience a diet sometimes or do not diet with a level of significance 5%.

Table 7. Log-rank test based on diet

Log-rank	df	*P-value*
3.3	2	0.2

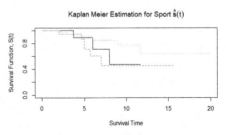

Fig. 5. Kaplan Meier sport curves

Sports Factors (X4).

Figure 5 shows that the graph of survival function is not sports (green line) higher compared to having exercise sometimes (red line) and having regular exercise (black line). So that means there is a tendency of diabetes mellitus patients who experience no exercise to have a higher chance of surviving longer than patients who experience occasional exercise and regular exercise. Then, to support the guess on the Kaplan-Meier curve, a log-rank test will be conducted strengthen the Kaplan Meier curve analysis as follows.

Table 8. Log-rank test by sport

Log-rank	df	*P-value*
3.2	2	0.2

Table 8 shows the result that in the variable exercise obtained p-value (0.2) > α (0.05), then the decision is failed to reject null hypothesis. Therefore, it can be concluded that there is no significant difference in survival time between patients with diabetes mellitus who experience no exercise and patients with diabetes mellitus who experience exercise sometimes or regular exercise with a degree of significance 5%.

4.3   Cox Proportional Hazard Modeling

Simultaneous Parameter Significance Testing. Simultaneous parameter significance testing is used to determine whether predictor variables that are suspected to affect the survival time of diabetes mellitus patients simultaneously have a significant effect on the survival time of diabetes mellitus patients. The following is the output of testing the significance of the Cox PH regression model parameters simultaneously using R.

Table 9. Likelihood ratio

Likelihood ratio	df	*P-value*
25.21	5	0,0001

Table 9 shows testing the significance of cox PH regression model parameters simultaneously obtained p-value (0.0001) $< \alpha$ (0.05), the decision is reject null hypothesis. Thus, it can be concluded that there is at least one predictor variable that has a significant effect on the survival time of diabetes mellitus patients with a significance level of 5%.

Partial Parameter Significance Test. Partial parameter significance testing is used to determine which predictor variables have a significant effect on the survival time of diabetes mellitus patients. The following is the output of testing the significance of the Cox PH regression model parameters partially using R.

Table 10. Significance of Cox PH model parameters

Variables	Coefficient	P-value
Genetics (X1)	1.61032	0.02342*
Age (X2)	−2.62841	0.00113**
Diet (X3)	0.92360	0.01354*
Sports (X4)	−0.16102	0.64454
Weight (X5)	−0.02860	0.59986

Table 10 obtained the results of testing the significance of the parameters of the partial cox PH regression model, and it is known that two variables have a p-value of less than α (0.05) namely genetic variables, age variables, and dietary variables, so that the decision of Reject H_0 so that it can be concluded that the type of diabetes mellitus and the type of event experienced by patients significantly affect the survival time of diabetes mellitus patients with a level of significance 5%.

4.4   Cox Proportional Hazard

After testing the parameter significance of all variables and obtained three variables that meet, then Cox PH modeling can be done. The following are the results of the estimated parameters of the Cox PH model.

Based on Table 11, the Cox PH model can be formed as follows:

$$\hat{h}(t) = \hat{h}_0(t)\exp(1.566\text{Genetics} - 2.799\text{Age} + 0.885\text{Diet})$$

It is known that diabetes mellitus patients who have diabetes offspring have a higher risk of death compared to diabetes mellitus patients who do not have diabetes. It is also

Table 11. Estimated parameters of Cox PH model

Variables	Coefficient	P-value
Genetics (X1)	1.56586	0.017824*
Age (X2)	−2.79935	0.000324*
Diet (X3)	0.88458	0.014730*

*significant at $\alpha = 0.05$

known that patients with diabetes mellitus aged ≤49 years old have a higher risk of death than diabetes mellitus patients who have ≥50 years of age. As well as diabetes mellitus patients who have a regular diet have a higher risk of death than diabetes mellitus patients who have a diet sometimes or not a diet. How influential diabetes mellitus and the type of factors experienced by patients affect the survival time of diabetes mellitus patients can be seen from the hazard ratio value as follows (Table 12).

Table 12. Hazard ratio value

Variables	Exp(coef)
Genetics (X1)	4.78678
Age (X2)	0.06085
Diet (X3)	2.42197

The hazard ratio value for genetic variables is 4.78678. It means that patients with decendence of diabetes have a 4.78678 times higher risk of death than patients who have no hereditary diabetes. The hazard ratio value for the age variable is 0.06085, and this means that older patients will have higher risk of death than the younger one. The hazard ratio value for dietary variables is 2.42197. In other world, patients who have a regular diet have a 2.42197 times higher risk of death than patients who diet sometimes or do not do diet.

4.5 Proportional Hazard Assumption

After obtaining the cox PH regression model, the PH assumptions were made against all predictor variables using the Grambsch and Thernau Test. Here is the output of the Grambsch and Therneau Test using R.

Based on Table 13, it is known that global has a p-value $> \alpha$ (0.05), then the decision is failed to reject null hypothesis. Then, it can be concluded that the covariate does not depend on time or can be said that PH assumptions are fulfilled. Next, it can be concluded that the Cox PH regression method can be used in such cases. In addition to use statistical tests, PH assumption testing can use residual Schoenfeld. The following is the residual Schoenfeld output on genetic variables, age variables, and dietary variables using R.

Table 13. Proportional hazard assumption

Variables	Chisq	DF	P-value
Genetics	0.539267	1	0.46
Age	1.825127	1	0.18
Diet	0.000351	1	0.99
GLOBAL	2.401981	3	0.49

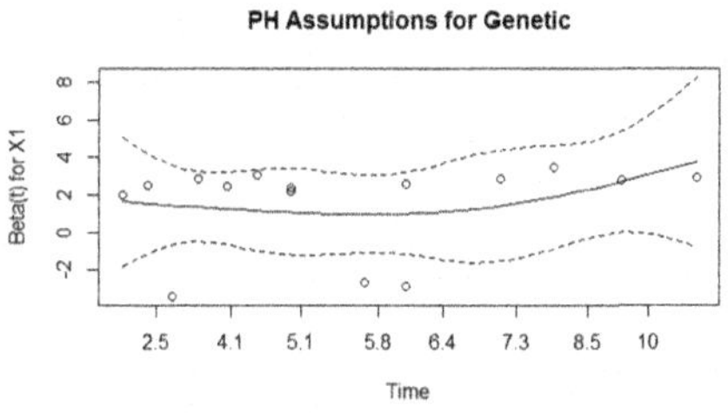

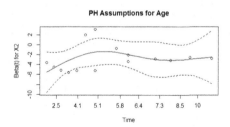

Fig. 6. Genetic PH assumptions

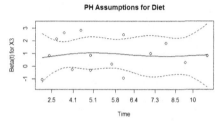

Fig. 7. Assumption of PH AGE

Fig. 8. Dietary PH assumption

Based on the picture above (Fig. 6, 7 and 8), we can find that residual Schoenfeld points in genetic, age, and diet variables tend to be around the interval limit, indicating that disease and event variables have met PH assumptions.

5 Conclusions

Based on the analysis, it can be concluded that the best model Cox Proportional Hazard can be formed by three variables, namely Genetic, Age, and Diet variables. These three variables that make up the Cox PH model obtained hazard ratio value for each variable. Diabetes mellitus patients with diabetes ancestry have a 4.78678 times higher risk of death than diabetes mellitus patients who do not have diabetes. The older patients tend to have higher risk of death than the younger patients. Last but not least, patients who have a regular diet had a 2.42197 times higher risk of death than patients with a diet sometimes or not dieting.

References

1. Ligita, T., Wicking, K., Francis, K., Harvey, N.: How people living with diabetes in Indonesia learn about their disease: a grounded theory study. PLoS One **14**(2), 1–19 (2019)
2. Hordofa, S.B., Debelo, O.: Statistical analysis of the survival of patients with diabetes mellitus: a case study at Nekemte Hospital, Wollega, Ethiopia. Am. J. Biom. Biostat. **4**(1), 6–12 (2020)
3. Kleinbaum, G.D., Klein, M.: Survival Analysis, 3rd edn. Springer, New York (2005)
4. Klein, J.P., Moeschberger, M.L.: Survival Analysis-Techniques for Censored and Truncated Data, 2nd edn. Springer, New York (2003). https://doi.org/10.1007/b97377
5. Hosmer, D.W., Lemeshow, S., May, S.: Applied Survival Analysis: Regression Modelling of Time to Event Data, 2nd edn. Wiley, Hoboken (2008)
6. Iddrisu, A.K., Alhassan, A., Amidu, N.: Survival analysis of birth defect infants and children with pneumonia mortality in Ghana. Adv. Public Health **2019**, 1–8 (2019)
7. Streib, F.E., Dehmer, M.: Introduction to survival analysis in practice. Mach. Learn. Knowl. Extr. **2019**(1), 1013–1038 (2019)
8. Kartsonaki, C.: Survival analysis. Diagn. Histopathol. **22**(7), 263–270 (2016)
9. George, B., Seals, S., Aban, I.: Survival analysis and regression model. J. Nucl. Cardiol. **21**(4), 686–694 (2014). https://doi.org/10.1007/s12350-014-9908-2
10. Collet, D.: Modelling Survival Data in Medical Research, 2nd edn. Chapman and Hall, New York (2003)
11. Ajagbe, O.B., Kabair, Z., O'Connor, T.: Survival analysis of adult tuberculosis disease. PLoS One **9**(11), 1–10 (2014)
12. Lee, E.T., Wang, J.W.: Statistical Methods for Survival Data Analysis. Lifetime Learning Publication, Bermount (2003)
13. Katz, M.H.: Multivariable Analysis: A practical Guide for Clinicians and Public Health Researchers, 3rd edn. Cambridge University Press, New York (2011)
14. Stevenson, M.: An Introduction to Survival Analysis. Massey University, New Zealand (2009)
15. Therneau, T.M., Grambsch, P.M., Fleming, T.R.: Martingale-based residuals for survival models. Biometrika **77**(1), 147–160 (1990)
16. Barlow, W.E., Prentice, R.L.: Residuals for relative risk regression. Biometrika **75**(1), 65–74 (1988)
17. Cox, D.R.: Regression model and life-tables. J. Roy. Stat. Soc. B **34**(2), 187–202 (1972)

Application of Machine Learning in Credit Risk Scorecard

Choon Yi Lee[1,3]([✉]), Siew Khew Koh[2], Min Cherng Lee[2], and Wei Yeing Pan[2]

[1] LKC Faculty of Engineering and Science, Universiti Tunku Abdul Rahman, 43000 Kajang, Selangor, Malaysia
choonyi@1utar.my
[2] Department of Mathematics and Actuarial Sciences, LKC Faculty of Engineering and Science, University Tunku Abdul Rahman, 43000 Kajang, Selangor, Malaysia
[3] No 30, Jalan 6, Taman Sri Ukay, 68000 Ampang, Selangor, Malaysia

Abstract. Machine Learning models have been extensively researched in the area of credit scoring. Banks have put in substantial resources into improving the credit risk model performance as improvement in accuracy by a fraction could translate into significant future savings. Given the lack of interpretability in machine learning models, it is often not used for capital provisioning in banks. This paper uses the Taiwan Credit Card dataset and illustrates the use of machine learning techniques to improve construction of credit scoring models. In factor transformation for a credit scorecard, Decision Tree technique showed the ability to produce quick and predictive transformation rule as compared to traditional approach. The resulting scorecard also showed high predictive power on Test sample. Given the ability of machine learning to produce predictive result, banks should explore on the techniques to improve their overall credit risk management framework. Credit underwriting scorecard could be built using higher discriminatory power techniques, as more good customers are likely to be better than score cut-off and thus accepted by banks.

Keywords: Data mining · Credit risk scorecard · Machine classifier · Bank credit underwriting scorecard · Predictive modeling

1 Introduction

1.1 Background

Credit lending has been one of the main driving forces behind the economies of most leading industrial countries. According to Thomas (2009), the founding of Bank of England in 1694 was one of the first signs of the financial revolution which would allow mass lending. Over the years, bank began lending to the noble, and slowly lending begun to be offered by manufacturer in the form of hire purchase, where they would sell machines to client in the form of credit lending. Banks and financial institutions today offer credit facilities range from Sovereign Bond, Corporate Lending, specific project

© Springer Nature Singapore Pte Ltd. 2021
A. Mohamed et al. (Eds.): SCDS 2021, CCIS 1489, pp. 395–410, 2021.
https://doi.org/10.1007/978-981-16-7334-4_29

financing, to consumer credit such as mortgages, hire purchase, personal loan, credit cards, overdrafts and many other financial products.

Today, the world has entered into dawn of the fourth industrial evolution, which differs in speed, scale, complexity, and transformative power compared to previous revolutions. Xu et al. (2018) examined the opportunities and challenges that are likely to arise as a result of the revolution. Rapid development of Machine Learning tools in the recent year have solved challenges in many areas, including the banking and finance industry. Gan et al. (2020) attempted to predict Asian option prices using deep learning model, and showed that the speed of the trained deep learning model is extremely fast, with high accuracy. Other than finance industry, Wang et al. (2020) used machine learning methods to forecast binary New Product Development (NPD) strategy for Chinese automotive industry, which is crucial for decision making to ensure the scarce resources are allocated effectively.

Machine Learning in Banking Risk Management has gained significant amount of attention from academia and industry. Digitalization of risk processes in bank and financial institutions have become increasingly important, for example, conduct risk. By combining machine learning and transaction data, financial institutions are able to automate conduct monitoring for mortgage underwriting (Oliver Wyman 2017). Leo et al. (2019) and Abdou and Pointon (2011) had reviewed a number of available literature and evaluated machine-learning techniques that have been researched in the context of banking risk management, and identified problems in risk management that have been inadequately explored and are potential areas for further research.

1.2 Objectives

In this research, application of machine learning technique which could improve traditional credit scoring framework will be presented. Decision Tree technique will be applied to improve feature selection and variable transformation process in credit scorecard development. The effectiveness in terms of speed and accuracy will be measured and compared to traditional approaches. The resulting variables will then be used to construct a credit scorecard.

2 Literature Review

2.1 Credit Scoring

Increased competition in the financial lending sector have led banks and financial institutions to search for more effective ways to attract creditworthy customers. Optimizing scorecard's approval decision to ensure credible customers are offered with credit facilities while keeping capital charge low has become more challenging to most banks and financial institutions. Excessive lending to risky customers would result in very high provision and capital charge, which would affect the profitability of the business. In general, banks should focus on improving the book quality through better credit underwriting as well as portfolio review process. Application scorecard refers to the credit scoring model that rates customer upon the credit facility application. Behavioural scorecard refs to the credit scoring model that tracks customer payment, delinquent behaviour in order to estimate PD as of reporting date. Siddiqi (2005) highlighted that while there are various mathematical techniques available to build prediction scorecards, the most appropriate technique to be used can depend on various issues. For example, data quality, target variable type, sample size, implementation platforms, interpretability of results, and legal compliance on methodology as usually required to be transparent and explainable. Besides, the ability to track and diagnose scorecard performance is also key to selecting the most suitable technique. Siddiqi (2005) also outlined steps and methodology in traditional scorecard development, including exploring data, identifying missing values and outliers, correlation, initial characteristic analysis, multiple factor analysis, preliminary scorecard, reject inference, final scorecard production, scorecard scaling, and scorecard validation. The methodology for traditional scorecard approach presented in this paper is based on similar methodology, which is greatly practice by many banks and financial institutions for scorecard development. Sun and Wang (2005) highlighted that the validity of a rating model should be discriminative, homogeneous, and stable. They also proposed Kolmogorov-Smirnov Test (K-S test), Gini Coefficient and Receiver Operating Characteristic (ROC) as possible ways to validate credit rating model.

2.2 Data Mining

Sharma (2009) presented a useful Guide to credit scoring in R that uses the German Credit dataset to demonstrate traditional credit scoring using logistic regression, and also cutting edge techniques available in R. Yap et al. (2011) compared traditional credit scorecard approach (similar to the approach taken in this paper), with Decision Tree and Logistic Regression approaches. They found that the final selected model is credit scorecard approach and that Decision Tree approach has shown lower misclassification rate in Training Dataset, but higher misclassification rate in Validation Dataset. Wang et al. (2012) proposed use of ensemble techniques bagging and random subspace to improve accuracy rate of Decision Tree model, by reducing the influence of the noise data and redundant attributes. Barboza et al. (2017) compared machine learning approaches against the traditional approaches which are Multivariate Discriminate Analysis (MDA) and Logistic Regression, and found that machine learning models show improved bankruptcy prediction accuracy over traditional models.

Zhao et al. (2014) compared the Average Random Choosing method to Pure Random Choosing method when sampling Training, Validation, and Test set. They found that Average Random Choosing method has positive impact towards performance of the machine learning model. This algorithm will maintain a similar event rate across training, validation and test data set. Through the study on the results of other researchers, machine learning algorithm is worth noting that it can be more predictive than traditional scoring method. They also reported under severe class imbalance dataset, for example 99% vs 1% event rate, the Average Random Choosing method does not improve the performance of model, and approach such as oversampling is more preferred. They also highlighted MLP neural network computation time increases with the increase number of neurons, and such scenario is also observed in this research. Khashman (2011) applied an input normalization technique to transform all input variables into range between 0 and 1, by dividing them against the maximum value of each variable, before training neural network models. Kuhn and Johnson (2013) highlighted that to improve the effectiveness of neural networks model, various data transformation methods were evaluated. They found that the spatial sign transformation method on variable showed significant improvement on the performance of neural networks model. Öğüt et al. (2012) used Support Vector Machine and Artificial Neural Network to compare against traditional approaches (Multiple Discriminant Analysis and Ordered Logistic Regression) in predicting the financial strength rating of Turkish Banks. They found that both Multiple Discriminant Analysis and Support Vector Machine achieve the highest accuracy rate when pre-transformed variables are used as input variables. Whereas Ordered Logistic Regression performed the best when transformed factors scores are used as input variables. Desai et al. (1996) compared the performance of neural networks such as multilayer perceptron and modular neural networks, as well as some traditional techniques such as linear discriminant analysis and logistic regression. The finding reported that neural networks offers good improvement in percentage of bad loans correctly classified (sensitivity). However, on the measure of accuracy rate, logistic regression models are comparable to neural networks approach. Oreski et al. (2012) proposed a feature selection method by using combination of genetic algorithm with neural networks to improve accuracy rate of neural network classifier. They found that the approach is better than other techniques such as Forward selection, Information gain, Gain ratio, Gini index, and Correlation. Between the bank's internal behavioral scoring model and the external credit bureau scoring model, Chi and Hsu (2012) found that combining the two models is more predictive than by looking only at one of the model alone.

Yeh and Lien (2009) examined six major classification methods – Artificial Neural Networks (ANN), K-nearest neighbor, Logistic regression, Discriminant Analysis, Naïve Bayesian, and Classification trees. From their finding, ANN performs the best among all the other methods in terms of Accuracy Ratio. They also suggested the use of Accuracy Ratio to compare the model performance, instead of error rate. This is because in the credit card dataset used (similar dataset is used in this research), most records are non-risky, therefore the error rate is insensitive to classification accuracy of models. The ANN is also compared to the other five classification methods for default probability produced, and it is reported that ANN performs the best in presenting real probability of default.

3   Methodology

3.1   Overview

This paper presents machine learning techniques that can be used to improve traditional scorecard construction, as well as compare machine learning model performance to traditional scorecard approach. Development of a robust scorecard is a multi-step process that involves not just statistical analysis but also expert judgement (Fig. 1).

In general, construction of credit risk scorecard or predictive model begins with data collection, cleansing, and sampling. Next, model design is discussed and agreed among the modeler and all other stakeholders. Then, each variable is univariately analysed for potential shortlisting in final model. Transformation is also applied in the process to align variable in terms of intuition as well as support the subsequent modelling process.

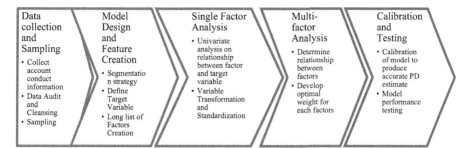

Fig. 1. Overview of Modelling Process

Multifactor analysis (MFA) is then performed to determine optimal weights for each factor and the appropriate rules to apply in order to arrive at final models. Calibration of model estimate is performed to ensure the prediction output matches underlying economic hypothesis. Finally, model performance is tested on both training and holdout samples.

3.2   Data and Sampling

This research uses the Default of Credit Card Clients dataset. The dataset contains information on default payments, demographic factors, credit data, history of payment, and bill statements of credit card clients in Taiwan from April 2005 to September 2005. The dataset source from a public repository, namely UCI Machine Learning Repository. We have downloaded the dataset from Kaggle website in UCI Machine Learning Repository. A brief description of the dataset is tabulated below.

Table 1. Description of dataset

No	Name	Data Type	Description
1	ID	Numeric	ID of each client
2	LIMIT_BAL	Numeric	Amount of given credit in NT dollars
3	SEX	Categorical	Gender
4	EDUCATION	Categorical	Education status
5	MARRIAGE	Categorical	Marital status
6	AGE	Numeric	Age in years
7	PAY_0	Numeric	Repayment status in September, 2005
8	PAY_2	Numeric	Repayment status in August, 2005
9	PAY_3	Numeric	Repayment status in July, 2005
10	PAY_4	Numeric	Repayment status in June, 2005

(continued)

Table 1. (*continued*)

No	Name	Data Type	Description
11	PAY_5	Numeric	Repayment status in May, 2005
12	PAY_6	Numeric	Repayment status in April, 2005
13	BILL_AMT1	Numeric	Bill statement in September, 2005
14	BILL_AMT2	Numeric	Bill statement in August, 2005
15	BILL_AMT3	Numeric	Bill statement in July, 2005
16	BILL_AMT4	Numeric	Bill statement in June, 2005
17	BILL_AMT5	Numeric	Bill statement in May, 2005
18	BILL_AMT6	Numeric	Bill statement in April, 2005
19	PAY_AMT1	Numeric	Previous payment in September, 2005
20	PAY_AMT2	Numeric	Previous payment in August, 2005
21	PAY_AMT3	Numeric	Previous payment in July, 2005
22	PAY_AMT4	Numeric	Previous payment in June, 2005
23	PAY_AMT5	Numeric	Previous payment in May, 2005
24	PAY_AMT6	Numeric	Previous payment in April, 2005
25	Default	Numeric	Default payment in next month

The dataset consists of 30,000 observations and 25 variables. The dataset is examined for potential issue such as missing values, outliers or any inconsistencies. No missing values is found in the dataset. Two exclusions were applied (Table 2):

Table 2. Sample Exclusion

No	Exclusion	Count
	Initial Dataset	30,000
1	Repayment status in September, 2005 is delinquent, with PAY_0 > 0, but however Amount of bill statement in September, 2005 (NT dollar) is ≤ 0. As the repayment status (delinquent) is not consistent with the amount of bill statement	1,689
2	Amount of bill statement in September, 2005 (NT dollar) ≤ 0, but however default payment next month. As the default event is not consistent with the amount owed	184
	Final Modeling Sample	28,127

To ensure the constructed scorecard is predictive and could generalize well on new sample, a portion of the modelling sample was separated out from model training. This approach is used to prevent overfitting of models. Modelling data sampling is the process of partitioning the modelling sample into training sample and holdout or testing sample. In this research, Stratified Random Sampling was performed to split modelling data into Train and Test sample. Strata variable was set to the target variable, to ensure the event rate across training and test dataset is similar. 80% vs 20% was used to split Train and Test sample.

3.3 Model Design and Feature Creation

Model design involves setting up scorecard to align with usage in scoring customers from specific segments. In this research, a single scorecard will be constructed using the variables transformed with machine learning method.

From the description of dataset in Table 1, the target variable is default payment next month, which can be modelled by inputs such as payment, usage, utilisation and some demographic factors. Bad is defined as defaulted within the next month, or performance period of 1 month. Under Basel Accord, default is commonly defined as delinquent for more than 90 days. In this research, since performance period is only 1 month, it might capture some good customers who forget to make payment before due date, and also "truly" bad customers.

In the process of constructing a robust credit scorecard, a list of factors is typically created to ensure the designed scorecard will fit well to the business strategy. The list of factors created can be driven by expert input or statistical analysis of the initial input field. In this research, a list of 65 factors is created from the original 23 input fields. The list of factors created is shown in Table 3.

Table 3. Summary of the long list of factors created

No	Name	Data Type	Description
1	UTIL1	Numeric	Utilization Current Month (Sep 2005)
2	UTIL2	Numeric	Utilization last 1 Month (Aug 2005)
3	UTIL3	Numeric	Utilization last 2 Month (July 2005)
...	...	...	...
65	Count_Pmt_GE_BAL_L1M	Numeric	Number of times Payment ≥ bill amount in last month

As shown in the table above, the list of factors can be summarized and categorized into broad categories such as payment, delinquent, utilisation information of the customers. The list covers few dimensions below:

1. Payment frequency, recency
2. Delinquent frequency, recency, severity
3. Utilisation frequency, recency

3.4 Machine Learning in Single Factor Analysis

Single Factor Analysis is a process by which a long list of potential factors is univariately analysed to arrive at a shorter list of candidate factors for inclusion in the credit scorecard. This process is also termed as Feature Selection process. The outcome of the process is to identify and remove low discriminatory power, too concentrated or poor distributed, unstable and redundant variables. During the process, variable transformation, standardization, variable shortlist decision was made for further analysis on the variables. Accuracy Ratio (AR), also known as Gini coefficient was used to determine the discriminatory power of variables and models. Information Value (IV) was also derived and used as secondary test on the variables. AR is the area ratio under the CAP curve, which is also known as the Gini curve, Power curve, or Lorenz curve. Gini coefficient ranges between 0 to 1, when it is equal to 1, it means the model output is fully able to differentiate non-defaulters and defaulters. When it is equal to 0, the rating model cannot discriminate between non-defaulters and defaulters. In reality, CAP curve of a rating model would run between the perfect curve and random model curve. Information Value (IV), or total strength of the characteristic comes from information theory, and is measured using the formula:

$$IV = \sum_{i=1}^{n} (Distr\ Good_i - Distr\ Bad_i) * \ln\left(\frac{Distr\ Good_i}{Distr\ Bad_i}\right) \tag{1}$$

where,

– $Distr\ Good_i$ is the Distribution of Good observation in group i

$$Distr\ Good_i = \frac{Count\ Good_i}{Total\ Count\ of\ Good} \tag{2}$$

– $DistrBad_i$ is the Distribution of Bad observation in group i

$$Distr\ Bad_i = \frac{Count\ Bad_i}{Total\ Count\ of\ Bad} \tag{3}$$

High IV indicates high predictive power, and vice versa. In this research, Gini $\geq 10\%$ was used as shortlisting criteria. 69 variables were shortlisted for multi-factor analysis.

Variable transformation was performed as per common approach in credit scorecard construction. In this paper, Weight of Evidence (WOE) transformation was applied as per the methodology outlined in Siddiqi (2005). The transformation involves Binning or grouping of identical risk subpopulation, and assign the WOE measure as the score for the subpopulation. WOE is based on the log of odds calculation:

$$WoE = ln\left(\frac{\%Distribution\ of\ Goods}{\%Distribution\ of\ Bads}\right) \tag{4}$$

where,

– % Distribution of Goods represents percentage of good customers in a particular group; and

– % Distribution of Bads represents percentage of bad customers in a particular group.

The WOE measures the strength of each attribute class in discriminating good and bad accounts. It is a measure of the difference between the proportion of good and bad accounts in each attribute class. Positive number implies that the particular attribute class is isolating a higher proportion of good than bad, and vice versa. A higher WOE value implies lower risk while lower WOE value implies higher risk in that attribute class. Variable standardization was applied in this stage. All scores were normalized to mean, $\mu = 0$ and standard deviation, $\sigma = 1$. An average customer would receive a WOE score close to 0, and negative value implies higher than average risk, and vice versa. Binning involves a combination of statistical analysis and expert input to arrive at final binning. The common steps for interval and categorical variables is shown below:
For interval variable,

1. Factors are first being fine-classed into 20 bands, where each band consists of approximately 5% of the total population.
2. To combine groups with similar bad rate.
3. Ensure there is sufficient observation ($\geq$ 5% of population).
4. Fine classed result may not produce monotonic risk trend across bands, thus further combine bands to produce monotonic (increasing/decreasing) risk trend.

For nominal variables,

1. Start by combining attribute with small sample size into group "Others".
2. For the remaining attribute, group similar bad rate attribute.
3. Further group until it has met guideline similar to those set upon interval variables.

The approach above has been applied and shows good performance on the transformed variable. However, it requires a lot of time due to the need to combine groups that may or may not result in monotonic risk trend. Besides, some of the characteristic, for example delinquent, is by nature do not comprise 5% of total population in each attribute, but having significantly different risk than no delinquent. Fine classing the factor into 20 bands with approximately 5% in each band would have left out any group that carries significant higher bad rate but lower than 5% distribution. As a result, a substantially higher risk group could be diluted into the lower risk group. Adjustment to include more bands, such as changing 20 bands to 50 bands with 2% in each band could be performed. However, this would introduce more complexity in combining similar bad rate groups as the potential merging scenarios of group has increased significantly. Besides, for nominal variable, it might be time consuming if there are too many attributes.

In this research, the effectiveness of Decision Tree approach in automating the above process will be explored. Decision Tree is built by repeatedly splitting training data into smaller and smaller samples. Beginning with the original segment, which is the entire data set, it is first partitioned into two or more segments by applying a series of simple rules. Each rule assigns an observation to a segment based on the value of an input for that observation. In a similar fashion, each resulting segment is further partitioned into sub-segments (segments within a segment); each sub-segment is further partitioned into more sub-segments, and so on. This process continues until no more partitioning is possible. This process of segmenting is called recursive partitioning, and it results in a hierarchy of segments within segments. Splitting observation using Decision Tree algorithm could produce binning that is optimal as the predictive power of the factor is maximized through the training process. Conceptually, the decision tree binning should provide best predictive power of the factor, while benefit from all the grouping advantages highlighted above.

There are factors which exhibit non-linear risk trend, which can also be captured by Decision Tree algorithm. However, if the non-linearity is counter intuitive and monotonic risk trend is necessary, we created a simplified monotonic binning algorithm to ensure monotonicity. The idea is to automate the process for better efficiency in credit scorecard construction. Monotonic binning scheme works as follows:

1. Factors are first being fine-classed into $n = 20$ bands, where each band consists of approximately 5% of the total population.
2. Examine the monotonicity of the risk across bands, if there is a break, repeat step 1 with $n = n - 1$.
3. Break if there are only 2 bands left.

4   Results

4.1   Variable Transformation in Factor Analysis

Applying Decision Tree algorithm has resulted in quick and predictive transformed variables. To compare between the common approach of classing and the proposed methodology on variable transformation process, we selected the first variable, credit limit, LIMIT_BAL to illustrate the difference. Firstly, credit limit is fine classed into 20 bands. Figure 2 shows the overview of the relationship between variable and the dependent variable.

It is observed that the bad rate decreases as the credit limit increases, indicating that customers who are having bigger limit have lower risk than the average customers. Combine groups with similar bad rates involve comparing bad rates in each band with the adjacent bands. In the LIMIT_BAL fine classed result, we compared the bad rates of each band, complexity arise when there is break in monotonicity of bad rate. It is observed that band [60000, 80000) (28.5% bad rate) could be merged with band [40000, 60000) (27.7% bad rate) or band [80000, 90000) (22.3% bad rate). As such, 2 bands should be merged and result in band [40000, 80000). The process will repeat using the resulting bin until all bands have achieved monotonicity in bad rate. Another complexity arises when the resulting bin breaks the monotonicity and hence require further grouping or re-perform

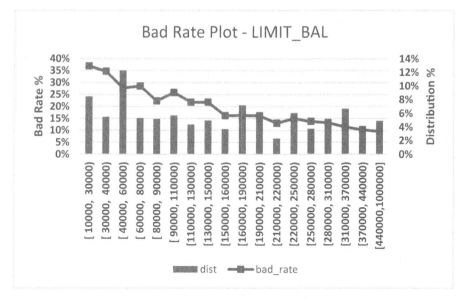

Fig. 2. Factor Bad Rate Plot

the grouping with another adjacent bin. The process is repeated for all variables. It is observed that with the decision tree and monotonic binning algorithm, we could generate the transformed variable automatically in an efficient way. The process provided much improvement to efficiency as the binning result also give optimal power on the variable. The final checking requires modeler to ensure the result matches underlying economic hypothesis and have intuitive connection with the dependent variable. Illustration of Decision Tree binning is shown in Fig. 3 below.

The above result implies that customers who have credit limit $\leq$ NTD 140,000 should receive WOE scores that are < 0 since it is higher risk compared to average population.

Figure 4 shows the result of performing monotonic binning strategy on credit limit. It can be observed that the number of resulting bins is more compared to decision tree binning and even though the risk remained its monotonicity across bins, however, some bins appear to have closer bad rate. We also compared the predictive strength of both the binning strategies, and observed that decision tree binning produced higher Gini result, i.e. 27.9% vs 27.6% in monotonic binning result. Hence, the decision tree binning will be used as the final classing result for the variable and assign WOE scores to the population.

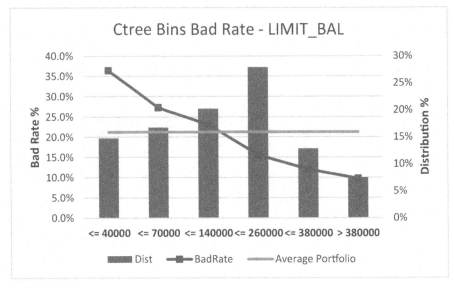

Fig. 3. Conditional Inference Tree Binning Scheme Bad Rate Plot

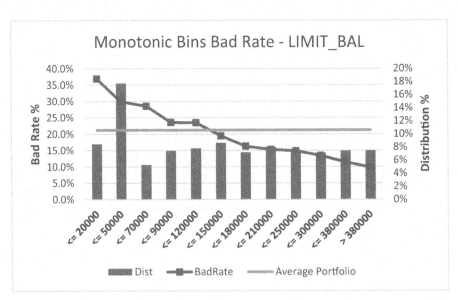

Fig. 4. Monotonic Binning Scheme Bad Rate Plot

4.2   Final Credit Scorecard

The final credit scorecard score assignment is given in Table 4 below.

Table 4. Credit scorecard scores assignment

Variable	Attribute	WOE	Standardized Score	Coefficient of regression	Final Score
woe_LIMIT_BAL	≤40000	−0.7527	−1.6169	−0.12406	−20
	≤70000	−0.3299	−0.7897	−0.12406	−10
	≤140000	−0.1086	−0.3567	−0.12406	−4
	≤260000	0.3802	0.5996	−0.12406	7
	≤380000	0.683	1.1920	−0.12406	15
	> 380000	0.9129	1.6418	−0.12406	20
woe_PAY_6	≤−2	0.5642	0.9345	−0.09586	9
	≤−1	0.3478	0.5253	−0.09586	5
	≤0	0.1353	0.1234	−0.09586	1
	>0	−1.3858	−2.7534	−0.09586	−26
woe_PAY_AMT3	≤0	−0.4625	−1.4748	−0.05782	−9
	≤2901	−0.0788	−0.3298	−0.05782	−2
	≤3912	0.1048	0.2181	−0.05782	1
	≤15587	0.3566	0.9695	−0.05782	6
	>15587	0.8617	2.4768	−0.05782	14
woe_PAY_AMT4	≤396	−0.2604	−1.0389	−0.07416	−8
	≤1668	−0.1809	−0.7454	−0.07416	−6
	≤4300	0.0955	0.2748	−0.07416	2
	>4300	0.4314	1.5148	−0.07416	11
woe_PAY_AMT5	≤0	−0.3227	−1.1729	−0.05639	−7
	≤2927	−0.0904	−0.3887	−0.05639	−2
	≤14100	0.2508	0.7633	−0.05639	4
	>14100	0.8606	2.8220	−0.05639	16

(*continued*)

Table 4. (*continued*)

Variable	Attribute	WOE	Standardized Score	Coefficient of regression	Final Score
woe_PAY_AMT6	≤910	−0.2452	−0.9305	−0.06904	−6
	≤2304	−0.0979	−0.4209	−0.06904	−3
	≤9794	0.2221	0.6861	−0.06904	5
	>9794	0.6679	2.2284	−0.06904	15
woe_MAX_UTIL_6m	≤0.4711	0.3408	0.9542	−0.06346	6
	≤0.6099	−0.0578	−0.2707	−0.06346	−2
	≤1.019	−0.2472	−0.8527	−0.06346	−5
	>1.019	−0.5861	−1.8942	−0.06346	−12
woe_MAX_BY_AVG_UTIL_3m	01 ≤ 1.1071	−0.343	−0.8067	−0.16056	−13
	02 ≤ 1.5385	0.2346	0.3352	−0.16056	5
	03 > 1.5385	0.5587	0.976	−0.16056	16
	04 Utilisation ≤ 0 in last 3 month	2.7401	5.2886	−0.16056	85
woe_Mths_since_status_GT1	01 ≤ 0	−2.0962	−2.3706	−0.81227	−193
	02 ≤ 1	−0.9778	−1.2288	−0.81227	−100
	03 ≤ 3	−0.3609	−0.5989	−0.81227	−49
	04 > 3	0.1667	−0.0603	−0.81227	−5
	05 All status ≤ 1 in last 6 month	0.7862	0.5722	−0.81227	46
woe_Cnt_Mth_With_pmt_L4M	≤3	−0.5643	−1.3228	−0.07905	−10
	>3	0.4245	0.7559	−0.07905	6

From the table above, the split point that was arrived using machine learning method could be of any continuous value from the variable. For example, the variable maximum utilisation in the past 6 months, "woe_MAX_UTIL_6m" can take any value from 0% (no utilisation) to more than 100% (over-limit plus unpaid interest). It is observed that the technique effectively searches for all possible split points for creating optimal rule. It can also be observed that the split points resulted from Decision Tree algorithm show clear differentiation of WoE across different bands. Final Credit Scorecard was trained using Logistic Regression approach on transformed variables. Multicollinearity issue was handled through setting pairwise correlation threshold at 50%. The resulting credit scorecard achieves Gini of 56.37% on Train sample and 57.43% on Test sample. Compared to traditional approach it is also easier to execute as there is less comparison of band merging scenario, thus reducing potential human error in merging the wrong bands.

5 Conclusion

In conclusion, machine learning can be viewed as a tool that allows for more effective credit scoring model construction. Decision tree splitting technique was very effective in factor transformation for traditional credit scorecard. The transformation using decision tree algorithm was quick and resulted in good predictive power variable. Performance of the traditional credit scorecard model is also improved by the transformation process. There are other machine learning techniques which could provide significant uplift to Banks credit underwriting scorecard. A "sharpened" cutoff strategy could be achieved through the use of better discriminatory power scorecard. Banks should look to explore on more potential techniques that give improvement to internal credit scoring methodology. Although it might come with higher cost on computing resources as well as implementation of machine learning algorithm, in the long run, the banks should benefit from accepting more good customers while rejecting the bad customers.

References

Abdou, H.A., Pointon, J.: Credit scoring, statistical techniques and evaluation criteria: a review of the literature. Intell. Syst. Account. Finance Manag. 18(2–3), 59–88 (2011)

Barboza, F., Kimura, H., Altman, E.: Machine learning models and bankruptcy prediction. Expert Syst. Appl. 83, 405–417 (2017)

Chi, B.W., Hsu, C.C.: A hybrid approach to integrate genetic algorithm into dual scoring model in enhancing the performance of credit scoring model. Expert Syst. Appl. 39(3), 2650–2661 (2012)

Desai, V.S., Crook, J.N., Overstreet, G.A., Jr.: A comparison of neural networks and linear scoring models in the credit union environment. Eur. J. Oper. Res. 95(1), 24–37 (1996)

Gan, L., Wang, H., Yang, Z.: Machine Learning solutions to challenges in finance: An application to the pricing of financial products. Technol. Forecast. Soc. Change 153, 119928 (2020)

Gunn, S.: Support Vector Machines for Classification and Regression. Image Speech & Intelligent Systems Group (1998)

Hayden, E., Porath, D.: The Basel II Risk Parameters: Estimation, Validation, Stress Testing - with Applications to Loan Risk Management. Engelmann, B., Rauhmeier, R. (eds.). Springer, Heidelberg (2011)

Hothorn, T., Hornik, K., Zeileis, A.: Unbiased recursive partitioning: a conditional inference framework. J. Comput. Graph. Stat. **15**(3), 651–674 (2006)

Khashman, A.: Credit risk evaluation using neural networks: emotional versus conventional models. Appl. Soft Comput. **11**(8), 5477–5484 (2011)

Kuhn, M., Johnson, K.: Applied Predictive Modeling. Springer, New York (2013). https://doi.org/10.1007/978-1-4614-6849-3

Leo, M., Sharma, S., Maddulety, K.: Machine learning in banking risk management: a literature review. Risks **7**(1), 29 (2019)

Öğüt, H., Doğanay, M.M., Ceylan, N.B., Aktaş, R.: Prediction of bank financial strength ratings: the case of Turkey. Econ. Model. **29**(3), 632–640 (2012)

Oliver Wyman: Next generation risk management (2017). https://www.oliverwyman.com/content/dam/oliver-wyman/v2/publications/2017/aug/Next_Generation_Risk_Management_Targeting_A-Technology_Dividend.pdf

Oreski, S., Oreski, D., Oreski, G.: Hybrid system with genetic algorithm and artificial neural networks and its application to retail credit risk assessment. Expert Syst. Appl. **39**(16), 12605–12617 (2012)

Sharma, D.: Guide to Credit Scoring in R (2009). https://cran.r-project.org/doc/contrib/Sharma-CreditScoring.pdf

Siddiqi, N.: Credit Risk Scorecards: Developing and Implementing Intelligent Credit Scoring. Wiley, Hoboken (2005)

Sun, M.Y., Wang, S.F.: Validation of credit rating models - a preliminary look at methodology and literature review. Rev. Financ. Risk Manag. **2**, 1–15 (2005)

Thomas, L.C.: Consumer Credit Models: Pricing, Profit and Portfolios. Oxford University Press, Oxford (2009)

Wang, G., Ma, J., Huang, L., Xu, K.: Two credit scoring models based on dual strategy ensemble trees. Knowl.-Based Syst. **26**, 61–68 (2012)

Wang, X., Zeng, D., Dai, H., Zhu, Y.: Making the right business decision: forecasting the binary NPD strategy in Chinese automotive industry with machine learning methods. Technol. Forecast. Soc. Change **155**, 120032 (2020)

Xu, M., David, J.M., Kim, S.H.: The fourth industrial revolution: opportunities and challenges. Int. J. Financ. Res. **9**, 2 (2018)

Yap, B.W., Ong, S.H., Nor Huselina, M.H.: Using data mining to improve assessment of credit worthiness via credit scoring models. Expert Syst. Appl. **38**(10), 13274–13283 (2011)

Yeh, I.C., Lien, C.H.: The comparisons of data mining techniques for the predictive accuracy of probability of default of credit card clients. Expert Syst. Appl. **36**(2), 2473–2480 (2009)

Zhao, Z., Xu, S., Kang, B.H., Kabir, M.Md.J., Liu, Y., Wasinger, R.: Investigation and improvement of multi-layer perceptron neural networks for credit scoring. Expert Syst. Appl. **42**(7), 3508–3516 (2014)

Multivariate Analysis to Evaluate the Impact of COVID-19 on the Hotel Industry in Indonesia

Prilyandari Dina Saputri[✉], Arin Berliana Angrenani, Dinda Galuh Guminta,
Fonda Leviany, Ika Nur Laily Fitriana, Santi Puteri Rahayu, and Hidayatul Khusna

Department of Statistics, Institut Teknologi Sepuluh Nopember, Kampus ITS Keputih Sukolilo,
Surabaya 60111, Indonesia
{prilyandari.206003,arinberliana.206003,dindaguminta.206003,
fonda.206003,ika.206003}@mhs.its.ac.id,
santi_pr@statistika.its.ac.id, hidayatul@its.ac.id

Abstract. Pandemic has a significant impact on many sectors, especially for the
hotel industry sector in Indonesia. To find out the impact of the pandemic on the
hotel industry sector, we conducted an inferential statistic using a nonparametric
location test to determine the significant differences between variables in 2019 and
2020. Then, we conducted cluster analysis using K-Means and Self-Organizing
Map (SOM) methods. We also create the perceptual mapping by Biplot. Using the
paired-fisher test for multivariate nonparametric location test, we found that the
differences between variables relating to the occupancy rate of hotel rooms in 2019
and 2020 have been significantly decreasing. According to the biplot analysis, in
2019, the characteristics between provinces were quite different. While, in 2020,
almost all provinces have identical characteristics. The result shows that SOM and
K-Means have the same performance. In 2019, there are 4 clusters, and in 2020
there are 3 clusters. There has been a change in cluster members before and during
the COVID-19 pandemic. Bali is the province that most affected by the COVID-19
incident because the tourism sector is the primary regional income. We found that
the small and medium hotel industry is severely affected by COVID-19 outbreaks.

Keywords: Biplot · Clustering · COVID-19 · Hotel · K-Means · SOM

1 Introduction

The COVID-19 pandemic has spread across the world. The number of cases shows an
increasing number [1]. As of 23 June 2021, over 2.03 million COVID-19 cases have
been reported in Indonesia [2]. The latest policies in Indonesia, known as Community
Activities Restrictions Enforcement "PPKM Darurat" is implemented to control the
spread of COVID-19 by restrict the society movement to go to public places such as
malls, cinemas, places of worship, and tourism places [3]. Since the beginning of this
global pandemic, the tourism sector has experienced a loss of potential income of Rp
90 T during January 2020 - April 2020. The reduced number of tourists has an impact
on the occupancy rate of hotel rooms, which at that time averaged only 49.2%. A drastic

© Springer Nature Singapore Pte Ltd. 2021
A. Mohamed et al. (Eds.): SCDS 2021, CCIS 1489, pp. 411–426, 2021.
https://doi.org/10.1007/978-981-16-7334-4_30

decline has occurred since the first case of the COVID-19 virus entered Indonesia until now. In 2019, there are 3516 star hotels in Indonesia. Moreover, the government through the Ministry of Tourism has closed 180 destinations and 232 tourist villages in Indonesia [4].

Clustering is the task of grouping data points so that they are more similar than data points from other groups [5]. Several previous studies applied the clustering method, Xiong et al. [6] explained that the K-Means method is susceptible to the selection of the initial cluster. Xiong et al. [6] optimizes the initial central cluster with grouping results that produce an average accuracy of 80%. A study using records of electricity consumption shows that SOM provides effective customer grouping results through graphical representations [7]. Meanwhile, previous research has analyzed hotel room occupancy rates. The LSTM model was used to forecast the occupancy rate of star hotel rooms [8]. The research conducted by Wu, Law, and Jiang [9] shows that swine flu has negatively affected the tourism and hospitality industries in many countries. Moreover, an infectious disease influencing factor was identified and can affect hotel occupancy rates.

Research on clustering the occupancy rate hotel rooms has never been done, especially during the COVID-19 pandemic. At the same time, the rate of hotel room occupancy is likely influenced by the tourism sector during this pandemic. Hence, the researchers are interested in analyzing the occupancy rate of star hotel rooms in Indonesia during the COVID-19 pandemic using K-means clustering and Self-Organizing Maps, as well as forming perceptual mapping using biplot. K-means clustering is used because it is relatively fast and straightforward for clustering data, while Self-Organizing Maps clustering is an effective method for visualizing high-dimensional data. The purpose of using a biplot is to show the correlation between variables that affect the occupancy rate and the proximity between provinces that have star hotels.

2 Methodology

2.1 Dataset Description

The data used in this study is secondary data obtained from the Central Bureau of Statistics Indonesia publication. The data is about occupancy rate of the hotel room in 2019 and 2020, which consist of the number of room nights occupied in star hotels, occupancy rate of hotel room in star hotels, number of staying guests in star hotels, the number of foreign guests in star hotels, the number of domestic guests in star hotels, and bed occupancy rate in star hotels. In biplot analysis, we used the aggregate number of all levels of star hotels, whereas in other analyses we disaggregated each level of star for each variable. There are five levels of stars, each variable is calculated for each type of star hotel, so the total of variables is 30 variables.

2.2 Multivariate Normality Test

QQ plot correlation coefficient test is performed to test the normality of the data. The linear relationship between the value of squared mahalanobis distance and its corresponding chi-square quantiles indicates normality in the population. The correlation coefficient can be calculated as

$$r_Q = \frac{\sum_{j=1}^{n} \left(X_{(j)} - \overline{X} \right) \left(q_{(j)} - \overline{q} \right)}{\sqrt{\sum_{j=1}^{n} \left(X_{(j)} - \overline{X} \right)^2} \sqrt{\sum_{j=1}^{n} \left(q_{(j)} - \overline{q} \right)^2}} \tag{1}$$

The null hypothesis that the data follow the multivariate normal distribution will be rejected if the value of coefficient correlation is less than the critical value [10].

2.3 Nonparametric Location Test

In case that the data do not follow the multivariate normal distribution, a nonparametric location test can be used to test the differences in location parameters. The null hypothesis used in this test is $P(X \leq Y) = \frac{1}{2}$, with the statistic test is described as

$$\tilde{U}_{m,n} = \frac{\frac{1}{mn} \sum_{i=1}^{m} \sum_{j=1}^{n} I\{X_i \leq Y_j\} - \frac{1}{2}}{\sqrt{\frac{1}{m}\hat{\xi}_x + \frac{1}{n}\hat{\xi}_y}} \tag{2}$$

where $\hat{\xi}_x = \frac{1}{m-1} \sum_{i=1}^{m} \left(\frac{1}{n}(S_i - i) - \frac{1}{m} \sum_{i=1}^{m} \left(\frac{1}{n}(S_i - i) \right) \right)^2$, and $S_1 < S_2 < \ldots < S_m$ are the ordered ranks of the X. $\hat{\xi}_y$ can be obtained using similar procedures [11]. The comparison between the observed statistics and the permutation distribution is used to test the global null hypothesis that contains all variables. This procedure will result in statistics and p-value for testing the global null hypothesis [12].

2.4 Biplot Analysis

Biplot similarity provides plots of the n observations and the positions of p variables in two dimensions [13]. The plots are based on the singular value decomposition (SVD). This state that the $(n \times p)$ matrices $\mathbf{X}$ consisting of n observations with p variables measured about their sample means can be written as:

$$\mathbf{X} = \mathbf{ULA}' \tag{3}$$

where $\mathbf{U}$ and $\mathbf{A}$ are $(n \times r)$ and $(p \times r)$ matrices respectively, each with orthonormal columns, L is an $(r \times r)$ diagonal matrix with elements $t_1^{1/2} \geq t_r^{1/2} \geq \cdots \geq t_r^{1/2}$, and r is the rank of $\mathbf{X}$. To include the information on the variables in this plot, we consider the pair of eigenvectors from the first two sample principal components.

2.5 K-Means

The K-Means method is a clustering method that uses distance as a metric so that there are k predefined clusters [14]. This method works by grouping each object into a cluster with the closest cluster center [10]. The first step of the K-means algorithm are divide the data into k clusters. Then, determine the value of the cluster center (centroid). The initial centroid is determined randomly while in the later stages using the following equation.

$$\bar{x}_{jh} = \frac{\sum_{i=1}^{n_h} x_{ij}}{n_h} = \frac{x_{1j} + x_{2j} + \ldots + x_{n_h j}}{n_h} \tag{4}$$

where $\bar{x}_{jh}$ is the mean of objects in the h-th cluster for the j-th variable and n_h is the number of objects in the h-th cluster. Calculate the distance of each object to each centroid using Euclidean distance:

$$D_{ih}^2 = \sum_{j=1}^{m} \left(x_{ij} - \bar{x}_{jh} \right)^2 \tag{5}$$

where D_{ih}^2 is the square of the distance between ith object and hth centroid, x_{ij} is the value of the ith objects in the jth variable, and m is the number of variables. Next, group objects to the nearest cluster center. Objects that have the closest distance to cluster h are included in cluster h. Return to step of determination the value of the cluster center until the centroid has not changed, meaning no cluster members have changed.

2.6 Self-organizing Maps

Self-Organizing Maps (SOM) is a type of neural network model. It is named "Self-Organizing" because it does not require supervise to learn the data and called "Maps" because SOM tries to map its weights to match the input data given. SOM allows the visualization and projection of high-dimensional data to lower dimensions, most often a 2-D plane while maintaining the topology of the data [15]. The steps of SOM are as follows [16]:

1. Neuron at input layer (input neuron) size of i ($x_1, x_2, x_3, \ldots x_i$) and neuron at output layer (output neuron) size j x l ($y_{11}, y_{12}, y_{13}, \ldots, y_{jl}$.). The weight of connection between input neuron and output neuron notated as W_{ijl}.
2. Initialization the connection weight (W_{ijl}) randomly from 0 to 1.
3. Repeat the step 4 to 7 until convergent (the change of weight is small relatively or smaller than the tolerance limit) or cycle (the step 4 to 7) has finished the iteration.
4. Choose randomly a vector of x (which is a random number from 0 to 1) that will be clustered and selected to input neuron.
5. Get the distance of input vector to weight of connection (d_{jl}) for each output neuron using Eq. (5)
6. Find the index $b = j$, $c = l$, where d_{jl} is minimum, the output neuron bc is called *Best Matching Unit* (BMU).
7. For every W_{ijl}, renew the weight connection using the formula:

$$w_{ijl}(t+1) = w_{ijl}(t) + \alpha \left(x_i(t) - w_{ijl}(t) \right) \tag{6}$$

where, $w_{ijl}(t+1)$ is the new weight vector of l-th at the component of i-th at input vector of j-th, $w_{ijl}(t)$ is the lastest weight vector of l-th at the component of i-th at input vector of j-th, α is the learning rate, and $x_i(t)$ is the component of i-th.

2.7 Cluster Evaluation (*The C-H Pseudo F Statistic*)

The C-H pseudo F statistic is the ratio of the mean squares for a given grouping divided by the mean squares of the residuals [17], which can be written as:

$$Pseudo - F = \frac{\left(\frac{R^2}{k-1}\right)}{\left(\frac{1-R^2}{n-k}\right)} \tag{7}$$

where $R^2 = \frac{SSB}{SST} = \frac{(SST-SSW)}{SST}$, SST is the total sum of squared, SBB is the sum of squared between groups, and SSW is the sum of squared within the group. The highest *pseudo F* indicates that the number of clusters formed has been optimal. The diversity within the group is very homogeneous, while the diversity between groups is very heterogeneous [18]. Assessment of the best cluster performance results can be determined using the internal cluster dispersion rate or commonly known as the ICD rate, which can be calculated using Eq. (8). The smaller the ICD rate, the better the grouping results [19].

$$ICD\,rate = 1 - \frac{SSB}{SST}. \tag{8}$$

3 Results

3.1 Data Exploration

There is no missing value in the data. The boxplot of data aggregation of 1 to 5-star hotels for each variable related to occupancy rate is shown in Fig. 2. From the boxplot, it can be obtained that there are outliers in the number of room nights occupied, the occupancy rate of a hotel room, number of staying guests, number of foreign guests, and sources of domestic guests. However, in this study, outliers are not handled because the purpose of the study is to cluster. The cluster results are expected to be genuinely homogeneous within-group and heterogeneous between groups by ignoring outliers.

Figure 1 shows that there was a significant decrease from 2019 to 2020 of hotel occupancy rate data in Indonesia. The biggest decrease was in the number of foreign guests by 80% from the previous year. The number of staying guests decreased by 55%, the number of room nights occupied decreased by 53%, and the number of domestic guests decreased by 51%. Meanwhile, the occupancy rate of hotel room decreased by 38%, and the bed occupancy rate decreased by 37%.

From the exploration data, it can be seen that pandemic has greatly impacted the hospitality sector in Indonesia. All variables are decreasing while the number of star-hotel in Indonesia are increasing from 2019 to 2020 as shown at Fig. 3. From the exploration data, it can be seen that the number of star-hotel in Indonesia had been increased from 2019 to 2020 but the rate of hotel occupancy and the guests had been decreased because of pandemic situation at the end of 1st quartal in 2019.

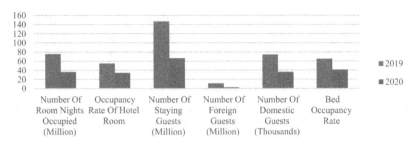

Fig. 1. Barchart of the occupancy rate of the hotel rooms for 2019 and 2020 in Indonesia.

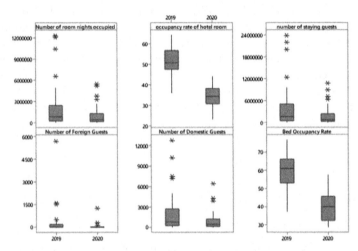

Fig. 2. Boxplot of the occupancy rate of the hotel rooms for 2019 and 2020 in Indonesia.

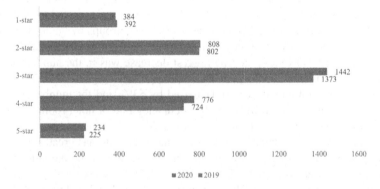

Fig. 3. The number of 1 to 5-star hotel for 2019 and 2020 in Indonesia

3.2 Differences for All Variables

In this section, we provide the differences of all variables in 2019 and 2020 using graph and hypothesis testing. The comparison for all variables can be plotted in the density of each variable in each year, as in the Fig. 4, which shows us that variables with a huge difference in the two years are the Occupancy Rate of Hotel rooms and Bed Occupancy Rate. The year 2020 (blue line) tends to be more lifted to the left than the year 2019 (black line). This implies that the parameter of the year 2020 is less than the parameter in the year 2019. In other words, the decreasing values for the 2020 year exist in these two variables. To make an inference with this conclusion, we provided the hypothesis testing for location parameters. The multivariate normality test for all variables in 2019 and 2020 results that the data does not follow the multivariate normal distribution. The result of the QQ plot correlation test for normality is provided in Table 1.

Table 1. Multivariate Normality Test for all variables in 2019 and 2020

Correlation Coefficient	Critical Value	Result
0.56236	0.96389	The null hypothesis is rejected

In this study, we used the nonparametric location test to determine the significant differences between variables in 2019 and 2020. The result of the Paired Fisher Sign Test for multivariate nonparametric location test can be found in Table 2, while the univariate nonparametric location test for all variables are provided in Table 3.

Table 2. Multivariate Nonparametric Location Test for 2019 and 2020

Statistics	p-value	Result
5.488	<0.001	The null hypothesis is rejected

Table 3. The p-value of Univariate Nonparametric Location Test for 2019 and 2020

Variables	5 star	4 star	3 star	2 star	1 star
Number of Room Night Occupied	1.000	0.005	<0.001	0.001	<0.001
Occupancy Rate of Room	1.000	0.005	<0.001	0.001	<0.001
Number of Guest Nights	1.000	0.001	<0.001	0.001	<0.001
Number of Foreign Guests	1.000	0.001	<0.001	0.022	0.319
Number of Domestic Guests	1.000	0.001	<0.001	0.001	<0.001
Bed Occupancy Rate	1.000	0.001	<0.001	0.001	<0.001

The multivariate nonparametric location test implied that there is a significant difference ($\alpha = 0.05$) between the median of the data in 2019 and 2020. The univariate nonparametric location tests provide interesting results. All the variables in 5-star hotels have the same median for 2019 and 2020 with p-value close to one. At the same time, most of the remaining types of hotels result in a positive estimate for the median with p-value less than 0.05. It can be concluded that there are significant decreasing values for all variables in 2020 compared to the same variables in 2019 in the remaining types of hotels.

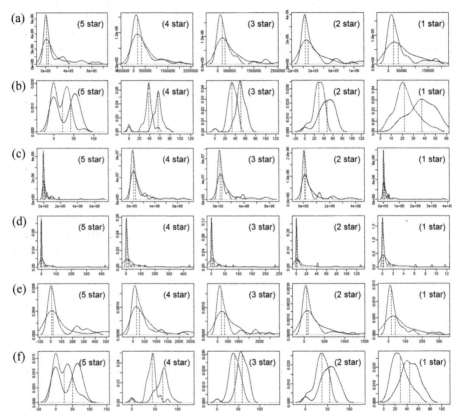

Fig. 4. (a) Number of Room Night Occupied, (b) Occupancy Rate of Hotel Room, (c) Number of Guest Nights, (d) Number of Foreign Guests, (e) Number of Domestic Guests, and (f) Bed Occupancy Rate for Each Type of Hotel in 2019 (black line) and 2020 (blue line) (Color figure online)

3.3 Biplot

Biplot is used to compare variables based on the occupancy rate in 2019 and 2020. Figure 5(a) shows that the biplot represents 99.79% of the total variance in the data, the first axis gives 95.08%, and the second axis gives 4.71% for total variance. Furthermore, Fig. 5(b) shows that the biplot represents 99.81% of the total variance in the data, the first axis gives 95.65%, and the second axis gives 4.16% for total variance. Concluded from the length of the vector that all variables have the similar variance. It can be seen that all variables have a relatively high and positive correlation, this is because the angle formed by the two vectors is less than $90°$, except for the room occupancy rate variable and the number of foreign guests facilities, also for the number of foreign guests facilities and bed usage rate which have a negative correlation because the angle formed by the two vectors is more than $90°$.

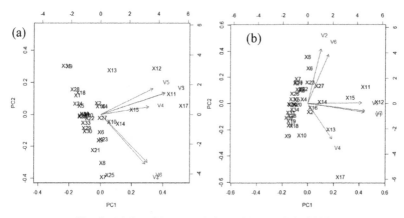

Fig. 5. Biplot of Star Hotels in (a) 2019, and (b) 2020

In 2019, the characteristics between provinces were quite different, as indicated by the points spread. Still, almost all provinces did not excel in all variables, except for South Sulawesi (X27) and Riau Islands (X10), which excel in occupancy rate of hotel room (V2) and bed occupancy rate (V6). While DIY (X14) only excels in bed occupancy rate and West Java (X12) in the number of domestic guests (V5). In addition, DKI Jakarta (X11) excels in the number of room night occupied (V1) and the number of guest nights (V3). Of course, this is a natural thing considering DKI Jakarta is a province that has the most star hotels. Provinces that are also different because they excel in the number of foreign guests (V4) are Bali (X17) and East Java (V15), considering that these provinces have a lot of tourism.

According to the biplot analysis, in 2020, almost all provinces have identical characteristics, as indicated by the close-together points. COVID-19 causes changes in the characteristics of star hotels to all variables, as shown by only two provinces that consistently excel on the same variables as the previous year, namely West Java and Bali. Banten (X16) replaces DKI Jakarta, which excels in the number of room nights used and the number of nights guests stay and South Sulawesi replaces DIY, which excels in the bed usage rate. Meanwhile, DKI Jakarta, East Java, and DIY now excel in the number of domestic guests, and Central Java (X13), North Sumatera (X2), follow Bali excel in the number of foreign guests. This can be caused by the arrival of foreign guests and domestic guests from abroad who must self-isolate in selected hotels around the international airport. However, Bali experienced a significant decrease in foreign guests considering that COVID-19 is a pandemic, so it can discourage foreign guests who want to visit Bali.

3.4 Clustering

First, we do a factor analysis technique that was carried out from 30 variables 3 factors (2019), and 2 factors (2020). Then using latest data from factor analysis, clusters were carried out using K-Means and SOM methods. Pseudo F-statistics can obtain a determination of the optimum number of clusters by using Pseudo F-statistics criteria.

Table 4. Evaluating K-Means using Pseudo F-Statistics

Number of Clusters	2019	2020
2	13.73172	22.76414
3	22.47794	73.86342
4	64.85457	71.91994

Table 4 shows that the highest Pseudo F-statistics obtained using K-Means in 2019 indicate that the formation of four clusters is optimal. When using the occupancy rate of hotel room data in 2020, the highest Pseudo F-statistics is obtained with two optimum clusters. The provinces in Indonesia based on hotel room occupancy rates are also grouped using SOM. The number of clusters is the multiplication of its grid dimension as the result of Table 5.

Table 5. Evaluating SOM using Pseudo F-Statistics.

Grid Size	2019	2020	Grid Size	2019	2020
2 × 1	13.732	19.655	4 × 1	64.854	71.517
1 × 2	13.732	22.696	2 × 2	64.854	71.639
3 × 1	21.076	73.863	1 × 4	64.854	71.639
1 × 3	21.076	73.863			

Table 5 shows that the highest Pseudo F-statistics obtained using SOM in 2019 data is 64.854, while in 2020, the highest Pseudo-F is 73.863. These results indicate that the number of four clusters is optimal in grouping the provinces in Indonesia based on hotel room occupancy rates in 2019 and three clusters for grouping 34 provinces in 2020. This study uses the ICD rate to determine the best method in cluster analysis to group provinces in Indonesia, as shown in Table 6.

The results of the ICD rate both using K-Means and SOM clustering are the same, so it can be concluded that the two methods have the same performance in grouping 34 provinces in 2019 and 2020 based on hotel room occupancy rates. Hence the member

Table 6. Method Evaluation using ICD Rate.

Method	2019			2020		
	Number of Clusters	ICD Rate	Pseudo F	Number of Clusters	ICD Rate	Pseudo F
K-Means	4	0.133592	64.854	3	0.173449	73.863
SOM	4	0.133592	64.854	3	0.173449	73.863

of the cluster using K-Means and SOM is equal. Table 7 shows the members formed in each cluster. The members of cluster 1 are provinces that have several tourist attractions. Meanwhile, Bali as the cluster 2 member is popular globally as the tourist destination for both foreign and domestic tourists. DKI Jakarta, West Java, Central Java, and East Java are business and industrial centers where the country's economy is centered to this region. Then, the remaining fifteen provinces are grouped into cluster 4, where the provinces in this cluster are lack the attractiveness that makes less people visit the related province. Table 7 also shows the changes of each cluster member in 2020. Provinces in cluster 4 in 2019 are joined the province in cluster 1 to form a new cluster as cluster 1 in 2020. Clusters 1, 2, and 3 respectively show members of the province with medium, low, and high room occupancy rates.

Table 7. Clustering Results Using 2019 Data.

Year	Cluster	Cluster Member	Number of Members
2019	1	Aceh, North Sumatera, West Sumatera, Riau, Jambi, South Sumatera, Riau Islands, DI Yogyakarta, Banten, West Nusa Tenggara, East Nusa Tenggara, East Kalimantan, North Sulawesi, South Sulawesi	14
	2	Bali	1
	3	DKI Jakarta, West Java, Central Java, East Java	4
	4	Bengkulu, Lampung, Bangka Belitung Islands, West Kalimantan, Central Kalimantan, South Kalimantan, North Kalimantan, Central Sulawesi, Southeast Sulawesi, Gorontalo, West Sulawesi, Maluku, North Maluku, West Papua, Papua	15
2020	1	Aceh, North Sumatera, West Sumatera, Riau, Jambi, South Sumatera, Riau Islands, DI Yogyakarta, Banten, West Nusa Tenggara, East Nusa Tenggara, East Kalimantan, North Sulawesi, South Sulawesi, Bengkulu, Lampung, Bangka Belitung Islands, West Kalimantan, Central Kalimantan, South Kalimantan, North Kalimantan, Central Sulawesi, Southeast Sulawesi, Gorontalo, West Sulawesi, Maluku, North Maluku, West Papua, Papua	29

(continued)

Table 7. (*continued*)

Year	Cluster	Cluster Member	Number of Members
	2	Bali	1
	3	DKI Jakarta, West Java, Central Java, East Java	4

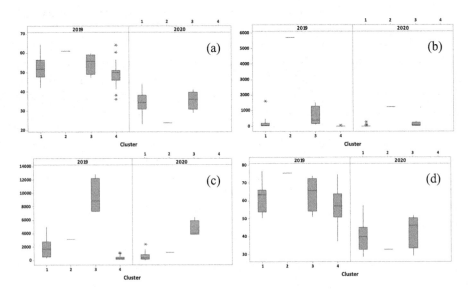

Fig. 6. Cluster Characteristics Based on (a) Occupancy Rate of Hotel Room in Star Hotels, (b) Number of Foreign Guests Staying in Star Hotels, (c) Number of Domestic Guests Staying in Star Hotels, (d) Bed Occupancy Rate in Star Hotels.

From the boxplot in Fig. 6 we know that the change of all variables in each cluster from 2019 to 2020 become decreasing. Figure 6 also shows that the Occupancy Rate of Hotel Room in Star Hotels in Fig. 6(a) and Bed Occupancy Rate in Star Hotels in Fig. 6(d) have a significant decrease from 2019 to 2021 in all clusters. So it can be concluded that the Occupancy Rate of Hotel Room in Star Hotels and Bed Occupancy Rate in Star Hotels experienced the most contraction between 2019 and 2020 in each cluster. Figure 7(a) shows that the proportion of province which has medium and low level of hotel room occupancy rates is almost the same. While the Fig. 7(b) explicitly shows the provinces which has medium and low hotel room occupancy rates in 2019 are joined into one cluster in 2020 which has the medium level of hotel room occupancy rates.

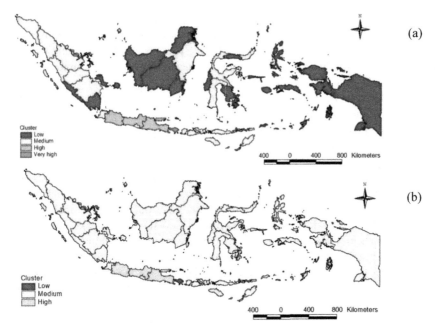

Fig. 7. Provincial mapping in (a) 2019 and (b) 2020.

4   Discussion

COVID-19 has a significant impact on hotel industry. One of the impacts is the hotel and restaurant sector, which has experienced a decline in hotel room occupancy rates. The analysis of grouping of provinces in Indonesia based on the occupancy rate of hotel rooms shows that SOM and K-Means have the same performance. There has been a change in cluster members before and during the COVID-19 pandemic. The clusters 1 and 4 from 2019 are joined into the same cluster in the next year, which means that 29 provinces have similar characteristics in 2020. Moreover, as a popular tourist destination, Bali has a higher room occupancy rate than other clusters. However, since the government's lockdown policy was implemented, Bali's hotel room occupancy rate has fallen below other clusters. This decline was due to the decrease in tourist who visits Bali, causing some hotels in Bali having difficulty in operating. In addition, several hotels in Bali have been changed functionally become isolation places for COVID-19 patients. Thus, Bali is the most affected province by the COVID-19 incident in 2020 since this province relies on the tourism sector as the primary regional income.

The other provinces in cluster 3, namely DKI Jakarta, West Java, Central Java, and East Java, have similar characteristics as economic centers in Indonesia with good infrastructure. So many businessmen dominated the hotel area in these provinces. However, during the pandemic, many office activities were carried out online or work from home, thus making the room occupancy rate in this cluster decreased. Considering that the highest spread of COVID-19 cases occurred on the island of Java, several hotels in cluster 3 were converted as places of isolation for COVID-19 patients.

Several members in cluster 1 were also affected, which can be seen from the decline occupancy rate of hotel room. Cluster 1 members such as DI Yogyakarta, West Nusa Tenggara, and East Nusa Tenggara in 2019 had a room occupancy rate that was not so low because this province has a tourist attraction beach. However, this number decreased drastically in 2020 due to the COVID-19 outbreak.

The differences between variables relating to the occupancy rate of hotel rooms in 2019 and 2020 have been significantly decreasing. The first Large-Scale Social Restrictions in Jakarta was applied in April 2020, followed by many other cities afterward [20]. This policy brought impacts in several fields, including the field of hotel and tourism. For the comparison, we used the data of hotel usage in 2019 for the data before COVID-19 arisen, and the data of hotel usage in 2020 for the data after COVID-19 existed. We already provided statistical inferences that in the 5-star hotels, the decrease of hotel usage is not significant, while in the other types of the hotel (medium and small), the hotel usage was significantly decreasing as the effect of COVID-19 outbreaks. These findings are consistent with the previous research conducted by Lee and Phi [21] explains that the small and medium hospitality in Vietnam would run out of resources caused by the pandemic. In Pakistan, small and medium enterprises are the most affected sector of the COVID-19 outbreaks, including the industrial fields such as restaurants and hotels [22]. Alonso et al. [23] studied 45 international small hospitality businesses and found that owners and managers must change and adapt their businesses to create industry resilience against COVID-19 outbreaks.

5 Conclusion

In this study, we found that there was a decrease in hotel room occupancy rates from 2019 to 2020 in Indonesia. The occupancy Rate of Hotel rooms and Bed Occupancy Rate are the variables that have the most significant difference for two years. The result of the multivariate nonparametric location test shows a significant decrease in 2020. Based on a perceptual map using a biplot, the results show that in 2019, the characteristics between provinces were quite different, whereas in 2020, all provinces had identical characteristics. The two provincial grouping methods used in this study, namely SOM and K-Means, provide the same performance. In 2019, there were four optimum clusters: very high, high, medium, and low; in 2020, there were three: high, medium, and low. Bali is the province most affected by COVID-19, with lower hotel room occupancy rates and fewer tourists. There are two clusters from 2019 are joined into the same cluster in the next year, which means that the more provinces have similar characteristics in 2020. During the COVID-19 outbreak, the small and medium hotel industry was the most affected sector. Several hotels in Indonesia are open, but only for COVID-19 victims. In conclusion, the incident at the beginning of 2020 was indeed very unpredictable due to the COVID-19 that shook the world because the virus had a significant impact on the hotel industry. Gradually, the Government makes regulations regarding the reopening of tourism objects while still paying attention to health protocols. Limiting the number of visitors who visit tourist attractions in an area by considering the level of spread of COVID-19 in the area so that it can have an impact on the level of the economy. Further research in investigating the effect of COVID-19 for each province is necessary to be conducted such as analyzing the impact of the opening of tourism objects during the pandemic.

Acknowledgements. We would like to thank the anonymous reviewers whose comments and suggestions helped improve and clarify this manuscript.

References

1. World Health Organization. Coronavirus Disease 2019 (COVID-19) Situation Report - 1. WHO Indonesia Situation Report vol. 1 (2020)
2. World Health Organization. Coronavirus Disease 2019 (COVID-19) Situation Report - 60. WHO Indonesia Situation Report vol. 60 (2021)
3. Menteri Dalam Negeri RI. INSTRUKSI MENTERI DALAM NEGERI NOMOR 15 TAHUN 2021 (2021)
4. Andriani, D.: Industri Pariwisata Kehilangan Potensi Pendapatan Hingga Rp90 Triliun (2020). https://ekonomi.bisnis.com/read/20200416/12/1228446/industri-pariwisata-keh ilangan-potensi-pendapatan-hingga-rp90-triliun
5. Kaushik, S.: An Introduction to Clustering and Different Methods of Clustering. MAIT (2016)
6. Xiong, C., Hua, Z., Lv, K., Li, X.: An improved K-means text clustering algorithm by optimizing initial cluster centers. In: International Conference on Cloud Computing and Big Data, pp. 265–268. IEEE (2016)
7. Azaza, M., Wallin, F.: Smart meter data clustering using consumption indicators: responsibility factor and consumption variability. Energy Procedia **142**, 2236–2242 (2017)
8. Zheng, T., Liu, S., Chen, Z., Qiao, Y., Law, R.: Forecasting daily room rates on the basis of an LSTM model in difficult times of Hong Kong: evidence from online distribution channels on the hotel industry. Sustainability **12**, 7334 (2020)
9. Wu, E.H.C., Law, R., Jiang Brianda, B.: The impact of infectious diseases on hotel occupancy rate based on independent component analysis. Int. J. Hosp. Manag. **29**, 751–753 (2010)
10. Johnson, R.A., Winchern, D.W.: Applied Multivariate Statistical Analysis. Pearson Education, London (2007)
11. Chung, E.Y., Romano, J.P.: Asymptotically valid and exact permutation tests based on two-sample U-statistics. J. Stat. Plan. Inference **168**, 97–105 (2016)
12. Helwig, N.E.: Statistical nonparametric mapping: Multivariate permutation tests for location, correlation, and regression problems in neuroimaging. Wiley Interdiscip. Rev. Comput. Stat. **11**, 1–24 (2019)
13. Jolliffe, I.T.: Principal Component Analysis. Springer, Heidelberg (2002). https://doi.org/10. 1007/b98835
14. Yuan, C., Yang, H.: Research on K-value selection method of K-means clustering algorithm. J **2**, 226–235 (2019)
15. Feldman, R., Sanger, J.: Text Mining Handbook: Advanced Approaches in Analyzing Unstructured Data. Cambridge University Press, Cambridge (2007)
16. Setiawan, K.: Paradigma Sistem Cerdas. Bayumedia Publishing, Malang (2003)
17. Orpin, A.R., Kostylev, V.E.: Towards a statistically valid method of textural sea floor characterization of benthic habitats. Mar. Geol. **225**, 209–222 (2006)
18. Vogel, M.A., Wong, A.K.C.: PFS clustering method. IEEE Trans. Pattern Anal. Mach. Intell. **PAMI-1**, 237–245 (1979)
19. Mingoti, S.A., Lima, J.O.: Comparing SOM neural network with fuzzy c-means, K-means and traditional hierarchical clustering algorithms. Eur. J. Oper. Res. **174**, 1742–1759 (2006)

20. Menteri Kesehatan RI. Keputusan Menteri Kesehatan Nomor HK.01.07/Menkes/239/2020 (2020)
21. Le, D., Phi, G.: Strategic responses of the hotel sector to COVID-19: Toward a refined pandemic crisis management framework. Int. J. Hosp. Manag. **94**, 102808 (2021)
22. Shafi, M., Liu, J., Ren, W.: Impact of COVID-19 pandemic on micro, small, and medium-sized Enterprises operating in Pakistan. Res. Glob. **2**, 100018 (2020)
23. Alonso, A.D., et al.: COVID-19, aftermath, impacts, and hospitality firms: an international perspective. Int. J. Hosp. Manag. **91**, 102654 (2020)

Identifying Sequential Influence in Predicting Engagement of Online Social Marketing for Video Games

Joseph Chia Wei Chen$^{(\boxtimes)}$ and Nurulhuda Firdaus Binti Mohd Azmi Ais

Razak Faculty of Technology and Informatics, University Teknologi Malaysia Kuala Lumpur (UTM KL), 54100 Kuala Lumpur, Malaysia
huda@utm.my

Abstract. Advancement of online social networks has seen digital marketing use platforms like YouTube and Twitch as key levers for video games marketing. Identifying key influencer factors in these emerging platforms can both deliver better understanding of user behavior in consumption and engagement towards marketing on social platforms and deliver great business value towards video game makers. However, data sparsity and topic maturity has made it difficult to identify user behavior over a sequence of different marketing videos, with a key challenge being identifying key features and distinguishing their contribution to the measure that defines sustained engagement over sequential marketing. This paper presents a method to understand sequential behavioral patterns by extracting features from marketing frameworks and develop a supervised model that takes all the features into consideration to identify the best contributing features to predicting engagement that delivers sustained interest for the next video in a series of marketing videos on YouTube. Experiment results on dataset demonstrate the proposed model is effective within constraint.

Keywords: Online social marketing · Regression prediction · Video game trailers · Machine learning · Sequential pattern

1 Introduction

Social media has become a major component of modern digital marketing today as technology play a front-and-center role in everyday life, driving the success of platforms like Twitch and YouTube in any successful marketing outreach today, as conventional marketing paradigms are being democratized with the expansion of technology-driven marketing such as influencers and memes [1]. Among the industries that are seeing this rapid adoption of this paradigm shift is the video games industry, which historically share a strong, symbiotic relationship with technology due to the critical role technology plays for the medium [2]. Today, marketing for video games is actively exploring the measure of engagement as a means of success. For example, number of likes for a message on Facebook is seen as an endorsement of messaging effectiveness for a company's social marketing strategies [3].

© Springer Nature Singapore Pte Ltd. 2021
A. Mohamed et al. (Eds.): SCDS 2021, CCIS 1489, pp. 427–437, 2021.
https://doi.org/10.1007/978-981-16-7334-4_31

However, identifying the key features that drive engagement has been challenging due to a range of factors, such as the nascent state of marketing analytics for video games and entrenched marketing fundamentals that are still prevalent in much of video games marketing [4, 5]. This also contributes to existing work that does not adequately address this problem. Study of popularity features measuring of critical and commercial ratings of video games are common precursor studies debunked by Zhu and Zhang's nominal work identifying that pre-existing brand popularity have a strong correlation for popularity and commercial impact against conventional assumption that critical ratings play a strong role in video games engagement [6]. This is further supported by Maeyer's work, arguing that in relation to consumer engagement, unobserved factors and time-varying factors needs to be a key element of any studies exploring engagement [7].

This paper studies the discussed issues and proposes a solution, with two key contributions.

- We propose relevant engagement features, extracted from the dataset as well as development of new features modelled from relevant studies on user engagement and marketing analytics.
- We propose a supervised machine learning model to offer prediction of proposed engagement features generated from YouTube dataset of video game trailers.

The rest of the paper is structured as follows. Section 2 reviews related work for feature identification and machine learning prediction methodologies. Section 3 discusses the methodology involved including data pre-processing steps as well as feature development considerations. Experiments and results are presented and discussed in Sect. 4, with Sect. 5 concluding the paper.

2   Related Work

Studies of marketing analytics and consumer engagement via digital marketing has an extensive history of study, primarily centered around development of marketing frameworks oriented around identifying different stages of a customer conversion stage in marketing. What researchers have identified is that it has become ever more challenging to contextualize the new dimensions of measuring engagement for digital marketing, as the number of touchpoints and interaction avenues available expand [8]. The prevalent works exploring engagement in digital marketing have tried exploration of Brackett's attitude model [9, 10] and hierarchical effect models [11] to orient the engagement metrics around the multi-stage customer experience journey of product awareness to intent-of-purchase. This paper goes deeper to analyze the features discussed in these papers, which are like, dislike, popularity, product, and sentiment.

Duffet [12] proposes the development of a model that estimates a post-awareness stage of marketing communication, which expands to identify actions that demonstrate the like or dislike of a product. His work was instrumental in a successful application of attitudinal stages to hierarchy-of-effect model to identify correlation with positive second-stage attitudes to marketing content on YouTube. In addition, [12] also notes that if we're able to extrapolate proof of a positive response in the dimensions associated to a consumer's condition in the first stage of YouTube content consumption to compel them to either meet one of these three post-awareness stage, these are significant features that can be used in consideration for YouTube related metrics, applying multi-stage engagement level flow of actions in a sequential pattern.

There are also work that explores the dimensions associated with pre-existing popularity, as the component of a product's brand prowess and unobserved features. Ahmad's [13] highlighted that feature reflecting of pre-existing popularity can be observed with budget, quantity of reviews, views and additionally defined with weighted adjustments to like and dislike features as per Eq. 1:

$$weighted \frac{like}{dislike} ratio = total \frac{likes}{dislikes} * g \qquad (1)$$

Although many studies are devoted to the exploration of engagement in marketing, it remains a highly abstract concept, with there being no unified definition or metric that can take into account the elements of emotional congruence and relevance, which are unobserved features.[14] As per the works of [6] and [10] the risk of variables or features that were not considered in the initial study like popularity and bias factor of sarcasm in body language could misinterpret polarity data substantially. And the same challenge applies to the non-qualitative assumptions that goes into the data analytics, which has a wide range of options in modelling which has demonstrated success across classification and regression modes of prediction in digital media.

This paper synergizes the key learnings of previous work [9–13] and proposes a prediction framework that leverages previously identified features and develops new features from the dataset to derive a range of engagement traits that delivers the best performance across a range of previously effective machine learning algorithm.

3   Methodology

To deliver the best performing model, a comprehensive process that ensures robust data selection and model testing is required. That process is operationalized into the Operational Framework Structure. (see Table. 1).

3.1   Data Collection and Pre-processing

This section discusses the data selection consideration and all activities related to cleaning the selected data for model readiness. The data collection process was divided into two stages, representing different datasets considerations. For the first dataset, the core video games marketing data that is used in this study is obtained from a web-scrap of YouTube's public data [15]. Using a criteria-based sampling procedure, the data scrap was focused on official marketing trailers for video games that was showcased at the Electronic Entertainment Expo, a premier annual event used to video games marketing, generating a dataset consisting of 277 records over a 36-month period. The second dataset involved a selection of essential product information related to video games that wasn't available in the first dataset, obtained from VGChartz [16], a repository of video game information, serving the role to expand on feature development based on

Table 1. Operational framework structure.

Process	Summary
Preliminary study	Understanding of problem background and thorough research via literature review
Data collection and preprocessing	End-to-end data collection, integration, and cleaning process
Data analysis	Understanding of data collected and building of initial hypothesis regarding features
Model preparation and evaluation	Development and evaluation of model trained with completed dataset
Conclusion	Comparison and selection of best model

popularity and product features discussed in [11] and [14]. The data was cleaned and pre-processed with the primary consideration of establishing a dataset that demonstrated sequential relationship across three trailers for video game marketing.

3.2 Feature Extraction

This section explores extraction and development features from the dataset. Effective feature study is critical in work applying machine learning to increase the precision of the model developed, which requires a thorough review and analysis of all available data collected and identify the appropriate feature set which demonstrates meaningful correlation with the target parameter.

Product Features in Context of Content Marketing. Identifying product features relating to video games can be established with existing lexicons of product terminology, but that is different for the context of "content marketing" on YouTube without established typology. However, by applying historical knowledge base of marketing techniques for video games explored in [5] and thematic analysis of shared use cases in language in [3], it is possible to apply natural language processing to contextualize and represent trailers into categories of features that provide a meaningful representation for product in the context of YouTube marketing. Application of Latent Dirichlet Allocation (LDA) algorithms [17] to the features related to description of YouTube trailers, alongside keyword analysis allows the product features to be contextualized in view of familiarity with marketing content. By describing the keyword profiles obtained from the natural language analysis, qualitative as-assessment can be applied with a small test group to pose relevant questions relative to meaning of keywords in the context of marketing. When observing words such as gameplay, debut, or teaser – what are the context of a marketing strategy offered to observed group? Using this combination of language processing and a sample keyword review with a test group, we can convert the keywords identified in the dataset into product differentiation features (see Table.2).

Popularity Feature within YouTube. Despite the prevalence of popularity as a feature in marketing analytics, popularity's definition as a feature varies as different research

Table 2. Keyword profile and associated product feature in YouTube trailers

Keywords	Product features
'teaser', 'trailer 1', 'reveal', 'CGi', 'first look',	Teaser trailer
'reveal', in-game reveal', trailer 1/2', >5m'	Reveal trailer
'reveal', 'with gameplay', 'trailer 1',	Reveal demo
'gameplay', 'not reveal', 'trailer 2/3', 'trailer'	Gameplay trailer
'demo', ' > 8m', 'deep-dive', 'gameplay' 'trailer 2/3'	Gameplay demo
'trailer', 'cutscene', 'trailer 2/3', 'no gameplay'	Regular trailer
'launch', 'trailer 3', 'launch trailer'	Launch trailer

explores popularity with different models, and suited algorithms. [19] For YouTube, prior work shows that a traditional logarithmic model and online media can be applied to predict con-tent popularity within videos. [2, 3, 18] An application of that model to this study would be represented as the following equation, reflected in a developed feature of popularity score as per Eq. 2:

$$\frac{comments}{views} * (likes - 1.5 * dislikes) \tag{2}$$

Applying the fundamentals of Eq. 1 developed from the findings of [13], the popularity features can also be expanded to consider the following:

$$weighted \frac{views}{likes} * (100\%) \tag{3}$$

$$weighted \frac{views}{comments} * (100\%) \tag{4}$$

Sequential Features for Marketing Engagement Continuity. The key objective of the study is the prediction of marketing engagement status with the assumption of a time-series relationship between trailers, based off existing data to extrapolate the infor-mation in a way where t = trailer sequential order, to find out the continued engage-ment of t + 1. Marketing research and analytics apply data manipulation and customer segmentation or labeling approach with the goal of dividing information into subsets to distinguish different groups. [20] Adopting that approach, the dataset is divided, labeled, and reintegrated **divide** the trailer data into two subsets of sequential relationship, which are trail-er one-to-trailer-two and trailer two-to-trailer three, where t + 1 is reflected in the relation between two sequential trailers The labeling for trailer sequence is taken through application of distinguishing the sequence with the timestamp or trailer date to reflect time-series information, with the earliest trailer in the sequence labeled one, two and three in chronological order, with all records that do not have a sequence of three trailers removed. Table 3 shows the shape of the dataset applying the sequential features of the products.

With the assignment of a sequence-based labeling to the dataset, it develops a predic-tion opportunity for the data, where model performance is considered across two distinct segments of t1 + 1 or t2 + 1 as well.

Table 3. Sequencing of trailer by timestamp

Product genre	God of war	Call of duty	
Timestamp	Trailer sequence	Timestamp	Trailer sequence
201702	Trailer one	201902	Trailer three
201802	Trailer two	201802	Trailer two
201809	Trailer three	201609	Trailer one

4   Experiment and Results

One of the key objectives of this paper is the inference of contributing elements to marketing engagement through analysis of patterns in the relationship between features. The experiments for this paper are oriented around application of regression-based analysis to detect the casual correlation between features that are identified as engagement factor and build a model that maximizes the capability in prediction of said engagement factor in relation to the effects of the features.

As elaborated in Sect. 3.2, the experiment design involves development and extraction of several key features, such as: product trait for both game and marketing such as genre of game and type of trailer, popularity score, numerical YouTube metrics like views, likes and developed features such weighted engagement features. The engagement features are tested for their correlation coefficient contributor to both $t + 1$ factors with Pearson's coefficient to identify best performing features, suitable for machine learning modeling. [21] The features are converted, where necessary to support regression-based analysis. Applying the best performing supervised learning algorithms from the prior work in [2, 3, 6, 7, 10–13], Random Forest Regressor, Linear Regression and Support Vector Machine Regression are used in the experiments. The experiments are conducted in Python with their default learning parameters, with a 5-fold cross validation method, with 80:20 split in training and testing for the dataset. Normality testing was also conducted for the dataset to test for suitability of correlation analysis methodology with the quantile-quantile plot approach.

4.1   Feature Analysis

For this paper, given the external extrapolation and development of features onto the dataset, feature analysis is a critical part of the experiments. We adopt the use of the Pearson product-moment correlation coefficient onto the dataset to study what are consumer behavior elements that demonstrate statistically relevant relationship between product, popularity or sequential features extrapolated from the research methodology. [21] The outcome of that correlation analysis shown in Table 4.

From Table 4, we observe that product features related to the video games, that being 'franchise' and 'game genre' themselves do not contribute meaningful correlation coefficient from the regression analysis, but product features related to marketing indicate that the type of trailer debuting in trailer one is positively related to the likelihood that trailer two will retain a degree of engagement across views, likes and comments, though

Table 4. Top performing features with highest r^2 value based on sequential traits for $t + 1$

Features	R^2 score for trailer one-to-trailer two correlation			R^2 score for trailer two-to-trailer three correlation		
	Views	Likes	Comments	Views	Likes	Comments
Franchise	0.089	0.170	0.280	0.110	0.230	0.220
Game genre	0.016	0.056	0.017	0.045	0.077	0.100
Trailer type	0.570	0.590	0.420	0.120	0.120	0.190
Trailer views	0.530	0.470	0.460	0.480	0.440	0.350
Trailer likes	0.560	0.590	0.480	0.610	0.660	0.570
Trailer dislikes	0.160	0.250	0.140	0.480	0.150	0.120
Trailer comments	0.660	0.540	0.470	0.480	0.670	0.590
Trailer popularity	0.540	0.510	0.410	0.380	0.670	0.590

it performs best with likes at 0.59 coefficient score. We also note that for both trailer one and trailer two, numerical metrics related to views, likes, and comments do contribute to sustained engagement on their own. Trailer one-to-two demonstrates the broader range of engagement continuity, with trailer type, views, likes, comments, and popularity score showing a positive correlation coefficient with engagement compared to trailer two-to-three.

4.2 Experiment Results

Through Sect. 4.1, we have developed a better understanding for the potential in the predictive power of the identified features. Applying the findings of that analysis, we propose a regression-based prediction model to derive the prediction for 'number of likes' generated by trailer two, by applying trailer one assumptions of trailer product feature 'trailer type' and three assumptions of trailer metrics in 'number of views, likes and comments' generated by trailer one, applying Random Forest Regressor, Linear Regression and Support Vector Machine Regression. [2, 3, 6, 7, 10, 12, 13] The accuracy evaluation metrics for the model is measured by the r^2 score of the model, which is correlation coefficient between the input and output features selected and mean absolute error, which measures the average magnitude of errors comparing the predicted and actual data. The result of the experiment is summarized in Table 5.

Based off Table 5, we observe that among the models tested, only Linear Regression and Random Forest prove to be sufficient in delivering higher than 50% r^2 score, with Random Forest Regressor demonstrating the strongest performance with the correlation coefficient at approximately 77% strong positive relationship correlation between the selected features, as well as the best error score, with the lowest mean absolute error of the tested models at 26,154.

To better understand the actual scenario application of the model as well as any limitations of the model in its ability to predict likes, the best performing model, Random

Table 5. Performance comparison between algorithms for predicting like for trailer two

Category	R^2 score	Mean absolute error
Linear regression	0.544	39,803
Random forest regressor	0.774	26,154
Support vector machine	−0.139	53,637

Forest Regressor is applied the full original dataset for a performance prediction across a scatter plot in Fig. 1 and box plot in Fig. 2.

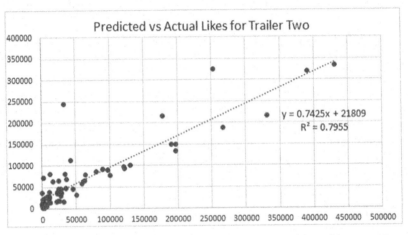

Fig. 1. Comparison on predicted likes in original dataset and prediction with scatter plot graph

From Fig. 1, we can see that model outperformed the metrics from the evaluation, with an r^2 value of 0.795 and a range of 21,809, shown in the table in the formula of $y = mx + b$. This demonstrates the model's overall capability in predicting the value in line of the actual likes value. However, the scatter plot does also highlight several situations where the prediction is demonstrably higher or lower than the trend, demonstrated by the b-value of 21,809 which is still substantial given that the average likes for the dataset range in the 60,000 likes range, which places the error range of the like prediction at close to 30%. That is supported by Fig. 2's illustration of the boxplot, where the maximum for deviation from the quartile value is close to 50,000 higher than the actual, which increases the deviation value.

These results show that even though the model is capable of a prediction analysis that is broadly capable of predicting the likes with a reasonable level of accuracy across a large set of data, the model bears risk of a relatively high deviation value that would skew the interpretation of the data.

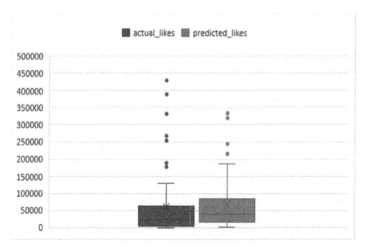

Fig. 2. Comparison on predicted likes in original dataset and prediction with box plot graph

5   Conclusion and Future Work

This paper proposes a novel approach towards regression-based prediction of marketing engagement for video game trailers on YouTube through analysis and development of sequential features and models from the dataset. Utilizing a wide breadth of perspectives that has in prior work demonstrated sustained interest, characteristics like popularity, number of likes, views, comments are investigated and repurposed using natural language processing, text mining and statistical formulations to generate a wide range of features, in which the features are tested and experimented to validate the value of features in the prediction of sequential engagement for video game trailers on YouTube.

The experiment results show that there are four core features contributing to the prediction of engagement, and they are reflected in number of views, likes, and comments for the first trailer to the second in its ability in predicting the number of likes for the second trailer, with the marketing product feature of trailer type being the sole product feature contributor when modeling prediction of said engagement. The best performing regression model in the prediction would be the Random Forest Regressor, which delivers a 77 to 79% prediction accuracy.

The paper has several limitations for consideration of future works. The error range of the prediction is not sufficiently controlled, limiting the scope and flexibility for the model's predictive ability. Secondly, though the paper can demonstrate the sequential correlation for a small number of features, despite many features tested. Future work can expand the depth and complexity of the product-related features across a wider range of models, including other approaches like classification-based prediction to further identify features that contribute to sustained engagement on YouTube [9]. Thirdly, due to the sequential feature development of the study, the dataset for the final modeling was not substantial in size. The quality of model could be further expanded if scaled to a larger dataset, though time-factor still needs to be considered due to organic growth of active userbase on YouTube over multiple years [12].

Despite limitations discussed, given that much of content-based marketing in video games are still approached traditionally with limited utilization of big data, [5] we believe the findings of the paper would serve to benefit marketing and content development analytics for video game developers in retrofitting existing popularity metrics relative to the type of marketing that they plan to debut with to predict the effectiveness of any marketing plan in terms of sustained engagement for subsequent trailers.

Acknowledgements. This work was funded by the Ministry of Higher Education under Fundamental Research Grant Scheme (FRGS/1/2019/ICT04/UTM/02/11).

References

1. Decoder. The Verge. https://www.theverge.com/22174582/decoder-podcast-interview-cad illac-cmo-melissa-grady-advertising. Accessed 15 Dec 2020
2. Zhang, X.: Improving cloud gaming experience through mobile edge computing. IEEE Wirel. Commun. **26**, 178–183 (2019)
3. Yu, B., Chen, M., Kwok, L.: Toward predicting popularity of social marketing messages. In: Salerno, J., Yang, S.J., Nau, D., Chai, S.-K. (eds.) SBP 2011. LNCS, vol. 6589, pp. 317–324. Springer, Heidelberg (2011). https://doi.org/10.1007/978-3-642-19656-0_44
4. Leeflang, P., Peter, C., Dahlström, P., Tjark, F.: Challenges and solutions for marketing in a digital era. Eur. Manag. J. **32**(1), 1–12 (2014)
5. Zackariasson, P., Dymek, M.: Video Game Marketing: A Student Textbook, 1st edn. Routledge, London (2016)
6. Zhu, F., Zhang, X.: Impact of online consumer reviews on sales: the moderating role of product and consumer characteristics. J. Mark. **74**(2), 133–148 (2010)
7. De Maeyer, P.: Impact of online consumer reviews on sales and price strategies: a review and directions for future research. J. Prod. Brand Manage. **21**, 132–139 (2012)
8. McKinsey. https://www.mckinsey.com/industries/consumer-packaged-goods/our-insights/. Accessed 20 June 2021
9. Brackett, L.K., Carr, B.N.: Cyberspace advertising vs other media: consumer vs mature student attitudes. J. Advert. Res. **41**(5), 23–32 (2001)
10. Yang, K.C., Huang, C.H., Yang, S.Y.: Consumer attitudes toward online video advertisement: YouTube as a platform. Kybernetes **46**(5), 840–853 (2017)
11. Gauzente, C.: Information search and paid results—proposition and test of a hierarchy-of-effect model. Electron Markets **19**, 163–177 (2009). https://doi.org/10.1007/s12525-009-0015-1
12. Duffett, R.: The YouTube marketing communication effect on cognitive, affective and behavioural attitudes among generation Z consumers. Sustainability **12**(12), 1–25 (2020)
13. Ahmad, I., et al.: Movie revenue prediction based on purchase intention mining using YouTube trailer reviews. Inf. Process. Manag. **57**, 5 (2020)
14. Mollen, A., Wilson, H.: Engagement, telepresence, and interactivity in online consumer experience: reconciling scholastic and managerial perspectives. J. Bus. Res. **63**, 919–925 (2010)
15. YouTube scraper. https://apify.com/bernardo/youtube-scraper. Accessed 20 July 2021
16. VGChartz. https://www.vgchartz.com/. Accessed 20 July 2021
17. Blei, D., Ng, A.Y., Jordan, M.I.: Latent dirichlet allocation. J. Mach. Learn. Res. **3**, 993–1022 (2003)

18. Szabo, G., Huberman, B.A.: Predicting the popularity of online content. Commun. ACM **53**(8), 80–88 (2010)
19. Semantic Scholar. https://www.semanticscholar.org/. Accessed 25 July 2021
20. Wu, L., Zhu, Y., Yuan, N.J., Chen, E., Xie, X., Rui, Y.: Predicting smartphone adoption in social networks. In: Cao, T., Lim, E.-P., Zhou, Z.-H., Ho, T.-B., Cheung, D., Motoda, H. (eds.) PAKDD 2015. LNCS (LNAI), vol. 9077, pp. 472–485. Springer, Cham (2015). https://doi.org/10.1007/978-3-319-18038-0_37
21. Hwang, T.-G., Park, C.-S., Hong, J.-H., Kim, S.K.: An algorithm for movie classification and recommendation using genre correlation. Multimedia Tools and Applications **75**(20), 12843–12858 (2016). https://doi.org/10.1007/s11042-016-3526-8

Author Index

Printed in the United States
by Baker & Taylor Publisher Services